FILTERING AND CONTROL OF MACROECONOMIC SYSTEMS

A Control System Incorporating the Kalman Filter for the Indian Economy

CONTRIBUTIONS TO ECONOMIC ANALYSIS

160

Honorary Editor
J. TINBERGEN

Editors
D. W. JORGENSON
J. WAELBROECK

NORTH-HOLLAND
AMSTERDAM · NEW YORK · OXFORD · TOKYO

FILTERING AND CONTROL OF MACROECONOMIC SYSTEMS

A Control System Incorporating the Kalman Filter for the Indian Economy

M. J. Manohar RAO
University of Bombay
India

1987

NORTH-HOLLAND
AMSTERDAM · NEW YORK · OXFORD · TOKYO

ISBN: 0 444 70188 5

Publishers:
ELSEVIER SCIENCE PUBLISHERS B.V.
P.O. Box 1991
1000 BZ Amsterdam
The Netherlands

Sole distributors for the U.S.A. and Canada:
ELSEVIER SCIENCE PUBLISHING COMPANY. INC.
52 Vanderbilt Avenue
New York, N.Y. 10017
U.S.A.

Library of Congress Cataloging-in-Publication Data

Manohar Rao, M. J.
Filtering and control of macroeconomic systems.

(Contributions to economic analysis ; 160)
Bibliography: p.
Includes indexes.
1. India--Economic policy--1980- --Econometric models. 2. Macroeconomics. 3. Filters (Mathematics) 4. Control theory. I. Title. II. Series.
HC435.2.M34 1987 339.5'0954 86-32928
ISBN 0-444-70188-5 (Elsevier Science Pub. Co.)

PRINTED IN THE NETHERLANDS

Introduction to the series

This series consists of a number of hitherto unpublished studies, which are introduced by the editors in the belief that they represent fresh contributions to economic science.

The term 'economic analysis' as used in the title of the series has been adopted because it covers both the activities of the theoretical economist and the research worker.

Although the analytical methods used by the various contributors are not the same, they are nevertheless conditioned by the common origin of their studies, namely theoretical problems encountered in practical research. Since for this reason, business cycle research and national accounting, research work on behalf of economic policy, and problems of planning are the main sources of the subjects dealt with, they necessarily determine the manner of approach adopted by the authors. Their methods tend to be 'practical' in the sense of not being too far remote from application to actual economic conditions. In addition they are quantitative.

It is the hope of the editors that the publication of these studies will help to stimulate the exchange of scientific information and to reinforce international cooperation in the field of economics.

The Editors

To my parents

PREFACE

The history of attempts to introduce the theory of optimal control into the field of economics goes back at least thirty five years. It is, however, only since the early seventies that the explosion in computer technology, coupled with the increasing sophistication in macroeconometric modelling, has paved the way for rapid progress in the formulation and solution of optimal control and filtering programmes, especially in the sphere of macroeconomic policy designing.

It is these recent advances in systems methodology that have prompted the need for an interface between optimal control theory and dynamic macroeconomic analysis. While a tremendous amount of research has been carried out elsewhere to examine the implications of this 'close encounter', it is a matter of speculation whether policy planners in most developing countries have even considered the implementation of these techniques at any level of planning.

We feel that control and systems theory can provide immense assistance in the stabilization of the economies of these countries, most of which are plagued by problems of a cyclical as well as a structural nature. It is our hope that this effort in demonstrating the applicability of control and filter theory to short-term macroeconomic planning does, in fact, encourage its implementation, since its externalities can easily span different contexts. At the same time, we hope that this book will help to narrow down the communications gap that exists in India, and indeed elsewhere, between economists and control engineers by illustrating the extent to which the existing concepts in control theory can help in the solution of diverse economic phenomena.

The study is the result of my postdoctoral research carried out over the period May 1984 to January 1985, during which time I was a Fulbright scholar at the Department of Economics, UCLA. It would have been impossible to complete it without the active assistance and cooperation of several individuals at that institution and elsewhere. I would particularly like to express my heartfelt appreciation and gratitude to Professor Michael Intriligator who agreed to be my faculty advisor. Besides providing me with all his technical expertise on the subject, he helped me out in every possible way to ensure that, as far as 'system performance' was concerned, all 'noise' was optimally 'filtered' out from the study. For this, and much more, I shall always be indebted to him.

I also wish to acknowledge my indebtedness to Professor Masanao Aoki whose lectures I had the good fortune of attending and who helped clarify many of the doubts I had on the topic of optimal filtering. I am also deeply obliged to Professor Jean-Paul Fitoussi whose expertise in the construction of French macroeconometric models led him to provide extremely valuable comments and suggestions regarding my modelling efforts.

Special thanks are due to Ramalinga Kannan who provided me with much of the data I needed, without which it would have been very difficult to have estimated the model in its present form. To Sandra Rosenhouse for her programming assistance which made possible many of the results in this work, my heartfelt gratitude. I owe a great deal to Zaki Eusufzai who introduced me to the TROLL system which was used for the simulation of the econometric model. He and Kishore Gawande were extremely kind enough to help me in several areas concerning the construction and testing of the model. All their suggestions and comments were most appreciated.

I owe an almost unrepayable debt to Anne Bodenheimer who, as Fulbright Coordinator, helped me out in more ways than I can ever imagine or remember. Without her help in persuading the Council for the International Exchange of Scholars (CIES) at Washington, D.C., to grant me an extension of my fellowship, this book would never have been completed.

I am deeply grateful to Sharada Nayak, Director, United States Educational Foundation in India, for the immense trouble she took on my behalf. It was at her insistence that my period of research was optimally rescheduled so that I was able to derive the maximal advantage of my stay in the U.S.

I am also very thankful to Professor P. R. Brahmananda, Director, Department of Economics, University of Bombay, who was more than willing to grant me the necessary extension of leave despite the considerable hardship I imposed upon him by my protracted absence.

More than ever, I wish to thank, knowing full well that it would be a hopelessly inadequate gesture, Theresa De Maria, Ruth Imperial and Karen Williams for their constant affection, support and encouragement, without which I can honestly say that my stay at UCLA would have been quite bereft of any really fond memories. To them I say, "Partir, c'est mourir un peu".

Bombay
November 22, 1986

M. J. Manohar Rao

CONTENTS

CHAPTER 1

INTRODUCTION

1.1 Preface

The objective of this study is to consider short-term Indian economic policy from the viewpoint of modern systems and control theory. The principal concern of policy planners in India has been to regulate the economy such that it progresses in a satisfactory manner. However, thirty-five years of macroeconomic planning has indicated, beyond the shadow of a doubt, that large scale perturbations have occurred very frequently in the Indian economy and, as our knowledge about the functioning of the economy and the effects of instruments has been far from perfect, it has been found impossible to prescribe precise compensatory action. In the absence of such countercyclical policy, the spontaneous regaining of equilibrium has been ruled out and the economy has thereby suffered swings of considerable amplitudes at great costs. This study is principally an attempt to try and use stochastic control theory for macroeconomic regulation, so that the inherent pitfalls of adopting policies of an intuitive nature are ruled out.

1.2 Short-Term Economic Policy And Optimal Control Theory

The principal objective of control theory has been to try and improve system performance through the regulator concept, especially when uncertainty is involved. The feedback control policy leads to a simple method for determining optimal control actions, given appropriate statistics based on available information. The determination of these statistics, namely, the conditional mean and the error covariance matrix of the system state, takes place separately. The relationship between the system state and the information data is explicitly kept in view. The observations of economic activity are assumed to contain observation errors, including changing or fragmentary information, and the incorporation of such indicators into the system is achieved through the Kalman filter. The filter provides minimum-variance, unbiased estimates of the system state, conditional on the available information.

The strategy of, what has been termed in the literature, feedback control is adopted in order to obtain the optimal control actions. This type of control has been defined as the policy where controls are 'some deterministic function of the current and past observations on the system state variables and of past employed controls' (Aoki 1967). In such a policy, the information comprising previous control actions and system observations available uptil the present moment when control action is to be specified is utilized in computing the control actions. Such a determination of the optimal control trajectory is achieved in two separate and sequential steps. In the first, the estimation of the system state based on available information is obtained, and subsequently, the optimal control action is determined from a deterministic system, which is obtained from the corresponding stochastic system by invoking the Certainty Equivalence Principle (Simon 1956, Theil 1964), implying thereby the replacing of all the random quantities by their conditional expectations.

As more information accumulates, the conditional expectations need to be updated in order to apply the Certainty Equivalence Principle. The Kalman filter (Kalman 1960, Kalman and Bucy 1961) is a very convenient technique to revise optimally the estimates based on past information in the light of new information alone. In effect, the conditional mean and the error covariance of the system state summarize all the accumulated information, as far as the determination of the optimal control is concerned. The Kalman filter enables their updating recursively and past information need not be used again nor stored, since its effect is summarized in the earlier estimates. Similarly, the results of Meditch (1967) facilitate the recursive revision of the past conditional statistics with the coming of fresh information. The optimal control actions resulting from the closed-loop policy depend not only on the latest observation but also on the past observations. The policy attaches appropriate weights to the sequence of observations, and the outcome is that the optimal control action is a weighted average of the latest and past observed errors in prediction. This feature is technically called the proportional -plus-integral control, and it results in smoother control actions and adds to overall stability.

1.3 Scope Of The Study

Under the present Indian practice, economic regulatory policies tend to be heavily biased towards current rather than current and past observations of the state of the economy. Moreover, they have a tendency to overlook the fact that more decisions will have to be taken later on in the light of new information. These drawbacks are overcome under the present framework. Moreover, while the usual economic literature assumes, rather naively, that the current state of the system is completely known, so that all uncertainty is concentrated in the future, the use of stochastic control theory allows for the systematic treatment of the more realistic case when information is scarce, contradictory and inexact. It is here that the Kalman filter comes in to make the most of the available information. Vishwakarma (1974) was one of the first to apply these elegant prediction and control algorithms developed by Kalman and Bucy to a macroeconomic regulatory problem, within a sort of quasi-Monte Carlo framework. This study is an attempt to apply it in a more rigourous and extensive manner well suited for the Indian context.

The mathematical analysis of linear closed-loop control leads to important results such as the separability of prediction from policy determination and certainty equivalence, even if the poor quality of the data is taken into account. This formally justifies the similar separation carried out by Tinbergen (1956) and his concentration on a deterministic analysis.

The application of optimal control techniques to solve short-term economic problems presupposes that it is possible to construct and operate a plausible mathematical model of the economy. Based on the analogy between the structures of certain economic and physical problems, there is a *prima facie* case for applying optimal control to such a model of the economy to help analyze, probe and ultimately control the dynamic behaviour of the economic system (Ball 1978). The very essence of the optimal, as opposed to the automatic, control problem is that one does not know in advance what exactly one would like to happen; and since economic analysis is concerned for the most part with the exercise of choice, given constraints, the possibility of transferring optimal choice technology from the physical to economic systems via optimal control theory appears to be very appealing. This is the basic essence of the study.

1.4 Optimal Control Theory

1.4.1 Problem formulation

A macroeconomic model attempts to describe the dynamics of an economy over T periods. The model consists of a system of n difference equations relating; n endogenous variables $x(i,t)$, denoted by the vector $x(t) \in R^n$ and describing the state of the economy; s control variables $u1(j,t)$, denoted by the vector $u1(t) \in R^s$ and describing the instruments of economic policy at the disposal of the planner to guide the trajectory of the economy towards a specified optimal state; m exogenous variables $u2(k,t)$, denoted by the vector $u2(t) \in R^m$ and describing those variables whose values are uncontrollable but are assumed to be known (the existence of such uncontrollable exogenous variables is invariably the case with most econometric models) and which are expected to affect the values of the endogenous variables; parameters describing the structure of the economy and its relationship with the environment; and residual random variables (which from now on we assume specified and incorporated in the functional form of the equations).

To explain the value $x(i,t)$ taken by the *ith* endogenous variable at period t from the series of past historical data, econometricians estimate equations taking into account the values taken by other variables at period t as well as earlier periods. The foremost past period taken into account in the equation explaining the *ith* endogenous variable is called the lag of the *ith* equation. In general, a macroeconomic model comprises behavioural (or reaction) equations explaining endogenous variables and accounting identities expressing *ex-post* equilibrium conditions enforced at each period. The latter are not explicitly solved for one endogenous variable in terms of the others. Thus, a model in structural form can be described by the system of equations

$$h(i)x(i,t) = f(i,t)(x(t),x(t-1),...,x(t-p);\ u1(t),u1(t-1),...,u1(t-q);\ u2(t),u2(t-1),...,u2(t-r))$$

$$i=1,2,...,n \quad \text{and} \quad t=1,2,...,T \qquad \text{---(1.1)}$$

$$x(i,t) \text{ given for } i=1,2,...,n \text{ and } t= -p+1,-p+2,...,0 \qquad \text{--(1.2a)}$$
$$u1(j,t) \text{ given for } j=1,2,...,s \text{ and } t= -q+1,-q+2,...,0 \qquad \text{--(1.2b)}$$
$$u2(k,t) \text{ given for } k=1,2,...,m \text{ and } t= -r+1,-r+2,...,0 \qquad \text{--(1.2c)}$$

where p,q and r are the maximal lags with respect to the endogenous, control and exogenous variables, respectively; and h(i) are given parameters such that

$$h(i)=\begin{cases}1, & \text{if the } \underline{ith} \text{ equation is solved with respect to } x(i,t)\\ 0, & \text{otherwise.}\end{cases}$$

We also assume that eq.(1.1) satisfies the following regularity conditions:

(i) The function f(i,t) is continually differentiable with respect to x, u1 and u2, and

(ii) The function f(i,t) and all its partial derivatives with respect to x, u1 and u2 are bounded.

Note that h(i) = 1 implies that $\partial f(i,t)/\partial x(i,t) = 0$. We introduce an nxn diagonal matrix H with ith entry h(i):

$$H = \text{diag}\,(h(i)) \qquad \text{---(1.3)}$$

and an a' vector d(t) of available information at the beginning of period t:

$$d(t) = (x(t-1),...,x(t-p);\ u1(t),...,u1(t-q),\ u2(t),..,u2(t-r)) \qquad \text{---(1.4)}$$

with a' = pn+(q+1)s+(r+1)m, called the vector of predetermined variables.

The simulation of the model (1.1) over the T periods consists therefore, once the sequences (u1(t)) and (u2(t)), t=1,2,..,T, are initiated, in the successive solutions for t=1,2,..,T in x(t) of the n-dimensional nonlinear system

$$Hx(t) - f(t)\,(x(t),d(t)) = 0 \qquad \text{---(1.5)}$$

where d(t) is given by eq.(1.4).

The determination of the optimal economic policy consists in finding a sequence $u1^* = (u1^*(1),u1^*(2),...,u1^*(T)) \in R^S$, with S=sT, of admissible values for the control variables, given the sequence $u2 = (u2(1),u2(2),...,u2(T)) \in R^M$, with M=mT, so as to minimize an objective function

$$j(x,u) = \sum_{t=1}^{T} j(t)\,(x(t),...,x(t-p);u1(t),...,u1(t-q)) \qquad \text{---(1.6)}$$

where $x = (x(1),x(2),...,x(T)) \in R^N$, with N=nT, is the sequence of states of the economy resulting from the choice of the sequence ul = (ul(1),ul(2),...,ul(T)) of controls. It amounts to the problem of minimizing the functional j given by eq.(1.6) subject to the equality constraints of the model given by eq.(1.1). The resulting optimal control sequence ul* gives rise to a corresponding x* = (x*(1),x*(2),...,x*(T)) which is known as the optimal state trajectory. In practice however, the constraint that $ul(t) \in R^S$ is not invoked explicitly.Rather, restrictions on the variations of the controllers are imposed through the control costs in the objective functional.

1.4.2 Technical assumptions

Of late, several researchers have found the optimality principle of Bellman (Bellman 1957) as well as the maximum principle of Pontryagin (Pontryagin 1961, Pontryagin et al 1962) eminently suited for application to such a type of economic control problem. The results in this study are obtained by invoking the maximum principle (also known in the literature as the minimum principle) within the framework of a 'Linear--Quadratic-Gaussian' (LQG) model of the Indian economy, and applying the necessary conditions for optimality to the discrete-time control problem. The optimal control that results is seen to depend on the solution of two difference equations; a 'Riccati' equation which depends on both the system itself and the weighting matrices in the objective function, and a 'tracking' equation which depends on the solution of the Riccati equation as well as the nominal state and control trajectories that are to be tracked.

The optimal control model is given by the following pair of equations:

State Equation: $x(t) = Ax(t-1) + B1u1(t) + B2u2(t)$
Observation Equation: $z(t) = Cx(t)$ ---(1.7)

While the first equation provides information on the way the system moves over time, the second one recognizes the empirical fact that the system state may not be observed as such; only certain linear combinations of it are measured as observation variables. As mentioned in the earlier section, the residual random variables contaminating each of the above equations are assumed to be incorporated in the functional form of the equation. A, B1 and B2 are nxn, nxs and nxm matrices, respectively, while C is an n*xn matrix, where n* is the dimension of the observation vector (In most applied research, it is assumed that $n^* < n$).

The feedback control rule which is obtained as a result of the application of the discrete maximum principle to the above system (see Appendix B) is given by

$$ul^*(t) = G(t)x^*(t-1) + g(t) \qquad \text{---(1.8)}$$

where:
x*(t) is the nxl optimal state trajectory vector at time t,
ul*(t) is the sxl optimal control vector at time t,
G(t) is the sxn time-varying state variable feedback matrix and
g(t) is the sxl time-varying intercept vector.

When a feedback control rule of this kind is used, the optimal values of the policy variables in the future will depend upon future observations of the optimal values of the state variables. Therefore, such a type of control philosophy is referred to as closed-loop control (see Dreyfus 1968).

As the control rule is of the form of eq.(1.8), the determination of the optimal control $u^*(t)$ will depend upon the optimal state in the previous period, i.e., $x^*(t-1)$. The optimal state is obtained in two steps. In the first step, we estimate the one-period ahead conditional prediction of the state vector which is given by

$$\hat{x}(t) = Ax^*(t-1) + B1u1^*(t) + B2u2(t) \qquad \text{---(1.9)}$$

where $x^*(t-1)$ is the optimal state in the previous period, $u1^*(t)$ is the optimal control in period $\underline{t}$ derived by invoking eq.(1.8), and $u2(t)$ is the value taken by the uncontrollable exogenous variable in period $\underline{t}$ which, by assumption, is known.

In the second step, we determine the conditional mean of the optimal state by the relation

$$x^*(t) = \hat{x}(t) + K(z(t) - C\hat{x}(t)) \qquad \text{---(1.10)}$$

where the quantity in brackets is the error in prediction based on information available upto one interval before, and K is the steady-state solution of the Kalman gain obtained by recursive application of the Kalman filter algorithm (see Appendix A). The optimal control for the next period $u^*(t+1)$ is now obtained from the estimate of the optimal state of the earlier period $x^*(t)$ and the solution procedure continues sequentially in this manner. We thereby end up with the optimal control as well as the optimal state trajectories.

1.4.3 Economic implications of the problem

In general, it has been noticed that economic situations very closely correspond to closed-loop control situations. There is a 'noisy' system which is imperfectly observed and these contaminated observations of the state are used to derive optimal policies that are intended to improve the economic situation, measured by some planning criteria. In the regulation case, this objective function can be the weighted sum of squares of differences between the predicted and desired values of certain target variables. This is in line with the quadratic preference functions popularized by Theil (1964), although many economists, including Friedman (1973), Chow (1975) and others, have argued against it. The controls are the 'instruments' as defined by Tinbergen (1956) and the model of the system can be a Tinbergen-type linear policy model too. It now seems apparent that an application of optimal control techniques will be most successful in the field of short-term policy planning, due to the fact that short-term econometric models are essentially linear (or easily capable of linearization), and this property makes them amenable to numerical solutions by recourse to linear control theory techniques. Thus, while our model was intrinsically nonlinear, we obtained its linearized approximation by using control engineering techniques.

1.5 Outline Of The Study

India may well be a uniquely suitable testing ground for refined economic regulation of the type indicated in this study. This is because the Indian economy is a heavily planned one and under such a framework the open-loop type of control actions adopted has implied a specification of the time-paths of all the policy variables right at the beginning of the planning period itself. However, if we take into consideration the fact that nearly forty percent of Indian gross domestic product, which stems from the agricultural sector, depends upon the vagaries of weather, then such a prior specification without regard to possible future events could well be sub-optimal. This has been proved repeatedly, time and again, within the Indian context if we realize that in over thirty-five years of national macroeconomic planning not once have we been able to achieve all our pre-designated targets.

It is our belief that this could have been avoided if the concept of uncertainty, both with regard to transition as well as observation errors, had been explicitly incorporated into the Indian plan framework. This study is an attempt to do precisely this by constructing an econometric control system incorporating the Kalman filter for the Indian economy.

The Kalman filter apparatus requires not only a linear model of the functioning of the economy, but also a model of the observation sub-system that converts the system state vector into published statistics, an objective function and covariance matrices of the noises existing in the system. The model constructed was nonlinear which was subsequently linearized into the LQG version by invoking the so-called 'operating point' concept. A resolution of the model was then carried out in order to isolate the loop variables and render the model recursive. The observation sub-model constructed was a novel one and fitted very well into the concept of errors in information signalling, which is currently quite prevalent within the Indian data base context. Its incorporation into the system provided a rapid convergence of the Kalman gain to a unique steady-state solution.

All the optimal control experiments which were then carried out on an *ex-post* basis were primarily designed to determine what optimal policies ought to have been adopted over the period of the original Indian Sixth Plan: (1977-82), in the light of events that transpired later on. The results prove that such counterfactual experimentation using optimal control theory in conjunction with the Kalman filter can be of invaluable assistance from the viewpoint of Indian macroeconomic planning, because they provide a basis on which the optimal values of the instruments can be predicted under alternative stochastic scenarios.

CHAPTER 2

FILTERING AND CONTROL OF STOCHASTIC SYSTEMS

2.1 Prologue

While applications of control theory to economic regulation problems may be said to have started with Tustin (1953), it was Tinbergen (1956) who was actually responsible for defining the concept of a 'policy model' in economic literature which predicted the effect of 'instruments' on 'targets'. The modern theory of linear stabilization policy stems from his fundamental proposition that if all relationships are linear, static and non-stochastic, and if all policy instruments can be freely adjusted at zero cost, then policy makers need use only as many instruments as they have targets in order to achieve the desired values for all the target variables, and any extra policy variables are redundant and may be set at arbitrary levels. This proposition provided a cornerstone for much of the subsequent development in this area.

However, it was Phillips (1957) who was the first to recognize that economic stabilization policy could be viewed as a problem in engineering feedback control. His basic contribution was to propose a set of alternative stabilization policies (proportional, integral and derivative policies) and to consider the stability properties of a dynamic multiplier -accelerator model when these policies were introduced. The three policies he considered were shown to impose different properties, both desirable as well as undesirable, on the resulting dynamic time-path of income. He showed that the application of certain types of 'intuitive' policies could generate unwanted oscillations and even create an instability in income.

All these models dealt with the problem of economic regulation, that is, trying to bring the economy to a preferred state, sometimes called the 'equilibrium' or 'stable' state. This concept, however, had no connection with the 'general economic equilibrium' of classical Walrasian economics or to the 'stability' of dynamic systems in the Lyapunov sense. Most of the examples of features of a desired state were: a low and constant rate of unemployment, prices that did not diverge too much from world prices, a zero deficit in the balance of payments, a high rate of growth of national income, amongst others.

Once a reduction in deviations from a desired state was taken as the main aim of regulatory policy, it became natural to try and formalize this goal in the form of an objective function consisting of a weighted sum of squares of these deviations. Probably the very first practical application of this concept was by Simon (1956) who developed a computationally feasible technique for solving the control problem when the economic model was linear with additive errors and a quadratic objective function was to be minimized. He invoked the Certainty Equivalence principle by means of which, under the assumptions specified above, the additive stochastic disturbances were replaced by their expected means for the purpose of determining the optimal policy. This concept was later on generalized, because of its obvious great advantages, by Theil (1957,1958) and Holt (1960,1962). It is these works, including the later ones by Theil (1964,1965) dealing with macroeconomic applications, which specifically considered the minimization of quadratic objective functions subject to the constraint of a linear econometric model in its reduced form, that can be considered as the true precursors of the application of optimal control theory to economic stabilization problems. As a result of these and several other analyses of macroeconomic policy, including Howrey (1966), Fox, Sengupta and Thorbecke (1966), Allen (1967), Chow (1967,1968), Aoki (1968), amongst others, it became apparent that the dynamic structure of the economy did not permit 'well-intentioned' or 'intuitive' policies, and more rigourous methods were needed. It is in this context that there arose the application of control and filtering theory in macroeconomics.

2.2 Control And Filter Theory In Economic Regulation

Wishart (1969) provided a schematic representation of a stochastic control system as described by Fel'dbaum (1953). This is presented in Figure 2.1 overleaf with designations fitting the regulation problem to be analyzed.

Applied to the economic regulation problem, all the various elements in it can be interpreted in the following way. The main system is the (national) economy, not including the sub-system to be described. Its state is observed by an observation sub-system, in particular, a statistical agency (in India, this is either the Central Statistical Organisation or the Reserve Bank of India or both), which produces observations $\underline{z}$, adding 'observation noise', $\underline{e2}$, in the process. These observations enter into the filtering and control subsystem, typically the government agency (in India, this is the Planning Commission) which evaluates the information and converts it into predictions $\underline{\hat{x}}$ of the system state $\underline{x}$. These predictions are then compared with the desired state $\underline{\tilde{x}}$ and in the light of the perceived deviations from it, there arises the need for regulatory policy $\underline{u}$, which is determined by the execution subsystem, that is, the executive branch of the government (in India, this is typically the Ministry of Finance). At this juncture, a specific 'noise', $\underline{e3}$, enters the model, which distorts the policy action (This could very well be in the form of inter-ministry rivalry). The actual policy followed is then shown to influence the main system, which itself is subject to 'noise', $\underline{e1}$. As the resulting observations of the main system (z) are taken into account for further regulatory action, Figure 2.1 represents a closed-loop control system in which there is an arrangement to measure the system state and to make these measurements available to the decision making subsystem for incorporation into subsequent policy. Thus, the decisions can be made not only in the light of information which is available on the past states, but also with the awareness that the current system is being imperfectly observed, and as such, any policy decisions based upon this information will need to be modified accordingly. The resulting closed-loop arrangement is a device to reduce the losses caused by the uncertainties in the system.

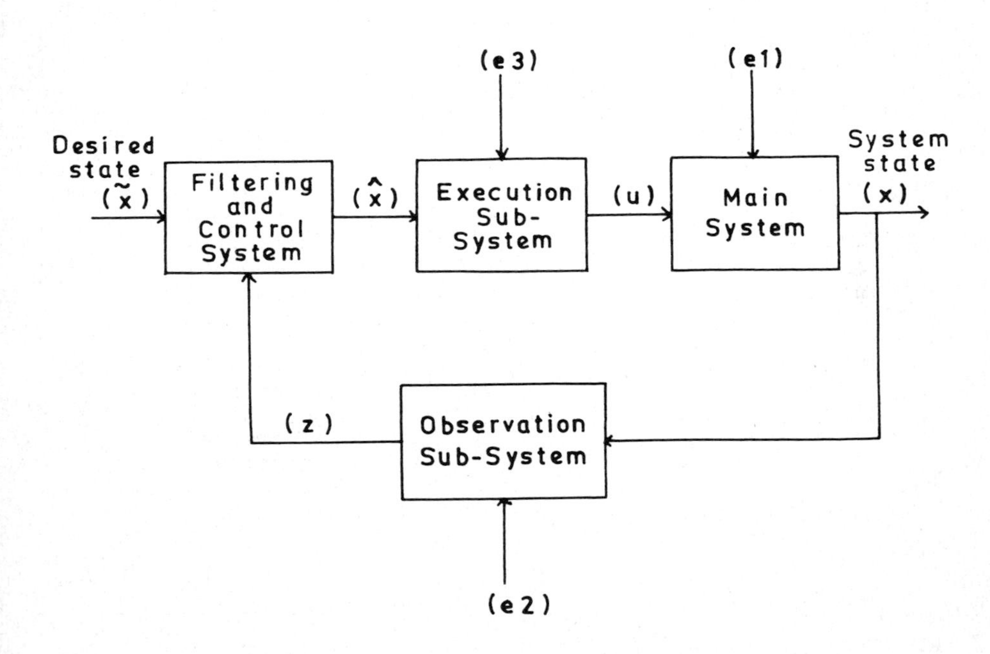

Figure 2.1
Schematic Representation Of A Stochastic Control System

The systematic approach to determine the values of the instruments for economic regulation is possible only if carried out in a number of sequential phases. Because the execution of policy takes time and information about the economic activity is available only with certain irreducible delays, forecasts about the state of economic activity are essential. Predictions are therefore made assuming a change in the instrument values which are optimally designed so as to bring about a closer allignment of the predicted state with the desired state. These changes in the policy instruments are effected also taking into account the likely developments in the exogenous variables and their possible repercussions upon the economy. On the tacit assumption that these predictions of the system state, based on the latest available information, will be incapable of perfect observation, any further changes in the future values of the policy instruments will depend upon and be indicated by the estimate of the reconstructed state.

It is seen that the overall control strategy for the determination of optimal regulatory policy comprises the following three steps:

Step 1. Obtaining the one-interval ahead conditional prediction of the system state.

Step 2. Updating the conditional mean of the system state based upon its conditional prediction.

Step 3. Determining the optimal instrument values based upon the conditional mean of the system state.

It is noted that the prediction of the future course of the economy is separated from the phase of specifying the optimal values of the instruments. In this study, the Kalman filter will be employed to update the conditional mean of the system state. The Pontryagin maximum principle will be employed to determine the optimal instrument values.

2.3 The Optimum Linear Filter

2.3.1 The Weiner filter

The publication by Weiner in 1949 of the report "Extrapolation, Interpolation and Smoothing Of Stationary Time Series" (Weiner 1949) marked the beginning of the engineering awareness of the problem of state estimation. The work of Weiner formed an important milestone because it posited the problem in the frequency domain framework, which was of immense practical use at that time. Unfortunately, because the results were expressed in the frequency domain, they could not be extended directly to nonstationary problems. Although the general formulation of the nonstationary problem was developed by the Weiner-Hopf equation, very few practical results emanated as a consequence. This was because the filter, which came to be known as the unrealizable Weiner filter, generally possessed poles in the right half s-plane. We must remember that since the solution procedure entailed using Fourier transforms, poles in the right half-plane did not indicate an unstable system, but rather a noncausal system requiring response before stimulus, which was physically unrealizable.

While Booton (1952) and Zadeh and Ragazzini (1950,1952) did make notable contributions in as far as achieving practical results using the Weiner filter were concerned, the matrix Weiner filter for the multivariable stationary estimation problem was first presented by Darlington (1958), Wong and Thomas (1961), and Davis (1963). However, this material never gained wide acceptance in engineering practice, because of the extremely severe computational problems associated with the required matrix spectrum factorization. Although Kalman and Ho (1966), Riddle and Anderson (1966), and Anderson (1967) did present computational procedures for spectrum factorization based on the solution of matrix Riccati-type equations, the utilization of the Kalman filter algorithms eventually superseded most of the uses of the matrix Weiner filter.

2.3.2 The Kalman filter

The works by Kalman (1960) and Kalman and Bucy (1961) were extensions of Weiner's classical work, where they focussed attention upon a class of linear minimum-error-variance sequential state estimation algorithms. These have come to be referred to in the literature as "Kalman-Bucy filters", or more commonly, just "Kalman filters", in recognition of their initial impetus to the theoretical developments in this area which has witnessed extensive applications in recent years. Kalman filter algorithms have been applied in numerous practical situations, including navigation, space guidance, orbit determination, and of late, economics.

While the problem of linear minimum variance sequential filtering had been solved by Weiner and others for single input-output systems, the principal contribution of Kalman was to extend the Weiner filter to nonconstant coefficient multivariable systems, with nonstationary finite-time noise records and to obtain a sequential form of the solution. In the derivation of the Kalman filter, it is assumed, as well as required that the observations be processed sequentially. Regardless of whether the estimator is sequential or nonsequential, the values of the resulting state estimates are unaltered. Moreover, the Weiner filter was expressed in frequency-domain transfer-function notation and its state-transition equation was specified by the spectral density of the signals. On the other hand, the Kalman filter was expressed in state-variable notation and its state-transition equation was specified in terms of a first-order vector differential (or difference) equation. Therefore, probably the most significant contribution of Kalman and Bucy was to put the linear minimum variance estimator into a state-variable sequential framework which was responsible for providing it with such a tremendous amount of computational feasibility.

2.3.3 Derivations of the Kalman filter

The treatment of the Kalman filter began historically with the discrete-time version of the problem, that is, discrete-time observations of a discrete dynamic system. This approach had a number of inherent theoretical advantages because for simple problems, the discrete algorithms could be manually manipulated. Moreover, this step-by-step processing of information led to the gaining of considerable insight which spurred further developments, particularly as far as obtaining certain theoretical properties of the problem were concerned.

Two basic approaches to the derivation of the discrete algorithm have been adopted in the literature. In the first case, the optimum linear minimum variance algorithms have been obtained via the maximum à posteriori (MAP) estimation approach (see Ho and Lee 1964). In the second, these have been obtained using the concept of orthogonal projections (see Deutsch 1965).

As far as the derivation of the Kalman filter for continuous processes was concerned, three basically different, but complimentary, methods have been considered in the literature. The first two derivations were approaches which applied the methods of variational calculus to the linear minimum--error-variance estimation problem. The former was a direct application of the matrix minimum principle to the problem of minimizing the variance of the estimation error in order to obtain the Kalman filter algorithm (see Athans and Tse 1967). The latter approach dealt with the transformation of the Weiner-Hopf integral equation into the sequential differential equation of the Kalman filter (see Sage 1968). The third case considered the continuous problem as the limiting case of the discrete problem, as the sampling interval was reduced to zero thereby letting the samples become dense, and then applying a limiting argument to the discrete algorithm (see Sage and Melsa 1970).

If the observation time becomes infinite, or at least long compared with the transient response of the system dynamics, and if the system parameters are time-invariant and the prior statistics are at least wide-sense stationary, then a stationary asymptotic form of the Kalman filter is obtained, which is a degenerate form of the general Kalman filter. This stationary Kalman filter is identical with the Weiner-Kolmogorov filter.

These derivations of the Kalman filter are by no means the only possible derivations. Ancillary derivations applicable to nonlinear filtering problems also exist, but shall not be dealt with here. The essential approach has been to obtain the extended nonlinear filter algorithms for continuous and discrete-time filtering by a sub-linearization of the nonlinear state transition and observation equations and subsequent application of linear filter theory (see Sage and Melsa 1971).

2.3.4 Asymptotic behaviour of the Kalman filter

Extensive research has been carried out to determine the asymptotic characteristics of the Kalman filter. The proofs of these properties, wherever they exist, make use of stability theorems broadly classified as the second method of Lyapunov (see Schultz and Melsa 1967). There are two related questions concerning the asymptotic behaviour of the Kalman filter. First, the conditions that must be satisfied for the existence of a steady-state solution of the variance equation. The answer to this problem is useful because it tells us when a solution to the stationary, or Weiner, filter estimation problem exists, and indicates conditions necessary for the estimation process to remain meaningful as the observation period becomes long.

The second, and by far the more important, question deals with the stability of the filter algorithm. If the filter is stable, then the effect of any small error in the prior statistics for the initial state variance will become negligible as time evolves. The practical value of a computationally stable filter algorithm cannot be over-emphasized, since we seldom have perfect knowledge of the initial distribution of the state vector. While an understanding of controllability and observability are essential for any study on the stability of the Kalman filter, these concepts will not be elaborated here because we have discussed them extensively in Appendix C. Suffice to say, these notions were introduced by Kalman (1960) to describe a set of conditions which were related to the existence of solutions for certain linear control problems. Exhaustive treatments of the subject are available in Kalman, Ho and Narendra (1962), Kalman (1963,1964), Sage (1968), amongst others.

As mentioned above, the stability of the Kalman filter is of overriding concern. If the algorithm is stable, then any error in the initial error variance will decay to zero as time tends to infinity, and optimal performance will result. The rate of convergence, which is often quite rapid provided all the preliminary conditions of the Kalman-Bucy theorem (see Appendix A on the statement of these conditions) are satisfied, can be estimated by the use of the second method of Lyapunov (see Kalman 1963).

2.3.5 Divergence in the Kalman filter

A lot of ongoing research is being carried out to examine the sensitivity of the filter algorithm to errors in the statistical moments, as well as the state transition and observation equations, of the model. These have established that implementation of the Kalman filter algorithm can run into practical problems in the form of a display of divergence or instability even though, theoretically, the algorithm is computationally stable.

In some applications, it has been found that the actual estimation errors have greatly exceeded the values which would have been theoretically predicted by the error variance matrix. In fact, it has been noticed that the actual error has become unbounded, even though the error variance in the Kalman filter algorithm has become vanishingly small. This phenomenon, referred to as divergence, or data saturation, can seriously vitiate the usefulness of the Kalman filter. The possibility of such unstable behaviour was first suggested by Kalman (1960), and later noted by Knoll and Edelstein (1965), amongst others, in the application of Kalman filter algorithms to space navigation and orbit determination.

While the fact that the Kalman filter diverges inspite of the satisfaction of the initial preconditions of the Kalman-Bucy theorem may seem in direct conflict with the stability properties of the filter, this, however, is not the case, since divergence is caused by errors which were not accounted for in the analysis. In particular, two of the major causes of divergence have been established to be the inaccuracies in the modelling process used to determine the state transition or observation equation and the failure of linearization. Errors in the statistical modelling of noise variances have also been found to lead to divergence. Another source has been traced to round-off errors, inherent in any digital implementation of the filter algorithm, which has often caused the error-variance matrix to lose its positive-definiteness or symmetry.

All these have basically implied a divergence because the Kalman gain in the Kalman filter algorithm has approached zero too rapidly. Hence the estimate becomes decoupled from the observation sequence and is not affected by the growing observation error. Several modifications to the Kalman filter have been postulated to prevent divergence (see Schlee _et al_ 1967). The basic concept of these approaches has been to _limit_ the decrease in the gain in order to avoid the decoupling of the measurements. The procedures can be placed into three broad classes. One of the most obvious has been to limit the Kalman gain so that it does not become smaller than some fixed pre-assigned value. In other words, after, say, _M_ observations, we simply hold the gain constant for all future observations. In this case, one may argue that the filter is no longer optimum, and this would be true. However, a few may be willing to pay this price, in order to preempt divergence.

Another approach to eliminating divergence in the above problem has been to model the unknown input as a state of some dynamic system which is driven by white noise (see Friedland 1969). Although this procedure has often worked well, it has the unfortunate feature of raising the order of the problem. Since the order of the entire filter and required variance algorithm for an _nth_ order model is $n(n+3)/2$, even a small increase in _n_ can _be_ bad. For example, if _n_ was originally equal to 20, and we merely add one new state variable to model an unknown input, then the effective order of the filter and variance algorithm increases from 230 to 252. Even though this order can be dismissed in these days of computational largesse, one should remember that this procedure can be an effective one provided that the nature of the unknown is well enough understood to be effectively modelled.

The third approach is a more theoretical one. Divergence occurs because too little weight is being placed on new data, or conversely, too much weight is being placed on old data; hence the alternative term, data saturation. When the input noise is small as compared with the measurement noise, then the error variance, as computed by the Kalman filter algorithm, and hence the gain, tends to become very small as time evolves. In the case where there is no input noise, the gain approaches zero asymptotically. This result, which is to be expected since each sample contains (on the average) much more unwanted measurement noise than information about the input noise, tends to decouple the filter from the observation sequence and hence causes the divergence. As these problems of divergence are caused by the fact that the input noise is small as compared to the measurement noise, the solution has been to artificially increase the input noise variance.

Another approach which has been proposed for the control of divergence involves computing the square root of the error covariance matrix. This procedure, initially developed by

Potter (1967), and later extended by Bellantoni and Dodge(1967) and Andrews (1968), is particularly useful when round-off errors are the cause of divergence. By treating the square root of the error variance rather than the variance directly, the number of significant figures required in the computer is cut in half. For the square-root formulation, it is assumed that the input noise is zero; and this has proved to be no major restriction since, as we just examined, divergence is most often encountered when the input noise is small. Dyer and McReynolds (1969) have, however, developed an algorithm for nonzero input noise, as well.

Jazwinski (1968) presented another approach in which the estimate is based on only the most previous set of M observations, where M is some fixed number large enough to allow the system to be observable. Unfortunately, the method requires instrumentation of both the filtering and prediction algorithms, and hence becomes impractical if the filtering solution diverges, which is the exact reason for the method in the first place.

An application of adaptive filtering algorithms with an idea of eliminating divergence was initially carried out by Smith (1967), Jazwinski (1969), amongst others. Sage and Melsa (1971) believe that this is one of the most fruitful uses of the adaptive filtering algorithms since, when divergence occurs, the errors may become large, which provides us with more information for adaptation.

2.4 Optimal Control Theory

2.4.1 Time-optimal control problems

We now revert back to the post-Weiner but pre-Kalman era that existed during the fifties. This was a period when extensive research in the area of time-optimal control problems was carried out by mathematicians, both, in the United States as well as in the Soviet Union. By 1957, Bellman, Gamkrelidze, Krasovskii and LaSalle had developed the basic theory of minimum-time problems and had presented results concerning the existence, uniqueness and general properties of the time-optimal control problem. The recognition of the fact that control problems were essentially problems in the calculus of variations soon followed. However, it was noticed that classical variational theory could not readily handle the 'hard' constraints usually imposed in a control problem. This difficulty led Pontryagin, a Soviet mathematician, to first conjecture his famous maximum principle and then, together with Boltyanskii and Gamkrelidze, to provide a proof for it. The maximum principle was first announced at the International Congress of Mathematicians held at Edinburgh in 1958.

2.4.2 The maximum principle

The maximum principle (Pontryagin 1961, Pontryagin *et al* 1962) is in a certain sense a generalization of the Euler-Lagrange multiplier theorem in which the multipliers are allowed to vary as functions of time. The maximum principle yields a certain coupled system of ordinary differential equations, involving both the extremum control functions and the time-varying multipliers, which can often be used to determine the optimum control functions. In certain control problems, the maximum principle provides direct information about the optimum number of switching times. The maximum principle is well suited to handle problems involving a wide range of different types of constraints, and it often provides certain limited information about the nature of the optimum control functions even when the given problem is too difficult to be solved completely in closed form.

While the maximum principle may be viewed as an outgrowth of the Hamiltonian approach to variational problems, the method of dynamic programming and the optimality principle, which was developed by Bellman of the United States (Bellman 1957) may be viewed as an outgrowth of the Hamilton-Jacobi approach to variational problems. This approach basically uses the recurrence relationship or the algorithm of partial differential equations which must, in any case, be solved numerically to effect the solution of the control problem.

Both these methods (the maximum principle and the optimality principle) basically followed a different theoretical approach, which were quite independent of each other. However, there are a few who contend that the approach of Pontryagin might have had its precursors in the work by Valentine (1937) who dealt with variational analysis and inequalities, and the one by Hestenes (1950) who wrote a report for the RAND Corporation based on an approach quite similar to the one adopted by Pontryagin.

At about this time, control engineers became painfully aware that there existed the need to develop a major approach to the treatment of both the analysis and synthesis of linear control systems. The commonly used block-diagram approach which involved, in essence, the determination of the transfer characteristics of the system components and the overall transfer characteristics was found insufficient to solve the exacting problems frequently encountered in scientific and engineering applications of control theory, because the designer had to select that controller which would satisfy both static and dynamic performance specifications, and this very often complicated the issue.

Thus, a second approach was developed and this was based primarily upon the characterization of a system by a number of simple first-order differential (or difference) equations describing the state variables of the system, with the initial conditions given by the state-transition equations. The state variables were like the generalized coordinates in classical mechanics. This approach, which has come to be referred to as the state-space approach, presently forms the core of modern control theory.

Although the introduction of these techniques into the control field is relatively new, the basic concepts underlying these methods had long been used in classical dynamics, quantum mechanics, thermodynamics and other fields. The idea of state as a basic concept in the representation of systems was initially introduced by Turing in 1936. Later, the concept was employed by Shannon in his basic work on information theory. The application of the state-space concept in the field of control theory was initiated by the Soviet scientists, Aizerman, Fel'dbaum, Letov, Lur'e, amongst others. However, it was only in the early sixties, with the introduction of the concept of state and related techniques into the optimum design of linear as well as nonlinear systems by Bellman (1961) and the application of state-space techniques to automatic control by Kalman (1960) did the state-space approach come to play a really prominent role in advancing the state of the art of the rapidly growing field of control theory.

Simultaneous with the development of these revolutionary techniques in control theory was an almost continuous breakthrough in computer technology, which provided the control engineer with a vast array of expanded computational facilities and simulation aids at his disposal. Thus, the ready availability of special- and general-purpose computers greatly reduced the need for closed-form solutions and the demand that controllers amount to simple network compensation.

Modern control theory can therefore be viewed as the confluence of three diverse streams: the calculus of variations, the state-space approach and the development of the computer.

2.4.3 Generalizations of the maximum principle

Because most of the early applications of control theory to engineering involved continuous-time systems, the theoretical foundations of the maximum principle were couched in a continuous-time framework, where the dynamics of the system to be controlled were represented by differential equations. The maximum principle, which provided a set of necessary conditions for the solution of the general continuous-time optimal control problem, has found wide acceptance and applications to engineering problems and, of late, to problems in economic analysis as well.

The proof of the maximum principle was provided by Pontryagin <u>et al</u> (1962), although a more heuristic proof was obtained by Athans and Falb (1966). Later on, Dorfman (1969) provided a thorough statement of the problem with more of an economic content. In contrast to the earlier bulk of control theory literature (in particular, its applications to engineering problems) which formulated the problem in continuous-time, the more recent applications (particularly to economic problems) have found discrete-time to be more appropriate. In view of this, several derivations of the discrete-time maximum principle have appeared over the past years (see Katz 1962, Athans 1968). However, the resulting necessary conditions for optimality were essentially similar, - the differences in the results being ones of generality and the assumptions placed on the optimization problem. Jordan and Polak (1964) derived a set of necessary conditions using geometric arguments similar to those used by Pontryagin for continuous-time systems. A set of necessary conditions of greater generality also using a geometric proof was derived by Halkin (1964,1966). Both these proofs, however, required a convexity condition on the constraint set of the problem. Holtzman (1966), as well as Kleinman and Athans (1966), later generalized the results of Halkin by replacing the convexity condition with a less stringent requirement of 'directional convexity'.

Unfortunately, all these geometric derivations of the discrete maximum principle limited the generality of the necessary conditions by placing a very severe restriction on the function f(i,t) in eq.(1.1) by requiring that the matrix

$$\partial f/\partial x$$

be nonsingular for all time. In the case of a linear system described by eq.(1.7), this implied that the matrix A be nonsingular. In view of the fact that most econometric models contain variables with various lag and difference structures, this nonsingularity restraint was often violated by the state space representation of most economic models. It was therefore desirable to derive a maximum principle without placing this restriction on the function f.

Pearson and Sridhar (1966), as well as Rosen (1967), showed that a dynamic optimal control problem could be expressed and treated as a larger static convex programming problem. The Kuhn-Tucker theorems (Kuhn and Tucker 1951) were applied to this static problem directly, and the necessary conditions that resulted were, in turn, translated back into dynamic form yielding the desired maximum principle, which was then invoked to solve the discrete-time optimal control problem.

The maximum principle was developed basically for solving deterministic control problems. Random disturbances were assumed to be absent in the treatment of Pontryagin, although there have been a few attempts to modify the maximum principle for application to stochastic control problems (see Kleinman 1969).

One very important point that needs to be highlighted is that the maximum principle conditions are, in general, not sufficient, nor do they necessarily yield a unique solution or a global maximum. However, it has been shown that the conditions are necessary and sufficient if the Hamiltonian is linear in the control variables (Rozonoer 1966) or if the maximized Hamiltonian is a concave function of the state variables (Mangasarian 1966).

2.4.4 Economic applications of the maximum principle

It was only in the second half of the sixties did an interest in applying control and system theory to economic problems emerge, especially in the field of modelling and control of macroeconomic systems. By around this time, the initial interest in the Walrasian general competitive analysis - so apparent in the early fifties - had been replaced by models in which incomplete information, such as search behaviour, formation of expectations, trading out of equilibrium, began to predominate. Apart from these theoretical areas of research, economists also attempted to develop empirical theories to explain the sequence of decisions of individual economic agents that generated the time-paths of all the various economic variables. Pontryagin's maximum principle was found to be of inestimable value in tackling problems in all these areas, as well as in the theory of growth that dealt with highly aggregative economic models (see Hadley and Kemp 1971, Intriligator 1971). Simultaneously, economists began to show great interest in describing or modelling macroeconomic systems with more disaggregated variables or incorporating imperfect information into the system. Economists, by now, had come to realize that it was patently unrealistic to be exclusively concerned with only the equilibrium states of the economy, and that it would be more fruitful to consider economic systems in disequilibrium that work with imperfect information and without the perfect coordination of all economic agents and activities. Out of the latter developments, an increasing interest was evinced in the performance of models out of equilibrium, - a concern long familiar to control and system engineers.

In short, as economists began to gather more experience with models of varying calibres of complexity, and began to realize that digital computers could make simulation a more powerful and exciting way of exploring the dynamic behaviour of complex economic systems, they started to believe in the absolute necessity of a 'close encounter' between control theory and economic analysis. The culmination of these beliefs was the IFAC/IFORS Conference on Dynamic Modelling and Control of National Economies which was held in 1973, exactly twenty years after Tustin's first abortive attempt to introduce control and system theory for the purpose of economic analysis.

2.5 Control Systems And Quantitative Economic Policy

In quite a brief span of time we have gained considerable expertise regarding the hurdles and pitfalls of short-term forecasting with macroeconometric models. The earlier optimism about the capability of these models to improve forecasting ability has, of late, given way to a considerable degree of 'enthusiastic scepticism' about the possibility of forecasters to churn out sufficiently accurate short-term forecasts that can act as guideposts to macroeconomic planning. This is due to the failure of these models to live up to their inflated expectations as a result of which the inevitable, and often counterproductive, reaction has tended to distract attention from the unquestionable gains.

With hindsight, it is possible to say that we have been asking too much from our economic models. The partial nature of economic theory as an explanation of human behaviour and the inferential character of our knowledge regarding the complex circuitry of the economy, coupled to the highly stochastic nature of macroeconomic events, seem to suggest that the predictive power of invariably all econometric models is inherently low. No matter how disaggregated the models, there are certain limits to the forecasting accuracy of the equations, and thus the possibility of eventually reducing the forecasting error to some acceptable and irreducible minimum is ruled out. Rather than frame economic policies such that their success or failure depends on the accuracy of economic forecasting, it is better that policies reflect the inevitability of forecasting errors and be framed accordingly. It is in this context that stochastic optimal control theory has been able to make some contribution (see Westcott *et al* 1979).

The literature relating optimal control theory to problems of applied macroeconomic policy making can be separated into two distinct, though not equally formed, areas. The first and the most widespread has been concerned with the exploration of the dynamic properties of an econometric model in terms of its multipliers and the implied effectiveness of alternative policy constellations. The second, far less popular but equally, if not possibly more, important, topic has been concerned with the problem of designing robust economic policies for the real world. The difference between these two approaches has been mainly in the extent to which concern has been shown regarding the possible structure of the actual system of the real world being synthesized in contrast to the particular model, as used by the policy planner.

All too often, alternative 'optimal' policies are extracted from a deterministic model merely by changing the weighting pattern of the target variables in the welfare function. Thus, little or no research has been directed towards assessing the sensitivity of these apparently optimal policies to model-misspecification, which can include the possibility of uncertainty contaminating both the state transition and the state observation equations of the control system. Yet it is the examination of these types of misspecifications that is most relevant to the second kind of control study; namely the generation of robust economic policies for the real world. For instance, it is quite possible that a deterministic model might indicate that fiscal measures are optimal, in terms of welfare loss, for achieving a particular target. However, a sub-optimal monetary policy, extracted from the same model, could be less sensitive to model misspecification and thus a safer bet in terms of robustness.

Thus, optimal solutions based in isolation only on the dynamic properties of an econometric model are largely meaningless if the system is badly defined and the model is prone to misspecification. As Salmon and Young (1979) have demonstrated, if the optimal strategies are to be seriously considered as an aid to policy formulation, then they must take cognizance of uncertainty and possible model misspecification. In particular, control strategy designs should be developed that can, to some extent, be 'robust' in the face of uncertainty.

The practical success of feedback control in engineering and physical systems can be largely attributed to the interface between theory and practice, and whenever this aspect has not been given its due share of attention the practical exploitation of new theoretical concepts has often tended to be slow. Indeed, many economists are unaware that, even in the case of engineering control problems where systems are relatively well defined, optimal control experiments have had, in many cases, only limited successes in terms of proven practical control system design. This has resulted due to the 'gap' which developed over the late sixties between the highly theoretical developments in optimal control theory and the practical implementation of control strategies in real world applications.

In this sense, the prime concern of econometricians should be with the distinction between uncertainty in the stochastic case and robustness to misspecification (see Leamer 1978). In the stochastic case, the extra information inbuilt into the model through probability specifications will be invaluable, but this information per se is unlikely to yield a fully satisfactory control system design if model misspecification still persists. This latter consequence of uncertainty is equally relevant to the economic policy planner.

It is worth noting that in the LQG case, the optimal control strategy is determined with no direct reference to the full implications of the positions assumed by the closed-loop eigen values of the system. This is one of the main reasons as to why unsatisfactory control strategies result from the LQG case because due care to analyze the dynamic properties of the controlled system is not taken. It is seen that state variable feedback (SVF) control is not only important because it happens to coincide with the LQG policy, but also because of the considerable degree of control it exerts over the closed-loop pole locations (see Rao 1984b). It is therefore not very surprising that the control literature on SVF has not limited itself to optimal interpretations only, but has also been concerned with nonoptimal approaches to control system design, including the decoupling of multivariable systems (Gilbert 1969), modal control (Porter and Crossley 1972) and model matching (Wang and Desoer 1972).

SVF control has important implications for the design of control systems which should not be ignored. In particular, the great degree of control over the closed-loop dynamics imparted by SVF, as well as the relative ease with which multiple objectives may be incorporated through the cost functional, has important ramifications which should always be taken into account during the construction of the model.

2.6 Conclusions

Therefore, what seems to be required is a general approach and design which incorporates the best aspects of both the conventional econometric modelling and modern control systems methodologies in a unified framework. While such a satisfactory unified approach to multivariable control is not yet fully available at this point, there are existing design procedures, such as the one adopted in this study, which seem to offer some hope that the solution to the problem of designing optimal as well as robust macroeconomic policies should soon become available in the near future.

CHAPTER 3

AN ANNUAL NONLINEAR MODEL OF THE INDIAN ECONOMY

3.1 Introduction

The model presented in this chapter is based, to a certain extent, on the common elements of earlier models constructed for the Indian economy. However, it contains many refreshingly new ideas about the way the Indian economic system operates. It will be used for counterfactual experimentation over the period of the original Indian Sixth Plan (1977-82), and is a revised variant of the earlier prototype version estimated by Roy and Rao (1980), which was successfully used in the late seventies by the Indian Planning Commission for short-term macroeconomic forecasting during their Annual Plan exercises. Its main merit is simplicity and a consequent robustness to model-misspecification. It incorporates the minimum of relationships that an economic analyst would wish to see in a short-term policy model and thereby lends itself easily to designing optimal control strategies.

While econometric modelling is expected to be as objective as possible, many econometricians (see Klein 1982) feel that an artistic element plays an important role in that science. As all data come from life experiments, it is not surprising that experience in observing economic life provides useful *à priori* information for econometric analysis.

Part of this experience is knowing something about the institutional structure of an economy. Institutionally-based equations along with behaviourally-based equations help in making up the mathematical structure of an economy. From an Indian viewpoint, some of the more prominent institutions are the

--- control of production and consumption
--- regulation of savings and investment
--- scheme of foreign trade and exchange control
--- mechanism of price formation
--- banking system
--- government operations.

The estimation of a policy-oriented model for the Indian economy then becomes a question of possessing the adequate facts about the functioning of the Indian economy *vis-à-vis* these institutions and having the ability to translate this knowledge into the design of an econometric structure. It is only then that we can proceed with the statistical estimation of the coefficients of the model.

While many of the facts of institutional knowledge may

come from 'casual empiricism', a majority of them however stem from systematic statistical observation. Thus, a knowledge of the various parameters of the Indian economy has come from the studies conducted by the Central Statistical Organisation(CSO), Perspective Planning Division (PPD) and Reserve Bank of India (RBI), all of which have had a highly commendable history of a systematic approach towards empirical investigations. Rise in the savings rates and the capital-output ratios, constancy in the wage share of national income and the rate of capital consumption, fall in the income velocity of money and the ratio of currency to deposits, and many other empirical facts of the Indian economy have been probed into and extensively researched. This systematic measurement of key variables has been of tremendous help to investigators involved in the specification and *à priori* estimation of macroeconometric models of the Indian economy.

3.2 Macroeconometric Research In India: A Synthesis

After more than three decades of intense experimentation, quantitative descriptions of the Indian economy - in terms of macroeconometric modelling - can be deemed to have matured, both in terms of statistical sophistication as well as actual application. They have most certainly emerged from the dormant embryonic phase of being purely academic set-piece exercises and, of late, more and more government and private organizations have begun to take a deeper look at these efforts in an attempt to try and understand their underlying implications.

Models of recent vintage like those constructed by Pandit (1980), Krishnamurty (1983), Bhattacharya (1984) and Pani (1984) have been extremely innovative in terms of incorporating special features or specific aspects of the Indian economy in their modelling efforts. These models have been somewhat removed from the earlier ones by Narasimham (1956), Choudhary (1963), Marwah (1964) and Mammen (1967) which in some respects were akin to models of Western economies based on the precepts of Tinbergen and Klein (see Desai 1973).

There are many underlying features that distinguish the current trend in modelling research for the Indian economy. We shall highlight a few of them that have been quite ubiquitous.

(1) Most of these models have used annual data and can thus be considered as annual models. The resulting lack of degrees of freedom has seriously hampered the possibility of carrying out any rigourous investigation on the impact of structural changes in the Indian economy.

(2) A common feature has been the characterization of the functioning of the economy in the form of a two-sector model - the agricultural sector whose output generation at the aggregate level is, by and large, exogenous; and the rest of the economy where growth is more-or-less determined by demand. The interaction between the two sectors has been postulated in several ways: (i) by the propagation of price changes from one sector to the other, either one way or both ways, (ii) by the transmission of demand for the other sector's product due to income generation and additional output in one of the sectors, and (iii) the common impact of monetary impulses on the prices

of both sectors.

(3) All the models have underscored the importance of the operations of the government sector although the perceived impact on the two sectors has varied. Government investment has been assumed to be crucial not only because it is a generator of additional demand for capital (manufactured) goods, but also because it augments infrastructural facilities and adds to the productive capacity of the economy. The crowding-out, if any, of private investment by this activity has, however, not been studied in any great detail.

(4) The important linkages between government fiscal operations and the generation of monetary impulses has been the focus of many models. While a few of them have made a thorough study of the role of the budgetary constraint on sources and uses of funds, and of the budget deficit in the creation of money; many others have postulated these links in a reduced-form fashion, recognizing the fact that the use of monetary policy is to a great extent constrained by the actions of the public sector with regard to their draft on commercial banks' resources as well as by way of resort to borrowing from the central monetary authority.

(5) The general formulation of the price mechanism has been fairly standard for all the models. Sectoral prices and the overall price level have been based on the standard classical framework of the Fisherian identity, implying thereby that they are affected by excess liquidity in the economy. However, relative prices have been stated to act as equilibrators of demand and supply in the agricultural sector, while cost mark-up relations have been the general rule for industry and services.

(6) Labour market descriptions have been conspicuous by their absence from most of the models. The trade-off between unemployment and inflation has not been tested for the Indian economy because of a complete lack of data on unemployment rates, a few benchmark estimates notwithstanding. However, wage-price dynamics have been postulated on the assumption that organized labour tries to maintain a real wage rate in the long run.

(7) Descriptions of banking operations have appeared in considerable detail basically to determine the level of money supply in the economy. The links between banking activity, credit and output generation have, however, rarely been formulated. Two basic reasons can be cited for this lacuna. For one thing, while a considerable amount of rigourous theory exists on the demand for money or on the asset-holding behaviour of the household, the theory on the demand for credit for production purposes or the demand for funds for investment has neither been well postulated nor has it been empirically well tested even for developed economies. This has rendered the adoption or modification of existing theory to suit Indian conditions extremely difficult, especially if we consider that these conditions include a dual credit system with an unorganized market for funds. Added to this is the system of an administered interest rate structure in the banking sector which preempts a proper analysis of the role of the interest rate as an allocative instrument of available funds. Thus, the

working of the monetary and financial system in its relation to economic growth has not been a well-explored area and, as a result, its treatment in the modelling of the Indian economy has tended to be weak.

(8) Yet another important field which has been underspecified in many models is the foreign sector. The links between the domestic economy and this sector have been only superficially articulated for the overriding reason that the role of trade and the balance of payments, which hitherto have accounted for only a very small proportion of Indian gross national product, has not yet been fully comprehended. That exports and adverse terms of trade affect aggregate demand and domestic prices is well recognized. However, of late, it has been noticed that large gaps in demand and domestic supply have arisen due to the presence of foreign exchange constraints, regulated imports and related trade restrictions, - all of which have had a negative impact on economic growth. In view of the considerable changes in the foreign exchange markets,international financial flows and growing interdependencies of national economies over the period of the seventies and eighties, it is of utmost concern that Indian macroeconometric models highlight this sector in a more disaggregated manner so as to enable the formulation of suitable policy actions.

3.3 Model Of The Indian Economy: Preliminaries

3.3.1 Salient features

The mathematical representation of the Indian economy described here is in the spirit of the econometric model constructed in the Perspective Planning Division of the Indian Planning Commission (Roy and Rao 1980). The earlier version was designed to assist the Government of India in their "rolling-plan" exercises, which was an effort to integrate their five-yearly plans through a set of mutually consistent annual plans. The choice of economic variables was, however, largely influenced by the econometric models developed by Rao (1984) and Pani (1984). There are, nevertheless, many significant differences from these models. The determination of industrial output as a function of man-days lost *inter alia* is practically unique in the history of Indian econometric modelling. The simultaneous specification of the tax-rate as a generator of inflation as well as a liquidator of budgetary deficits (and thereby potential inflation) is another case in point. The scheme of influences is, by and large, much simpler than its earlier counterparts. This is because the presentation here explicitly proceeds along a control-theoretic structure in an effort to provide robust policy guidelines from a short-term forecasting standpoint.

Although the accuracy of a mathematical model (of an economic system) depends a great deal on its dimension, it would hardly be of any use to policy planners if the model were to incorporate a few marginal effects which would drastically increase the dimension of the model without any significant improvement in its accuracy. Motivated by the fact that the complexity and cost of controlling any model depends upon its dimension, several methods have been devised to reduce the dimension of a macroeconomic system. One obvious method is to describe the system by a cruder mathematical model of a lower dimension while retaining only its 'significant modes'. This technique serves to reduce the dimension without seriously affecting the accuracy of the model (Aoki 1968). There are instances, however, in which it is not possible to achieve the reduction of the dimension in this manner, because by overlooking certain 'parasitic effects' we may end up with an inherently unstable system (Kwakernaak and Sivan 1972).

The approach currently in vogue to designing systems of lower order dimensions is to use mathematical models which are as accurate as possible, without hesitating to incorporate marginal effects that may or may not have significance. However, we limit the dimension of the controller (i.e., the vector of control variables) to some fixed number m, less than n, where n is the dimension of the state vector. The technique is to select the smallest m that still produces a satisfactory control system (Sims and Melsa 1970). This is the general approach adopted in the study.

In the model, the stochastic nature of the economy is explicitly kept in view by the presence of exogenous random disturbance processes. These random disturbances affecting the endogenous variables are distinguished from possible errors in observation which are related to the observation mechanism. The disturbance processes involved are thus meant to represent not only the approximations in the modelling of economic activities, but also to account for innumerable other incidental, and primarily stochastic, causes which are bound to affect the transitions as well as the observations of the economy.

3.3.2 The mechanism of the model

The object of this study is to analyze the factors determining output, demand, prices, money-supply and government fiscal operations in the Indian economy. For this purpose a disaggregated macroeconometric model of the economy is formulated. The model is specified so as to capture the main characteristics of the economic structure as well as the major linkages between the important sectors.

The incorporation of supply constraints, particularly agricultural and industrial output, and the transmission of the effect of fiscal operations onto money-supply, which in turn influences price behaviour, are two of the major elements of the model. The supply-side is disaggregated into four sectors: (i) agriculture, (ii) mining, manufacturing, construction and utilities, (iii) transport and communications, and (iv) other services. Individual components of aggregate demand are then analyzed. These include private consumption expenditure, sectoral investment and imports. The 'two-gap' formulation provides the critical link between aggregate investment, domestic savings and capital inflows. Money supply is endogenously determined by the link to the government deficit *via* its mediation through the stock of high-powered money. This deficit itself is endogenized through a built-in government receipt and expenditure system. Sectoral prices are determined by the supply of money circulating in the economy, apart from demand and supply considerations, with the tax-rate being incorporated in order to test whether the ensuing loop that results between the tax-rate, price level, nominal income, budget deficit and money supply is an 'overheating' feedback or a 'self-liquidating' feedforward. The link between the monetary sector and the real sector is postulated according to the Quantity Theory with money (in terms of excess liquidity) affecting the price level (national income deflator), given the initial supply constraints.

The model does not, however, study the interaction between the availability of funds and sectoral production. Similarly, the ramifications of the labour market and its operations are missing from the model. The effect of population growth has also been completely ignored. We have assumed that this variable has a neutral effect on the structural parameters of the model.

3.3.3 List of variables and units of measurement

The model comprises ten sectoral blocks, these being: (i) the production sector, (ii) the consumption sector, (iii) the savings sector, (iv) the investment sector, (v) the trade sector, (vi) the price sector, (vii) the income sector, (viii) the monetary sector, (ix) the government sector, and (x) the miscellaneous sector.

The model contains 40 endogenous variables which are determined by a set of as many equations. Of these, 27 can be technically termed as reaction or behavioural equations in view of the fact that they describe the adaptive behaviour of economic entities to changing causes. The remaining 13 are definitional equations or identities. Of the reaction equations, 20 are linear but the remaining 7 which are not, as well as the 5 identities which link value, quantity, price and ratio variables, introduce considerable nonlinearities into the model. These are later removed through linearization by invoking the 'operating point' concept. The model also comprises 13 predetermined variables, of which 8 are exogenous variables and the remaining 5 are instruments.

In the following list of variables, the economic phenomena are distinguished as endogenous, i.e., the variables governed by the mathematical model of the economy; exogenous, i.e., phenomena that affect the endogenous variables but the generation of which is not explained within the model; and control variables, i.e., those which are at the discretion of the policy makers, who by their use attempt to influence the economic activity characterized by the endogenous variables. In the literature, control variables are synonymously referred to as decision variables, policy variables or instruments. The exogenous and control variables are often grouped together under the label 'data' or 'predetermined variables'. The past (or lagged) values of the endogenous variables are also included under this grouping. All the flow variables refer to yearly estimates and they are all measured at annual rates. The stock variables are measured, not as averages over each year, but as existing point estimates at the end of each financial year which, for the Indian economy, runs from April to March.

Table 3.1

List Of Variables And Units Of Measurement

Endogenous Variables

No.	Symbol	Phenomena	Units
1.	QF	Output of foodgrains	index no.1970-71=100.0
2.	QNF	Output of non-foodgrains	-----do-----
3.	QA	Agricultural output	-----do-----
4.	YAR	NDP at factor cost: Agriculture	Rs. crores at 1970-71 prices
5.	YMR	NDP at factor cost: Mining, manufacturing, construction and utilities	-----do-----
6.	YTR	NDP at factor cost: Transport and communications	-----do-----
7.	YSR	NDP at factor cost: Services	-----do-----
8.	YNAR	NDP at factor cost: Non-agriculture	-----do-----
9.	YNFR	NDP at factor cost	-----do-----
10.	YGFR	GDP at factor cost	-----do-----
11.	YPDR	Personal disposable income	-----do-----
12.	CPR	Private consumption expenditure	-----do-----
13.	s	Savings rate	percentage points
14.	SNR	Net domestic savings	Rs. crores at 1970-71 prices
15.	INR	Net domestic capital formation (NDCF)	-----do-----
16.	IAR	NDCF: Agriculture	-----do-----
17.	ITR	NDCF: Transport and communications	-----do-----
18.	KR	Capital stock	-----do-----
19.	DR	Capital consumption	-----do-----
20.	FR	Current account deficit	-----do-----
21.	MR	Imports of goods	-----do-----
22.	WPF	Wholesale price:Foodgrains	index no.1970-71=1.000
23.	WPNF	Wholesale price: Non-foodgrains	-----do-----
24.	WP	Wholesale price: All commodities	-----do-----
25.	P	National income deflator	-----do-----
26.	YNFN	NDP at factor cost	Rs. crores at current prices
27.	YNMN	NDP at market prices	-----do-----
28.	CC	Stock of currency	-----do-----
29.	SCBR	Scheduled commercial banks' reserves	-----do-----
30.	TBR	Total bank reserves	-----do-----
31.	M1	Money supply (narrow money)	-----do-----
32.	MS	Money stock	Rs. crores at 1970-71 prices
33.	v	Income velocity of money	units

No.	Symbol	Phenomena	Units
34.	BD	Budget deficit	Rs. crores at current prices
35.	NDE	Non-developmental expenditure	-----do-----
36.	TR	Tax revenue	-----do-----
37.	NTR	Non-tax revenue	-----do-----
38.	CR	Capital receipts	-----do-----
39.	AF	Area under foodgrains	Million hectares
40.	ANF	Area under non-foodgrains	-----do-----

Exogenous Variables

No.	Symbol	Phenomena	Units
1.	R	Rainfall index	Normal=100.0
2.	t	Time index	Units
3.	MDL	Man-days lost	Million man-hours
4.	XR	Exports of goods (inclusive of net export of services)	Rs. crores at 1970-71 prices
5.	PM	Unit value index of imports	index no.1970-71=1.000
6.	DFEA	Change in foreign exchange assets	Rs. crores at current prices
7.	DBFA	Net domestic borrowing and aid	-----do-----
8.	DEF	Defence expenditure	-----do-----

Control Variables

No.	Symbol	Phenomena	Units
1.	IGAR	NDCF: Agriculture (Public sector)	Rs. crores at 1970-71 prices
2.	IGTR	NDCF: Transport and communications (Public sector)	-----do-----
3.	GDE	Government developmental expenditure	Rs. crores at current prices
4.	T	Tax rate	Ratio
5.	BR	Bank rate	Percentage

Note: 'Rs. crores' stands for ten million Indian rupees (Rs. 10,000,000). Taking into consideration the current exchange rate, which is approximately U.S. $1=Rs. 12.50, one crore rupees would be roughly equal to 0.8 million dollars.

3.3.4 The estimation procedure

The empirical estimates of the equations are obtained by using annual data over the 21-year period, 1961-62 to 1981-82. Data for the following years could not be included in the analysis as relevant information in respect of several macro-economic aggregates, especially sectoral investments, was not available. Thus, an important limitation was the restricted span of the time series data involved in the estimation of the coefficients of the model. This was further compounded by the fact that for the period under consideration the Indian economy underwent two major structural changes. The first was in 1966 when the Indian rupee was devalued considerably by almost fifty percent. The second was in 1969 when fourteen leading commercial banks were nationalized. Both these brought about substantial changes in the trade and monetary structures of the economy. Under the circumstances, it was felt prudent to confine the sample spaces for estimating the equations of the trade and monetary blocks to the periods 1966-82 and 1969-82, respectively. However, the equations of all the remaining eight blocks were estimated using the full-period data, i.e., 1961-82.

The basic data were drawn from the National Accounts Statistics of the CSO, the Reports on Currency and Finance of the RBI, the Annual Plan exercises of the PPD and the Economic Surveys of the Ministry of Finance, Government of India. All the data used in the estimation of the model are provided in the Annexure.

The format of most of the equations was essentially linear, although there was no hesitation to include nonlinear specifications as and when they were thought to be necessary.

We have resorted to the Ordinary Least Squares (OLS) method for estimating the parameter values of the model. While OLS yields estimators that are biased and inconsistent, these estimators exhibit insensitivity to specification error which is a highly desirable property from the viewpoint of designing robust economic policies. Moreover, Monte Carlo studies have indicated that OLS generally retains the Gauss-Markov property of minimum variance. More specifically, it has been shown (Smith 1973) that OLS is appropriate if the matrix of coefficients of endogenous variables is sparse as was the case with our model. Such sparseness frequently arises when the size of the model increases, since the total number of endogenous variables usually goes up much faster than the number of such variables appearing in a typical equation of the model (see Intriligator 1978).

For each reaction equation, the backward-elimination, forward-selection and stepwise regression procedures were employed and the alternative 'best' formulations, thus isolated, were individually scanned. In assessing the adequacy of the final selected regression equation, apart from the usual economic criteria like the appropriateness of the sign and relative magnitude of the coefficients, certain statistical criteria were strictly observed. In no case was a t-value of less than 1.5 tolerated, no matter what theoretical considerations indicated. We felt that it was mandatory for theory to be sufficiently backed by facts before any regressor could be included in a behavioural equation. Apart from this, all equations had to possess a high coefficient of determination ($\bar{R}^2$) exceeding 0.75; and a low coefficient of variation (C.V.) not exceeding 15%. Throughout the entire estimation process, there was absolutely no compromise whatsoever on any of these three scores (Only in the case of eq.(3.28) did we permit the C.V. to exceed 15%, and that too only because we were measuring the dependent variable in terms of its first-difference).

In order to simplify the notation, certain conventions have been adopted for the symbols which are employed to represent diverse phenomena. The time index t, which designates a year t, is understood to be implicit in each symbol, i.e., a symbol P for a variable in year t would be more completely represented by P(t). The lagged variable P(t-k) is simply denoted by P(-k). In all the equations that follow, the figures in parentheses indicate the t-statistic of the coefficient beneath which they appear. Figures below them are the elasticities. Also shown are the $\bar{R}^2$, F-statistic, D.W. statistic (or the corresponding h-statistic, whenever the independent variables include lagged dependent variables), the coefficient of first-order autocorrelation (ρ), the standard error of the estimate (S.E.E.), the mean of the dependent variable (Mean) and the coefficient of variation (C.V.).

3.4 Estimates Of Model Structure

The model contains, as indicated earlier, ten sectoral blocks. These are: (i) the production sector, (ii) the consumption sector, (iii) the savings sector, (iv) the investment sector, (v) the trade sector, (vi) the price sector, (vii) the income sector, (viii) the monetary sector, (ix) the government sector and (x) the miscellaneous sector.

3.4.1 The production sector

Production is divided into agricultural and non-agricultural production. The former comprises output of the agricultural sub-sector, i.e., foodgrains and non-foodgrains, while the latter comprises output of the non-agricultural sub-sectors which are mining, manufacturing, construction and utilities; transport and communications; and other services. The combined output of all these sub-sectors provides us with an estimate of net domestic product at factor cost.

3.4.1(i) Output of foodgrains (QF):

The production of foodgrains is specified as a function of area under foodgrains (AF) and rainfall (R). We have also included a trend factor (t) to account for the rise in productivity due to an increasing use of fertilizers, pesticides, the dissemination of agricultural know-how and the continuing impact of the 'Green Revolution'. The estimated equation is provided on the following page.

What needs to be noted is that the time index is very significant and understandably so if we take into consideration the indisputable evidence (see annexure) that although the levels of foodgrains production have been continually fluctuating under the impact of adverse weather conditions, they have done so around an ever increasing trend.

We have the following equation:

$$QF = -168.285 + 1.4650\ AF + 0.8031\ R + 1.9210\ t \qquad ---(3.1)$$
$$(-2.62) \quad (2.07) \quad (2.83) \quad (4.52)$$
$$0.32 \quad 0.23 \quad 0.58$$

$\bar{R}^2 = 0.9477$ D.W. $= 2.613$ S.E.E. $= 4.6799$
$F = 121.71$ $\rho = -.314$ Mean $= 106.976$
C.V. $= 4.3747$

Thus it is also seen that the rainfall index (denoting pre-sowing weather conditions, in terms of precipitation received) as well as the area under foodgrains are both extremely important determinants of this all-important endogenous variable. In this context, it is a matter of great concern that the acreage planted under foodgrains seems to have reached a plateau during the late seventies and the early eighties.

3.4.1(ii) Output of non-foodgrains (QNF):

The production of non-foodgrains is also specified as a function of area under non-foodgrains (ANF), rainfall (R) and a trend factor (t). The estimated equation was:

$$QNF = -300.412 + 11.5406\ ANF + 0.3454\ R + 1.2239\ t \qquad ---(3.2)$$
$$(-5.56) \quad (6.19) \quad (2.74) \quad (5.89)$$
$$0.51 \quad 0.12 \quad 0.46$$

$\bar{R}^2 = 0.9662$ D.W. $= 1.872$ S.E.E. $= 3.0033$
$F = 191.56$ $\rho = .048$ Mean $= 110.481$
C.V. $= 2.7184$

It needs to be noted that both rainfall and the trend factor seem to have a greater influence on the production of foodgrains rather than on non-foodgrains. It is also seen that the acreage planted under non-foodgrains is a powerful stimulant of this sector's output. On the whole, the explanatory power of both these production functions, taking into consideration the extremely volatile nature of the Indian agricultural sector, is quite satisfactory.

3.4.1(iii) Agricultural output (QA):

Agricultural output is then obtained as a weighted function of foodgrains and non-foodgrains production. It was given by

$$QA = 0.6812\ QF + 0.3188\ QNF \qquad ---(3.3)$$

These weights were provided by the Ministry of Agriculture on the basis of the performance (in terms of output) of these sub-sectors in the base year, i.e., 1970-71. They reflect their relative importance in as far as agricultural production in the Indian economy is concerned.

We can thus obtain the total weighted contribution of both rainfall and the trend factor to overall agricultural output.

3.4.1(iv) NDP at factor cost - Agriculture (YAR):

The net domestic product at factor cost originating from the agricultural sector (value added) is estimated as a function of the gross output of the agricultural sector, assuming that the input costs are proportional to the value of output. This procedure also implies that net output in allied activities like forestry, fishing and animal husbandry is directly related to agricultural production. The estimated equation was:

$$\underset{}{YAR} = \underset{(8.95)}{2466.38} + \underset{\substack{(51.07)\\1.00}}{128.361}\ QA \qquad ---(3.4)$$

$\bar{R}^2 = 0.9924$ D.W. = 0.858 S.E.E. = 212.876
F = 2608.25 ρ = .462 Mean = 16340.4
C.V. = 1.3028

This is one of the three equations in the model wherein we have statistically estimated an endogenous variable which could have been ideally expressed by a linear or nonlinear identity. We have preferred the adoption of statistical constructs basically due to the non-availability of data in respect of some components constituting the sector under study.

3.4.1(v) NDP at factor cost - Mining, manufacturing, construction and utilities (YMR):

The net domestic product at factor cost originating from this sector is specified as a function of availability of raw materials (QNF) and the performance of the transport sector (YTR), within a partial-adjustment framework. It was also decided to introduce the number of man-days lost (MDL) as another independent variable affecting industrial production, because its impact on this sector has been especially severe in the last few years. The estimated equation was:

$$YMR = \underset{(-0.53)}{-312.689} + \underset{\substack{(3.91)\\0.66}}{0.6836}\ YMR(-1) + \underset{\substack{(2.56)\\0.16}}{18.5186}\ QNF$$

$$+ \underset{\substack{(1.54)\\0.25}}{0.7642}\ YTR - \underset{\substack{(-3.06)\\-0.12}}{22.8235}\ MDL \qquad ---(3.5)$$

$\bar{R}^2 = 0.9887$ h = 0.024 S.E.E. = 162.196
F = 308.27 ρ = -.114 Mean = 8140.80
C.V. = 1.9924

The speed of adjustment of actual output to desired output is low (0.32). The time constant of the lag is nearly 2.63 years. It is very encouraging to note that almost 99 percent of the variation in output is accounted for by this specification. It is a well known fact that the output of the manufacturing sector can be severely hamstrung by a poor performance of the transport sector, and this has been quite evident in the Indian economy wherein the latter sector has quite persistently continued to create bottlenecks in as far as the delivery of raw materials and the off-take of finished goods have been concerned. Attempts at using either capital stock or investment estimates as regressors were considered as failures, inspite of the high quality nature of these data. This tacitly leads us to believe that the many Indian economists who have been patently advocating the use of a heavy industry strategy _à la_ Mahalanobis (Mahalanobis 1953,1955) ever since the Indian Second Five Year Plan (1955-60) have erred along the way, and what the Indian manufacturing sector currently most needs is not further investment; but better capacity utilization, an efficient transport network and a far less strained labour-management relationship (thereby resulting in fewer man-days lost). The coefficient of MDL is very significant, and it is noted that the loss of a million man-days can decrease production by as much as $ 18 million in the short-run, while the long-run impact can be as high as $ 58 million. A quick glance at the annexure reveals that since 1966 more than 355 million man-days have been lost. Without considering any indirect or long-run effects whatsoever via the cumulative interim multipliers of the model, this information alone suffices to indicate that, over the last 20 years, the Indian economy has been deprived of more than $ 6 billion worth of output due to industrial unrest.

3.4.1(vi) NDP at factor cost - Transport and communications (YTR):

The net output of this sector is considered as a function of the investment in this sector (ITR), in a partial-adjustment framework. The estimated equation was:

$$YTR = \underset{\substack{(-1.08)}}{-77.9406} + \underset{\substack{(8.71)\\0.92}}{0.9820}\ YTR(-1) + \underset{\substack{(1.92)\\0.09}}{0.3307}\ ITR \qquad \text{---(3.6)}$$

$\bar{R}^2$ = 0.9901	h = 0.404	S.E.E. = 50.3066
F = 600.03	ρ = .099	Mean = 2172.38
		C.V. = 2.3157

The speed of adjustment of actual output to desired output is very low (0.02). The time constant of the lag is nearly 55.1 years. The long-run investment-income multiplier is seen to be almost 18.4. The popularly postulated specification of treating the net output of this sector as a hybrid function of the capital stock in this sector and the income from the agricultural and the manufacturing sectors was tested, but failed to be satisfactorily vindicated. A quick glance at the annexure will reveal the reason for this failure. It is noticed that of all the sectors in the economy, the transport and communications sector has been the only one (apart from the ubiquitous 'other services' sector) to register a continuous increase in output right from 1961 onwards. This has been regardless of the performance of the agricultural and manufacturing sectors, which have exhibited ups and downs. The inability of these oscillations to have transmitted themselves to the transport sector serves to indicate that the latter is not prone to be affected by the performances of the primary and secondary sectors of the economy. Rather, the transport sector is an investment-hungry one whose output is almost entirely a function of the capacity created in it. This clearly serves to underscore the critical importance of creating a sufficiently adequate transport infrastructure in the Indian economy, by means of a high level of investment, to meet the demands of all the other sectors.

3.4.1(vii) NDP at factor cost - Services (YSR):

The value added from this residual sector is the sum of the incomes from the trade sector, the finance and real estate sector, and the community and personal services sector. The largest contributor amongst these is the trade sector, comprising hotels and restaurants, whose output critically hinges on the performance of the primary sector. The finance and the real estate sector, comprising banking and insurance amongst others, is dependent on the transport and communications sector for its efficient performance. The community and personal services sector, comprising public administration and other services, is heavily influenced by the real expenditure of the government, both on the current as well as on the capital account. As such, the total value added in the overall services sector is regarded as a function of the income generated in the agricultural sector (YAR), income from the transport sector (YTR) and the real developmental expenditure of the government (GDE/WP), within a partial-adjustment framework. The estimated equation was:

$$\begin{aligned} YSR = & \underset{(-1.60)}{-507.617} + \underset{\substack{(7.36)\\ 0.67}}{0.7138}\, YSR(-1) + \underset{\substack{(3.37)\\ 0.08}}{0.0909}\, YAR \\ & + \underset{\substack{(2.09)\\ 0.20}}{0.9368}\, YTR + \underset{\substack{(2.00)\\ 0.06}}{0.1987}\, (GDE/WP) \qquad ---(3.7) \end{aligned}$$

$\bar{R}^2$ = 0.9981 h = -0.317 S.E.E. = 122.915
F = 2622.94 ρ = -.160 Mean = 9650.52
C.V. = 1.2737

The speed of adjustment of actual output to desired output is low (0.29). The time constant of the lag is estimated at 2.97 years. As expected, the demand for services is seen to depend upon the tempo of the agricultural and transport sectors. The role of real government developmental expenditure (both current as well as capital accounts) in promoting the output of the services sector is established, albeit weakly. Its long-run impact multiplier is seen to be about 0.70.

3.4.1(viii) NDP at factor cost - Non-agriculture (YNAR):

The output emanating from the non-agricultural sector is specified to be the sum of the output levels of the manufacturing sector (YMR), transport sector (YTR) and the services sector (YSR). It was given by

$$YNAR = YMR + YTR + YSR \qquad ---(3.8)$$

This endogenous variable has an important role to play in

the model as it helps to determine the savings rate which is based upon the distribution of income between the agricultural and non-agricultural sectors.

3.4.1(ix) NDP at factor cost (YNFR):

The net domestic product at factor cost for the entire economy is specified as the sum of the output levels of the agricultural sector (YAR) and the non-agricultural sector(YNAR). It was given by

$$YNFR = YAR + YNAR \qquad ---(3.9)$$

This variable is the single most important one in the entire model and plays a pivotal role in as far as the determination of most of the other variables of the system is concerned. Its precise estimation is therefore one of central importance from the viewpoint of the structure of the model. It forms the supply side of the system and is heavily influenced by the level of rainfall and the extent of man-days lost. The simulation of these exogenous variables can thus provide a key to analyzing the growth patterns of NDP under alternative scenarios. Policy specification on the basis of such guidepost solutions can be of vital importance if the consistency planning exercises of the Indian Planning Commission are to have any semblance with reality.

3.4.1(x) GDP at factor cost (YGFR):

The gross domestic product at factor cost for the entire economy is specified as the sum of the net domestic product at factor cost (YNFR) and depreciation (DR). It was given by

$$YGFR = YNFR + DR \qquad ---(3.10)$$

The role of this variable, although considerably muted, is basically one of obtaining the income velocity of money which helps to determine the aggregate price level thereby establishing the critical link between the real and monetary sectors of the economy.

3.4.2 The consumption sector

This sector is concerned initially with the estimation of personal disposable income which then forms the pivot variable in the determination of private consumption expenditure.

3.4.2(i) Personal disposable income (YPDR):

Personal disposable income is specified as a function of net domestic product at factor cost (YNFR).The estimated equation was:

$$YPDR = \underset{\substack{(538.81)\\0.97}}{0.9608}\, YNFR \qquad ---(3.11)$$

$\bar{R}^2$ = 0.9999	D.W. = 0.730	S.E.E. = 282.343
F = 290318.21	ρ = .615	Mean = 33324.9
		C.V. = 0.8472

While we could have expressed this endogenous variable by means of a linear identity, the non-availability of data with respect to some important components of transfer payments forced us to adopt this statistical construct.

The equation indicates that the sum of direct taxes paid by households plus corporation tax plus savings of the private corporate sector less net factor income from abroad less national debt interest (all at constant prices) amounted to about 3.9 percent of net domestic product at factor cost over the sample period.

3.4.2(ii) Private consumption expenditure (CPR):

Private final consumption expenditure is specified as a function of personal disposable income (YPDR), in a partial-adjustment framework. The estimated equation was:

$$\underset{(5.00)}{CPR = 2830.86} + \underset{\substack{(2.14)\\0.16}}{0.1633}\ CPR(-1) + \underset{\substack{(11.44)\\0.84}}{0.6607}\ YPDR \qquad \text{---(3.12)}$$

$\bar{R}^2 = 0.9941$ $\quad h = 2.338$ $\quad S.E.E. = 422.783$

$F = 1595.08$ $\quad \rho = .442$ $\quad Mean = 29525.6$

$C.V. = 1.4319$

The speed of adjustment of actual consumption to desired consumption is quite high (0.84). The time constant of the lag is about 0.55 years. The short- and long-run marginal propensities to consume are estimated at 0.66 and 0.79, respectively. The long-run elasticity of private consumption expenditure with respect to personal disposable income is 1.0. Specifications involving an income distribution variable, YNAR/YNFR, i.e., the ratio of non-agricultural income to total income, were tried but failed to provide significant results.

3.4.3 The savings sector

In this sector, we initially estimate the savings rate which is then used to obtain a measure of net domestic savings at constant prices.

3.4.3(i) Savings rate (s):

The net domestic savings rate in India has shown a secular rise in the last three decades. Particularly notable has been the dramatic increase during the latter half of the seventies. It rose more than three-fold from an average of 6.5 percent in the First Plan period (1951-56) to about 20 percent in 1982. The uptrend in the savings rate has, however, been characterized by considerable year-to-year fluctuations. Many hypotheses have been advanced to explain savings behaviour in India. The principal ones relate to: (i) The role of shifts in income (between the agricultural and non-agricultural sectors) on the savings rate, the propensity to save being lower in the agricultural sector (Raj 1962, Chakravarty 1973); (ii) Variations in the savings rate on account of lags in the response of consumption to changes in income, the lags being due to the transitory nature of the income changes (Raj 1962, Rao 1980); (iii) The impact of inflation on the savings rate via its effect on income distribution (Mujumdar et al 1980); and

(iv) The rise in the savings rate due to a general drift in an income distribution over time in favour of the upper income groups (Rao 1980).

As each of these hypotheses by themselves offer only a partial explanation for the observed savings behaviour, it was found desirable to construct an equation to explain the behaviour of the savings rate by incorporating all the hypotheses stated above. A significant feature of this equation is that it helps to estimate the difference in the propensities to save out of agricultural and non-agricultural incomes using aggregate time-series data. This is important in view of the fact that independent panel estimates on savings from the agricultural and non-agricultural sectors are not available for the Indian economy.

We postulate, initially, two separate savings functions - one for the agricultural sector and the other for the non-agricultural sector - within the framework of the current income hypotheses. Real savings of each sector is a linear function of real income originating from that sector. The two savings functions are given by

$$S(a) = A(a) + B(a)\ YAR \qquad \text{---(3.13i)}$$
$$S(na) = A(na) + B(na)\ YNAR \qquad \text{---(3.13ii)}$$

where B(a) and B(na) represent the marginal propensities to save of the agricultural and non-agricultural sectors, respectively. Adding eqs.(3.13i) and (3.13ii) together, we obtain total savings at constant prices as

$$SNR = A(a) + A(na) + B(a)\ YAR + B(na)\ YNAR \qquad \text{---(3.13iii)}$$

Making a substitution for YAR in terms of (YNFR-YNAR), since YNFR=YAR+YNAR, and dividing both sides of the equation by YNFR, we can rewrite eq.(3.13iii) as

$$s = B(a) + \frac{A(a)+A(na)}{YNFR} + \frac{B(na)-B(a)}{YNFR/YNAR} \qquad \text{---(3.13iv)}$$

where s stands for the savings rate, i.e., SNR/YNFR. Treating the term (A(a)+A(na)) as a constant, say C, we can rewrite eq.(3.13iv) as

$$s = B(a) + C(1/YNFR) + (B(na)-B(a))(YNAR/YNFR) \qquad \text{---(3.13v)}$$

The sign of the coefficient C will be negative if the sum of the intercept terms A(a) and A(na) are negative. Negative intercepts in the savings functions, eqs.(3.13i) and (3.13ii), imply that positive savings would emerge only after a certain level of income, and that the average savings rate would increase with income. There is substantial evidence from Indian survey data to show that savings are indeed negative in the lower ranges of income distribution. Referring to eq.(3.13v), we see that the constant term in the equation is the marginal propensity to save of the agricultural sector, while the coefficient of the share of non-agricultural output in national product, i.e., YNAR/YNFR, is the difference between the marginal propensities to save of the non-agricultural and agricultural sectors. If the marginal propensity to save of the agricultural sector is lower than that of the non-agricultural sector, then shifts in income distribution in favour of (against) agriculture would imply a decline (rise) in the savings rate. As there is considerable evidence from time-series and cross-section studies in India to show that the marginal propensity to save in the agricultural sector is lower than that in the non-agricultural sector, the coefficient of the share of non-agricultural output in total national output is expected to be positive and the magnitude of the coefficient should reflect the difference between the two propensities.

The second hypothesis relates to lags in the response of consumption to changes in income due to the transitory character of these income changes, and the consequent variation in savings. In this context, particular emphasis is laid on sharp changes in income arising as a result of bumper harvests or failures in agricultural production due to the vagaries of weather. A simple test of this 'normal income' hypothesis would be to include the growth rate of income (g) measured by the rate of change of NDP, as an explanatory variable in the determination of the savings rate. A positive and significant coefficient for the growth rate of income would provide evidence in favour of the normal income hypothesis, because the result implies variations in the savings rate behaviour due to lags in the response of consumption to changes in income.

The third hypothesis relates to the impact of inflation on the savings rate. This has aroused considerable interest in India, particularly because of the high rates of inflation witnessed in the first half of the seventies. There is no clear answer to the question as to whether inflation promotes or discourages savings, as there are opposite effects of inflation on savings. While it could be argued that inflation, through an income redistribution in favour of those with a high propensity to save, can increase the savings rate; it is equally true to assume that inflation can reduce the savings rate, if consumers resist cuts into real consumption, particularly in a country like India with low per capita consumption levels, and try to maintain real consumption intact. Which of the two opposite forces would dominate would depend upon a number of interrelated factors, such as the extent of inflation, the composition of consumption (durables and non-durables) and savings (physical and financial assets), expectations and the interest rate, amongst others. It is ultimately an empirical question and the evidence hitherto has been ambiguous. To test for the impact of inflation on the savings rate, we introduce the inflation rate (p) measured by the rate of change of the wholesale price index, as an explanatory variable in the savings rate function. A significant coefficient for the inflation rate would provide evidence in favour of this hypothesis, while the sign would be indicative of the prevailing force.

The last hypothesis suggests that there has been a secular rise in the household and domestic savings rates. A part of this increase could be attributed to slowly changing factors such as the gradual drift towards an income distribution in favour of higher income groups, spread of banking and credit institutions, etc. To test for the impact of these factors, we introduce the time trend (t) as an explanatory variable in the function.

Our savings rate equation, incorporating all these four hypotheses, was therefore given by the following linear specification:

$$s = a_0 + a_1(1/YNFR) + a_2(YNAR/YNFR) + a_3 g + a_4 p + a_5 t \quad \text{---(3.13vi)}$$

where it is assumed that a_0, a_2, a_3 and $a_5 > 0$ and $a_1 < 0$, there being no similar restrictions on a_4.

The final accepted form of the savings rate equation was:

$$\underset{(7.19)}{s = 0.1026} - \underset{\substack{(-3.94)\\ -0.66}}{4232.12\ (1/YNFR)} + \underset{\substack{(1.78)\\ 0.30}}{0.3155\ (YNAR/YNFR)} \quad \text{---(3.13)}$$

$\bar{R}^2 = 0.8472$ D.W. $= 1.489$ S.E.E. $= 0.0155$
F $= 56.44$ $\rho = .235$ Mean $= 0.1459$
C.V. $= 10.6533$

It needs to be mentioned that the alternative variants incorporating the second, third and fourth hypotheses failed to provide satisfactory results, and hence had to be rejected in favour of the first one. The growth rate of income invariably entered with a negative coefficient, while the rate of inflation as well as the time trend were insignificant throughout.

It is noticed that a_1 is indeed negative, bearing out the hypothesis that positive savings emerge only after a certain level of income. The marginal propensity to save of the agricultural sector (a_0) is about 0.10, while that of the non-agricultural sector (a_0+a_2) is more than four times higher at about 0.42. The equation explains almost 85 percent of the variation in the savings rate which is quite satisfactory considering its highly erratic nature.

3.4.3(ii) Net domestic savings (SNR):

The estimates of net domestic savings at constant prices were obtained by multiplying the savings rate (s) by net domestic product at factor cost (YNFR). It was given by

$$SNR = s \cdot YNFR \quad \text{---(3.14)}$$

The role of this variable is basically one of obtaining estimates of aggregate investment through the 'two-gap' model. It also helps in the determination of capital receipts which is a component of the budget deficit.

3.4.4 The investment sector

The equations in this sector are estimated at two levels. At the national level we initially obtain aggregate net domestic capital formation which is used to determine capital stock from which depreciation estimates are then extracted. At the sectoral level, we estimate investment functions to explain net domestic capital formation (both in the public and private sectors) in the primary and transport sectors.

3.4.4(i) Net domestic capital formation (INR):

The estimates of net domestic capital formation are obtained by invoking the 'two-gap' formulation, probably the most widely used approach for forecasting investment supply in developing countries. Foreign trade is brought into the model via the identity that capital inflows (i.e., the difference between imports and exports), add to investible resources, so that the savings-investment balance is given by

$$INR = SNR + FR \qquad ---(3.15)$$

where FR is the amount of real capital inflows. This equation is called the savings constraint, or 'gap', in development planning literature.

3.4.4(ii) Capital stock (KR):

The capital-stock generating forward recursion equation was given by

$$KR = KR(-1) + INR(-1) \qquad ---(3.16)$$

where we made the standard assumption, which all development planners are prone to make, that net investment is translated into capital stock only after a one-period lag.

3.4.4(iii) Capital consumption (DR):

The consumption of capital (depreciation) is linked up to the quantum of real capital stock in the economy (KR). It was assumed that the fraction of capital stock depreciated each period was a constant, and that the depreciation rate was the same on the average as well as on the margin. The estimated equation was:

$$DR = \underset{\substack{(101.41)\\ 1.08}}{0.0213}\ KR \qquad ---(3.17)$$

$\bar{R}^2$ = 0.9981	D.W. = 0.568	S.E.E. = 106.854
F = 10283.28	ρ = .707	Mean = 2279.86
		C.V. = 4.6869

The elasticity of capital consumption with respect to capital stock is very high (1.08), and the rate of depreciation obtained (about 2.1 percent of capital stock or about 29 percent of gross investment) is sufficiently indicative of the state of current Indian technology.

3.4.4(iv) Net domestic capital formation - Agriculture (IAR):

Net investment (inclusive of public and private sector investment) in this sector is specified to be a function of the total output of this sector (YAR), as well as net public sector investment in this sector (IGAR). It is assumed that government investment, including that for providing infrastructural facilities, is important in influencing the tempo of private sector investment, and thereby overall investment. Public sector investment is taken as exogenously given and is used as an instrument. The estimated equation was:

$$IAR = \underset{(-2.27)}{-545.055} + \underset{\substack{(2.00)\\0.25}}{0.0420}\, YAR + \underset{\substack{(5.85)\\0.74}}{2.1759}\, IGAR \qquad \text{---(3.18)}$$

$\bar{R}^2$ = 0.9009   D.W. = 1.737   S.E.E. = 128.644
F = 91.95 ρ = .112   Mean = 986.905
C.V. = 13.0351

The elasticity of total investment in agriculture with respect to agricultural output is quite low (0.25). However, it is much higher with respect to public sector investment (0.74). It is noticed that a unit increase in the value of the control variable would serve to swell up total investment by about 2.2 units. This implies that the private sector is coaxed into investing an additional 1.2 units for every unit invested by the public sector. This absence of crowding-out and the practically one-to-one matching of net investment by the public and private sectors augers well for agricultural activity and makes this instrument a feasible one

3.4.4(v) Net domestic capital formation - Transport and communications (ITR)

As before, and for the same reasons, net investment in this sector (inclusive of public and private sector investment) is specified to be a function of the total output of this sector (YTR), as well as net public sector investment in this sector (IGTR). Public sector investment is assumed to be given exogenously and enters as a control variable. The estimated equation was:

$$ITR = \underset{(2.76)}{97.0316} + \underset{\substack{(4.14)\\.29}}{0.0749}\, YTR + \underset{\substack{(11.18)\\0.77}}{0.9032}\, IGTR \qquad \text{---(3.19)}$$

$\bar{R}^2$ = 0.9368   D.W. = 1.628   S.E.E. = 37.9564
F = 149.32 ρ = .170   Mean = 686.762
C.V. = 5.5296

The results, as far as the elasticities of the variables are concerned, are qualitatively quite similar to the ones observed for the previous sector. The elasticity of total investment in transport and communications with respect to the performance of this sector is quite low (0.29). It is much higher with respect to public sector investment (0.77). However, as far as the coefficient of the controller is concerned, the difference between the two sectors is striking. It is noticed that a unit step-up in the value of the control variable would serve to increase total sectoral investment by only 0.9 units. This implies that not only is public sector investment incapable of coaxing additional private sector investment in this sector (as it did in the agricultural sector), it actually serves to crowd out the latter to the extent of 0.1 units for every extra unit invested by the former. This existence of crowding-out severely limits the viability of the instrument.

3.4.5 The trade sector

This sector is basically concerned with explaining the real current account deficit, as well as the quantum of imports of goods. The quantum of exports has been treated as exogenous, as per the standard procedure adopted in development planning literature. It needs to be noted that the imports equation is highly aggregative and is only expected to be broadly indicative in nature.

3.4.5(i) Current account deficit (FR):

The real current account deficit is defined in this study as the difference between the quantum of imports (MR) and the quantum of exports (XR). While the former comprises the volume of imports of goods alone; the latter is a catch-all and is the sum of the exports of goods, the net exports of services, the net factor income from abroad and other current transfers, all at constant prices. The identity was given by:

$$FR = MR - XR \qquad ---(3.20)$$

This variable is an important one because it is the link in the dynamic interaction provided by the feedback which runs from imports to the current account deficit to aggregate investment and back to imports. It also clues us to an idea of the extent of net real capital inflow and the drawdown of foreign exchange reserves envisaged under alternative scenarios which is an important indicator having a large bearing on the feasibility of the plan framework being adopted.

3.4.5(ii) Imports of goods (MR):

The volume of imports of goods, at the aggregative level, is basically expected to be influenced by the level of domestic activity and the relative price of imports. As such, it is expressed as a function of the sum of real consumption and investment (CPR+INR), which is taken as a proxy for domestic activity, and the ratio of the unit value of imports to the wholesale price (PM/WP), which is taken to be the relative price of imports, within a partial-adjustment framework. The estimated equation was:

$$\begin{aligned} MR = & \underset{(-1.69)}{-800.656} + \underset{\substack{(3.89)\\0.57}}{0.7742}\, MR(-1) + \underset{\substack{(3.16)\\0.52}}{0.0550}\,(CPR + INR) \\ & - \underset{\substack{(-1.68)\\-0.21}}{576.443}\,(PM/WP) \qquad ---(3.21) \end{aligned}$$

$\bar{R}^2 = 0.8241$ $\quad h = 0.073$ $\quad S.E.E. = 289.491$

$F = 24.42$ $\quad \rho = -.084$ $\quad Mean = 2262.06$

$C.V. = 12.7976$

The speed of adjustment of actual imports to desired imports is quite low (0.23). The time constant of the lag is

3.91 years. The elasticity of imports with respect to domestic activity is quite high at 0.52, while that with respect to relative import price is low at -0.21, although the latter does have a powerful negative influence on the quantum of imports. The long-run coefficient of the domestic activity variable is 0.24, while that of relative import price is -2552.89. The latter indicates that a one percent rise in the relative import price would result in a decline of approximately Rs. 25 crores ($ 20 million) in the quantum of imports over the long run. The price level of imports is treated as exogenous to the system.

3.4.6 The price sector

This sector is one of the most important ones in the model as it is concerned with the determination of prices which form the critical link between the real and nominal blocks of the macroeconometric system. Therefore, from the viewpoint of the inflation generating tendencies inherent in most Indian Five-Year plans it becomes the focussing point to determine the efficacy of alternative strategies. The estimation is carried out at two distinct levels. One, at the sectoral level wherein we estimate the price levels of foodgrains and non-foodgrains, separately. Two, at the aggregate level wherein we estimate the wholesale price level as well as the national income deflator. The scheme adopted will follow a causal-path recursive framework. The price level of food, which is the prime mover of all prices in the Indian economy, will be initially determined in a partial-adjustment set-up. The price level of non-foodgrains will then be obtained using food prices as a determinant. The wholesale price level of all commodities will thereafter be specified as a function of foodgrains and non-foodgrains prices *inter alia*. The wholesale price index will then be used (albeit *indirectly*) to help determine the national income deflator within a quantity-theory framework.

3.4.6(i) Wholesale price - Foodgrains (WPF):

The price level of food is determined within the context of a supply-demand mechanism. On the supply side it is expected to be affected by the availability of foodgrains, both current as well as lagged. On the demand side it is expected to be heavily influenced by the level of real consumption, because food *per se* absorbs almost two-thirds of private final consumption expenditure at constant prices. As such, the wholesale price level of foodgrains is specified to be a function of the sum of current as well as lagged levels of foodgrains output ($QF+QF_{-1}$) and the level of real consumption (CPR), within a partial-adjustment framework. The estimated equation is provided on the following page.

What needs to be noted is that the elasticity of the price level with respect to the combined supply of foodgrains is quite high at -0.66, while that with respect to real consumption is more than unity. The results indicate that an increase in the combined supply of foodgrains by one percentage point (as measured by index numbers, 1970-71=100.0) would serve to decrease the price level of food by almost one percentage point in the short run. They also indicate that increasing real consumption by Rs. 100 crores ($ 80 million) would have roughly the opposite effect to the extent of increasing the price level by the same amount.

We have the following equation:

$$WPF = \underset{(-1.22)}{-0.3937} + \underset{\substack{(3.35)\\0.60}}{0.6339}\ WPF(-1) - \underset{\substack{(-2.64)\\-0.66}}{0.0097}\ (QF + QF(-1))$$

$$+ \underset{\substack{(3.33)\\1.02}}{0.000098}\ CPR \qquad \text{---(3.22)}$$

$\bar{R}^2 = 0.9351$ $h = 0.356$ S.E.E. = 0.1411

$F = 92.19$ $\rho = .128$ Mean = 1.2988

C.V. = 10.8625

It is seen that the speed of adjustment of actual prices to desired prices is not very high(0.37). The time constant of the lag is 2.19 years.

3.4.6(ii) Wholesale price - Non-foodgrains (WPNF):

The price level of non-foodgrains is also determined within the context of a supply-demand framework. On the supply side it is assumed to be affected by the availability of non-foodgrains, both current as well as lagged. It is also assumed to be influenced by the output of the manufacturing sector, which represents a proxy for the demand for agricultural raw materials; as well as food prices which acts as the prime mover of all other prices in the economy. As such, the wholesale price index of non-foodgrains is specified to be a function of the sum of current as well as lagged levels of non-foodgrains output ($QNF+QNF_{-1}$), the income of the manufacturing sector(YMR) and the wholesale price index of foodgrains (WPF). The estimated equation was:

$$WPNF = \underset{(0.81)}{0.6100} - \underset{\substack{(-1.69)\\-0.48}}{0.0102}\ (QNF + QNF(-1))$$

$$+ \underset{\substack{(3.23)\\0.80}}{0.000250}\ YMR + \underset{\substack{(5.16)\\0.70}}{0.7266}\ WPF \qquad \text{---(3.23)}$$

$\bar{R}^2 = 0.9468$ D.W. = 2.920 S.E.E. = 0.1033

$F = 66.23$ $\rho = -.514$ Mean = 1.6031

C.V. = 6.4458

The elasticity of the prices of non-foodgrains with respect to its production (both current as well as lagged) is found to be -0.48. The results indicate that an increase in the supply of non-foodgrains by one percentage point (as measured by index numbers, 1970-71=100.0) would decrease its price level by slightly more than one percentage point in the short run. It is thus noticed that the supply impacts of foodgrains and non-foodgrains on their respective prices are quite similar. The elasticity of agricultural raw materials prices with respect to manufacturing output is very much higher at 0.80. An increase of Rs. 100 crores ($ 80 million) in industrial income would push up the price level of non-foodgrains by almost two and a half percentage points by raising the demand for agricultural raw materials. As expected, the impact of foodgrains prices on non-foodgrains prices is positive and very significant.

3.4.6 (iii) Wholesale price - All commodities (WP):

This is one of the key equations of the model and great care was paid to its proper specification. The wholesale price index of all commodities was sought to be determined by the liquidity in the economy relative to the production of agricultural and industrial output. Considering the fact that two of the major determinants of the wholesale price index, i.e., the price levels of foodgrains and non-foodgrains, had already been endogenized within the model, it was but natural to include them in the overall wholesale price determination equation, as prime movers of this index. These, in a manner of speaking, took care of the agricultural sector because they embodied in themselves the supply-demand aspects characterizing this block. As the price level of manufactured goods was not determined by the model, we had to include the output of the manufacturing sector as a regressor in the equation in order to represent the remaining portion of the supply position in the economy. But naturally, the liquidity aspect of the situation was represented by the level of money supply (narrow money comprising currency and demand deposits) circulating in the economy. One last variable that was included along with the others on both theoretical as well as control considerations was the indirect tax rate, i.e., the ratio of indirect taxes (less subsidies) to NDP at current prices. The increase in this variable was felt to be one of the motivating factors responsible for the high rate of inflation in India because of its inherent ability to directly impact itself upon the wholesale price level. Thus, the wholesale price index is specified to be a function of the price level of foodgrains (WPF), the price level of non-foodgrains (WPNF), the output of the manufacturing sector (YMR), the level of money supply in the economy (M1) and the tax rate (T). The estimated equation was:

$$\begin{aligned} WP = \underset{(1.06)}{0.1240} &+ \underset{\substack{(7.15)\\0.45}}{0.5287}\,WPF + \underset{\substack{(2.62)\\0.24}}{0.2963}\,WPNF - \underset{\substack{(-3.22)\\-0.25}}{0.000086}\,YMR \\ &+ \underset{\substack{(7.03)\\0.47}}{0.000039}\,M1 + \underset{\substack{(2.62)\\0.10}}{3.6484}\,T \qquad \text{---(3.24)} \end{aligned}$$

$\bar{R}^2 = 0.9960$ D.W. $= 1.277$ S.E.E. $= 0.0425$
F $= 990.92$ $\rho = .344$ Mean $= 1.3276$
C.V. $= 3.1998$

One of the most interesting features to be noticed is the rare occurrence of <u>each</u> of the five estimated coefficients in the equation (excluding, of course, the intercept term) being extremely significant. Between themselves, they help to explain more than 99 percent of the variations in the wholesale price index, thereby providing a very good description of its movement. The production of industrial output has the expected negative impact on the price level and its elasticity is estimated at -0.25. An increase in industrial production by Rs. 100 crores ($ 80 million) would serve to decrease the price level by little under 0.9 percentage points. As expected, the price levels of foodgrains and non-foodgrains are found to be significant and, in particular, food prices play a dominant role in the movement of the overall price level. Liquidity, as represented by the amount of money supply in the economy, is found to possess a positive and very significant coefficient. It is estimated that an increase in Rs. 100 crores ($ 80 million) in money supply would push up the price level by about 0.4 percentage points. The elasticity of the price level with respect to this factor is 0.47. The coefficient of the control variable, i.e., the tax rate, is significant although the elasticity of prices with respect to it is quite low at 0.10. It is estimated that increasing the tax/NDP ratio by one percentage point would push up the price level by more than 3.5 percentage points. This extremely high value renders this instrument a powerful one capable of inducing considerable changes in prices.

3.4.6(iv) National income deflator (P):

The aggregate price level (national income deflator) is formulated according to the Quantity Theory of money. The price equation implies that the price level varies to match the nominal money supply determined on the monetary side and the real demand for such balances by raising or lowering nominal incomes on the assumption of a stable demand for money function, as well as a given level of the income velocity of money.

In the tradition of the quantity theory, let us consider the following equation:

$$M_s \cdot v = y \cdot p \qquad \text{---(3.25i)}$$

where M_s is total nominal money supply in the economy, v is the income velocity of money, y is national income at constant prices and p is the aggregate price level. Rewriting eq.(3.25i), we obtain

$$p = (M_s/y) \cdot v \qquad \text{---(3.25ii)}$$

Assuming the existence of an equilibrium condition given by

$$M_d = M_s \qquad \text{---(3.25iii)}$$

where M_d and M_s are the demand and supply for nominal balances, respectively; and a stable demand for money function given by

$$\frac{M_d}{p} = \frac{M_d}{p}(y) \qquad \text{---(3.25iv)}$$

where (M_d/p) is the demand for real money balances. Then we can interpret eq.(3.25ii) in the following way: Assume that the velocity is given and is expected to remain constant. Now any change in real income (y) should, through eq.(3.25iv), change the demand for real money balances. Considering the fact that the equilibrium condition, given by eq.(3.25iii), must always be satisfied, this would imply that the price level in eq. (3.25ii) must vary such that the nominal money supply determined on the monetary side must be matched by the demand for such balances by the raising or lowering of nominal incomes.

Linearizing eq.(3.25ii) around each term and adopting the symbols used in the model, we obtain the following linear functional form of the model:

$$P = a_0 + a_1(M1/YNFR) + a_2 v \qquad ---(3.25v)$$

where a_1 and a_2 are both assumed to be positive.

The estimated equation was:

$$P = \underset{(-8.36)}{-0.9725} + \underset{\substack{(60.59)\\1.00}}{3.7468}\,(M1/YNFR) + \underset{\substack{(10.36)\\0.17}}{0.2349}\,v \qquad ---(3.25)$$

$\bar{R}^2$ = 0.9946	D.W. = 1.259	S.E.E. = 0.0410
F = 1841.88	ρ = .334	Mean = 1.2662
		C.V. = 3.2357

The estimated price equation accounts for more than 99 percent of the variations in the price level and catches almost all the important turning points in the movements of the national income deflator. Both the coefficients are found to be extremely significant, in particular the one relating to the excess liquidity in the economy. The elasticity of the implicit price deflator with respect to this variable is found to be unity. The results denote that an increase of one percentage point in the ratio of money supply to real output would lead to a rise of more than 3.5 percentage points in the overall price deflator.

3.4.7 The income sector

This sector comprises just two equations, both of which are definitional identities. Using the real income and price levels estimated earlier we obtain nominal income at factor cost. The tax rate is then introduced in order to determine nominal income at market prices. Both these estimates are then used as regressors while estimating the equations of the government sector in the model.

3.4.7(i) NDP at factor cost - Current prices (YNFN):

The estimates of net domestic product at factor cost at current prices are obtained by multiplying the estimates of net domestic product at factor cost at constant prices (YNFR) by the national income deflator (P). The identity was given by:

$$YNFN = P \,.\, YNFR \qquad ---(3.26)$$

The role of this variable is basically one of determining the tax and non-tax receipts of the government. Considering that an inflow of these two sources of revenue helps to reduce the budget deficit and consequently prices, proper budgetary planning to take advantage of this fact is indicated.

3.4.7(ii) NDP at market prices - Current prices (YNMN):

The estimates of net domestic product at market prices at current prices were obtained through a nonlinear identity rather than through the standard linear identity usually adopted in most macroeconomic models. This was because we specifically wanted to incorporate the indirect tax rate into the equation and not indirect tax receipts, which is what would have happened had we used the definitional identity provided below, which states that

$$YNMN = YNFN + IT \qquad ---(3.27i)$$

where IT stands for total indirect taxes (less subsidies). But having defined our tax rate (T) as the ratio of indirect taxes (less subsidies) to net domestic product at factor cost at current prices, i.e.,

$$T = IT/YNFN \qquad ---(3.27ii)$$

we can rewrite eq.(2.27ii) in terms of IT and substitute the result into eq.(2.27i) to obtain the following equation

$$YNMN = (1 + T)YNFN \qquad ---(3.27)$$

The role of this variable assumes great importance in view of the explosive feedback effect deliberately introduced into the model via nominal income at market prices and its cumulative and destabilizing effects upon itself. The loop is an important one from the policy implications of containing an induced inflation. The causal mechanism is initiated by an artificially high level of nominal income at market prices brought about by a high rate of taxation. This, through the behavioural assumption of a lagging (leading) system of government receipts (expenditures) invariably implies a high budget deficit. This brings in its wake an increase in the supply of money due to the impact of the burgeoning budget deficit on currency expansion. The rise in the inflation rate as a direct consequence of this would further increase the budget deficit in the next period. If, at this juncture, policy planners proposed a fresh round of taxation (in an effort to increase revenue receipts and consequently reduce the deficit), the explosive feedback generated could well destabilize the economy. Great restraint, under the circumstances, is therefore advocated and the model is well geared to provide proper policy guidelines in this situation.

3.4.8 The monetary sector

The standard approach to describe the monetary sector in a model of the Indian economy is as follows (see Pani 1984). The instruments associated with the Reserve Bank of India (RBI) are currency (CC), total bank reserves held with the RBI (TBR) and RBI credit to commercial banks (C). The instruments associated with commercial banks are deposits (D), bank reserves held by the RBI (TBR) and borrowings from the RBI (C); while those associated with the public are currency holdings (CC) and deposit holdings (D). High-powered money or base money (H) is defined as the sum of currency (CC) plus total bank reserves (TBR) less borrowed reserves (C), while (narrow) money supply (M1) is defined as the sum of currency (CC) plus deposits (D).

It must be noted that in the market for various instruments the transactor on the supply side is the RBI for CC, TBR and C; and commercial banks for D. The public, by and large, are the transactors on the demand side for CC and D; and commercial banks for TBR and C. High-powered money is considered as supply-determined by fiscal operations, trade transactions and general economic activity and can be denoted by $\bar{H}$. In the market for demand and supply of various instruments, the costs and prices that enter are deposit rates for bank deposits, zero price for currency (but not for currency substitutes) and the bank rate or discount rate for bank reserves and borrowings of banks from the RBI. Since all prices or costs (expressed in nominal terms) are either fixed or administered, commodity prices can act as equilibrating factors between the demand for and supply of various instruments.

In the case of bank reserves and borrowings from the RBI, commodity prices do not enter as these transactions are all reckoned or assessed only in nominal terms. Thus only the demand for and supply of currency or deposits is equated through price changes. As the same cannot be said of the demand for and supply of bank reserves or RBI lending to commercial banks, the composition of high-powered money in terms of TBR and C may change without affecting the level of high-powered money provided TBR and C are perfect substitutes. This assumes that whatever currency is supplied is absorbed by the system through variations in the price level, real income or interest rates which determine the demand for money. The problem is to then reconcile the demand for the components of base money with their supply so that, in the process, the money multiplier is suitably defined.

Writing suffixes 'd' and 's' to denote demand and supply aspects, the postulated set of economic relationships between all these variables can be written as follows:

$$\bar{H} = CC_s + TBR_s - C_s \qquad \text{...(1): By definition}$$

$$CC_s + D_s = M1 = m\bar{H} \qquad \text{...(2): By definition}$$

$$\left.\begin{aligned} CC_d &= k'(y,p,\bar{r}) \\ \frac{CC_d}{D_d} &= k(y,p,\bar{r}) \end{aligned}\right\} \qquad \text{...(3): Demand for currency by the public}$$

$$\left.\begin{aligned} TBR_d &= r'(D_s,CRR,\bar{r}) \\ \frac{TBR_d}{D_s} &= r(CRR,\bar{r}) \end{aligned}\right\} \qquad \text{...(4): Demand for reserves by the banking system}$$

$$\left.\begin{aligned} C_d &= g'(D_s, CRR, \bar{r}) \\ \frac{C_d}{D_s} &= g(CRR, \bar{r}) \end{aligned}\right\} \quad \text{...(5): Demand for borrowed reserves by the banking system}$$

Here, y represents real national income, p is the general price level, $\bar{r}$ is the vector of interest rates and CRR is the statutorily determined cash reserve ratio.

Four additional equations implicit in the system are

$$CC_d = CC_s \quad \text{...(6)}$$

$$D_d = D_s \quad \text{...(7)}$$

$$TBR_d = TBR_s \quad \text{...(8)}$$

$$C_d = C_s \quad \text{...(9)}$$

Eqs.(1)-(9) can now be used to define the money multiplier (m), which is given as

$$m = \frac{k(.) + 1}{k(.) + r(.) - g(.)} \quad \text{..(10): Equilibrating factor between the demand for and supply of money.}$$

The above analysis indicates that some assumptions need to be made in order to ensure that eqs.(6)-(9) are satisfied, although a few of them may not be compatible with reality. Eq.(1) connotes that, ignoring the feedback effect, the total $(CC_s+TBR_s-C_s)$ is given to the monetary system. If the assumption that $CC_d=CC_s$ through price or real income changes is true, then (TBR_s-C_s) is predetermined and this with eqs.(4),(5),(8) and (9) determine the four variables - the supply of and demand for bank reserves and borrowings from the RBI. If TBR_d and C_d are perfect substitutes, then either eq.(4) or (5) can be omitted. However if, as claimed by some, both TBR_s and C_s are predetermined given $\bar{H}$, then CC_s becomes a residual which is then determined by the budget deficit and other factors. This is the general scheme followed in the formulation of the monetary sector in our model.

The specification of the monetary sector therefore consists of a behavioural equation for currency expansion, an equation for commercial bank reserves, a statistical construct for total bank reserves and a money-supply determination equation using the money multiplier approach. The sector also contains two identities defining real money stock and the income velocity of money.

Although non-banking financial intermediaries are gaining importance in the economy, their ramifications were not considered in the model due to the lack of adequate data, the need for further disaggregation and also due to the absence of a well-defined theory on the impact of the operations of these institutions on output, demand and prices.

3.4.8(i) Stock of currency (CC):

The fiscal deficit of the government, i.e., its budget deficit, is determined by the government's total expenditure and revenue receipts. Given domestic borrowing and external borrowing, the deficit to be financed by the RBI is obtained. This, along with the increase in foreign exchange assets of the RBI and increase in the net absorbtion of government securities by the RBI, explains currency expansion. For the purpose of estimating the currency supply equation only the deficit out of the operations of the Central government, total borrowings (both domestic as well as external) and the change in foreign exchange reserves have been considered. This could be a limitation as RBI credit to State governments and the commercial sector is of an autonomous character. These factors also influence the changes in high-powered money to a substantial degree. We have considered the influence, on high-powered money, of private sector transactions with the rest of the world because this could only be through the balance of payments transactions and therefore show up as changes in foreign exchange reserves.

The change in the stock of currency is therefore expected to be dependent on the budget deficit (BD), the sum of net domestic borrowing and net foreign aid (DBFA) and the change in net foreign exchange assets (DFEA). The equation was specified in the following linear form:

$$\Delta CC = a_0 + a_1 BD + a_2 DBFA + a_3 DFEA \qquad \text{---(3.28i)}$$

where it is assumed that $a_1 > 0$, $a_2 < 0$ and $a_3 > 0$. While the first two restrictions are obvious, the third one needs some explanation. The reason is that these foreign exchange assets which are net holdings of the RBI represent Reserve Bank credit to the foreign sector, because they are the financial liabilities of the foreign sector. Most of these assets are held abroad in the form of foreign securities and cash balances. The RBI comes to acquire them as the sole custodian of the nation's foreign exchange reserves. As the controller of all foreign exchange transactions, whether on government or private account, it regularly buys and sells foreign exchange against Indian currency. All such transactions have a direct impact on high-powered money. When the RBI buys foreign exchange (and thereby increases its holdings of foreign exchange assets), it pays for it in terms of its own money and the supply of high-powered money in the economy increases. Since these transactions(buying and selling) in foreign exchange goes on all the time, what matters for a change in high-powered money is the change in net foreign exchange assets held by the RBI. Thus, a surplus (deficit) in the balance of payments increases (decreases) the supply of high-powered money in the economy, other things being the same. That is why the vast accumulation of foreign exchange reserves in India (mainly due to large inward remittances from Indians working abroad) over the recent years (after 1975) has led to substantial increases in high-powered money (and thereby in money supply) in the Indian economy.

Within the framework of our analysis, we ignored borrowed reserves (C_s) because they formed an insignificant component of high-powered money (H) which was now defined as the sum of currency supply (CC_s) and total bank reserves (TBR_s), i.e., $H = CC_s + TBR_s$. Therefore, any increase in high-powered money must reflect as an increase in currency or bank reserves or both. To keep the causal mechanism in the model simple, we postulated a recursive-path structure with currency affecting bank reserves and both of them (jointly) affecting high-powered money.Tracing these links backwards, we ascertain that any increase in high-powered money must therefore be due to a prior increase in currency. Thus, any increase (decrease) in foreign exchange assets implies an initial increase (decrease) in the supply of currency which is then translated into an increase (decrease) in the supply of high-powered money.

The estimated equation was:

$$\Delta CC = \underset{(3.75)}{452.772} + \underset{\substack{(5.83)\\1.53}}{1.0079}\,BD - \underset{\substack{(-3.90)\\-1.09}}{0.2550}\,DBFA + \underset{\substack{(4.76)\\0.88}}{0.6126}\,DFEA \qquad \text{---(3.28)}$$

$\bar{R}^2 = 0.8038$ D.W. $= 2.434$ S.E.E. $= 223.286$
$F = 14.66$ $\rho = -.297$ Mean $= 728.818$
C.V. $= 30.6367$

The equation indicates that about 80 percent of the variation in the change in currency holdings by the public is accounted for by the government deficit, domestic and foreign borrowings of the government and the change in foreign exchange assets of the RBI. Domestic and foreign borrowings have the expected negative impact on currency expansion. Given the government's fiscal deficit, an increase of 1 percent in domestic and foreign borrowings would lower currency expansion by about 1.1 percent. The elasticity of currency expansion with respect to the change in net foreign exchange assets is estimated at about 0.88. The budgetary deficit is found to be an extremely significant variable and it is seen that any increase in it would prompt an almost equal increase in the rate of currency expansion. Taking into consideration that this would serve to increase high-powered money (directly as well as indirectly via bank reserves), money supply (given an exogenously determined money-multiplier) and, thereby, the price level (assuming that the supply of output is relatively unresponsive to policy manipulations), great care needs to be exercised by the government to contain the budget deficit to reasonable levels as far as possible.

One point that needs to be emphasized is that we were not interested in the change in the stock of currency *per se*, because the estimation of the equations for scheduled commercial banks' reserves as well as for money supply involved using levels of currency as a regressor <u>inter alia</u>. We therefore invoked the definitional identity

$$CC = CC(-1) + \Delta CC \qquad \text{---(3.28ii)}$$

to estimate the level of currency at any given period. In the final form of the model, eqs.(3.28) and (3.28ii) were linked together into one equation in order to avoid an unnecessary increase in the dimension of the model.

3.4.8(ii) Scheduled commercial banks' reserves (SCBR):

The equation for determining the total bank reserves were, in the first instance, specified for scheduled commercial banks as they account for nearly 85 percent of aggregate reserves. Total reserves of the scheduled commercial banks are specified to be a function of the level of currency (CC) and the bank rate (BR). The former, used both as a proxy for government deficit and the demand for currency holdings by the public, is expected to play a dominant role in the equation. It is assumed that a high level of currency with the public would, in the context of branch expansion and development of banking habits, result in a larger inflow of reserves when the public changed its preferences towards the holding of deposits instead of cash. Thus this variable also acts as a proxy for the structural changes taking place in the economy. The latter was an exogenously given control variable and represented the tightness of monetary policy (supply of reserves to the banking system). We also included the level of net foreign exchange assets of the RBI in the belief that the multiplier or the rate at which the banking system as a whole can expand reserves is related to it. In periods when banks sell more foreign exchange than they purchase from the RBI, they can be confident of obtaining reserves with greater ease following this autonomous (from the banks' point of view) inflow of reserves. The estimated equation was:

$$\underset{(-3.97)}{SCBR = -1326.09} + \underset{\substack{(21.14)\\1.13}}{0.6102\ CC} - \underset{\substack{(-3.30)\\-0.18}}{199.441\ BR} \qquad \text{---(3.29)}$$

$\bar{R}^2$ = 0.9872 D.W. = 1.821 S.E.E. = 212.739
F = 462.35 ρ = .024 Mean = 1993.31
C.V. = 10.6727

The explanatory power of the equation is very high. It is seen that the level of currency has a significant effect on the reserves of the banking system. The elasticity of the supply of reserves with respect to this variable is 1.13. The impact of the bank rate on the reserves of commercial banks is negative and significant. As a proximate measure of the intentions of the monetary authorities, the supply of reserves may be expected to decline as the bank rate rises. According to the estimated equation, this decline for every percentage point jump in the bank rate is approximately to the tune of Rs. 200 crores ($ 160 million). Considering the fact that the mean of the dependent variable is about Rs. 2000 crores ($ 1.6 billion), the order of magnitude of the decline is slightly over 10 percent. It needs to be noted that alternative variants incorporating the level of foreign exchange assets of the RBI as an independent variable failed to yield significant estimates of the coefficient and hence had to be rejected.

3.4.8(iii) Total bank reserves (TBR):

Total reserves of all banks are obtained as a statistical function of the scheduled commercial banks' reserves (SCBR). The estimated equation was:

$$TBR = \underset{\substack{(84.32)\\ 1.00}}{1.1733}\ SCBR \qquad \text{---(3.30)}$$

$\bar{R}^2 = 0.9983$ D.W. $= 3.198$ S.E.E. $= 134.879$
F $= 7109.16$ $\rho = -.683$ Mean $= 2346.46$
C.V. $= 5.7482$

The equation indicates that the reserves of the scheduled commercial banks account for approximately 85 percent of total bank reserves. This implies that about 15 percent of aggregate reserves are other deposits (OD) with the RBI. While we could have treated OD as an exogenous variable and incorporated it into a linear identity of the form: TBR=SCBR+OD, this would have needlessly expanded the data requirements of the system. Thus, it was felt that the methodology adopted was a suitable one and the equation was justified being estimated in the form of a statistical construct.

3.4.8(iv) Money supply - Narrow money (M1):

Money supply rests upon the base (or high-powered) money produced by the RBI and the Government of India (small coins including one-rupee notes). The RBI is a quasi-independent organization which is charged with the responsibility of executing (although not determining) monetary policy through an appropriate adjustment of money supply.

The process by which money supply is determined is invariably via a money-multiplier and a monetary base. For this reason, the theory is often referred to as the 'money-multiplier theory of money supply' because the money supply is some multiple of the monetary base. Thus, we have

$$M1 = m \cdot H \qquad \text{---(3.31i)}$$

where m is the money-multiplier and H is the monetary base defined in our study as the sum of currency (CC) and total bank reserves (TBR). We therefore have

$$H = CC + TBR \qquad \text{---(3.31ii)}$$

Given this reformulation, eq.(10) on page 54 defining the money multiplier can be rewritten as

$$m = \left(\frac{CC_d}{D_d} + 1\right) / \left(\frac{CC_d}{D_d} + \frac{TBR_d}{D_d}\right) \qquad \text{---(3.31iii)}$$

Changes in the money supply occur because of changes in high-powered money or in the money-multiplier or in both. The money-multiplier can change only if the currency-deposit ratio or the reserve-deposit ratio changes. As far as the Indian economy has been concerned, the currency-deposit ratios have declined over the years. This tendency has been explained in terms of several factors including concentrated population, wealth of the community, development of banking habits, acceptance of cheques, amongst others. On the other hand, the reserve-deposit ratios have, of late, shown an inclination to increase. This rise in the ratio is essentially a reflection of the high level at which the statutory cash reserve ratio, including the incremental cash-reserve to aggregate deposits ratio, has been pegged. The combined effect of these two conflicting forces has been to gradually increase the money-multipliers. However, for the purpose of short-term forecasting we assume that the control exerted by the RBI over the monetary base is tantamount to controlling the supply of money in the economy.

We can substitute eq.(3.31ii) into eq.(3.31i) to obtain the following form of the money-supply determination equation:

$$M1 = m(CC + TBR) \qquad \text{---(3.31iv)}$$

where the money-multiplier (m) is assumed to be exogenously determined.

The estimated equation was:

$$M1 = \underset{\substack{(62.80)\\1.04}}{1.5073}(CC + TBR) \qquad \text{---(3.31)}$$

$\bar{R}^2$ = 0.9975	D.W. = 1.157	S.E.E. = 745.426
F = 3943.63	ρ = .199	Mean = 13053.8
		C.V. = 5.7104

The money-multiplier is estimated at approximately 1.5. The order of increase in money supply in the Indian economy has been mainly due to the huge quantum of rise in domestic credit, although the effect of the accretion of net foreign exchange assets has been considerable especially after 1975. Net bank credit to the government was, in terms of absolute outstanding amounts, as much as bank credit to the commercial sector until 1973. Net RBI credit to the government was consistently higher than the level of credit extended by commercial and cooperative banks to the government throughout the sample period, i.e., 1961-82. Thus, it appears that the creation of high-powered money has been made possible by the large level of government indebtedness to the RBI. This, together with the rising tendency of the money-multiplier, has triggered off an alarming acceleration in the rate of growth of money supply in the Indian economy.

3.4.8(v) Real money stock (MS):

The measure of money supply discussed so far was a measure of the nominal quantity of money since it was at current rupee amounts. However, it is sometimes useful to look instead at the quantity of money in terms of real goods and services which could be exchanged for it. This real quantity of money or real money stock, as it is alternatively called, is obtained by deflating the nominal quantity of money (M1) by an appropriate choice of a deflator. In our model, we opted for the wholesale price index (WP) on empirical considerations, instead of the generally used national income deflator. Our measure of real money stock was therefore given by

$$MS = \frac{M1}{WP} \qquad \text{---(3.32)}$$

Within the framework of our model we use this measure to compute the income velocity of money which, as seen earlier, helps to determine the aggregate price level in the economy.

3.4.8(vi) Income velocity of money (v):

The velocity of money is the ratio of income to money supply(at a given point in time) and is essentially a stock-flow relationship. For the Indian economy, these ratios are invariably computed on the basis of annual rates, i.e., the level of income pertains to its flow over a complete financial year. This ratio ought to be the same whether income and money supply are measured in nominal or real terms because, on purely theoretical considerations, the deflators for both of them are identical. However, as we deflated our estimates of national income and money supply by the national income deflator and the wholesale price level, respectively, within the framework of our model the ratios would have differed. We chose to define velocity in terms of real income and real money stock, and therefore we had

$$v = \frac{YGFR}{MS} \qquad \text{---(3.33)}$$

The variable represents the velocity with which the medium of exchange (in terms of base year rupees) circulates within the economy to generate final output in terms of real goods and services only. As far as the Indian economy is concerned, the velocity has remained fairly stable over a long period of time, showing outward signs of a slight fall in the recent years. One of the reasons put forward to explain this decline is the existence of a rapidly growing level of unearned incomes which has spawned an economy (popularly referred to as the 'parallel economy') within the economy. A lot of research in this area has been carried out and the evidence (*vis-à-vis* velocity) is highly conflicting, to say the least. However, these aberrations notwithstanding, the general trend has been to validate the results obtained by Friedman (1969) who hypothesized that velocity had a cyclical behaviour pattern and conformed positively to the business cycle, rising during expansions and falling during contractions. This cyclical effect has also shown up in a slower rate of decline in expansions than in contractions.

3.4.9 The government sector

This sector is entirely concerned with the fiscal operations of the Central government and provides the link in the feedback between nominal incomes, budget deficits and prices. Only Central government operations have been considered and separate revenue and expenditure functions for the State governments have not been specified. All the endogenous variables in this sector are measured in nominal terms.

3.4.9(i) Budget deficit (BD):

This is one of the central variables of the model which has, of late, assumed great importance in the light of investigations on the inflationary process in developing countries which have revealed that inflation and the fiscal deficit, with the concomitant increase in the money supply, are not independent. In many developing countries (including India), the fiscal deficit of the government is largely financed through credit from the central bank. Moreover, government expenditure is expected to increase at a faster rate in the face of mounting inflation due to the government's conscious desire to maintain real expenditure at a planned level. On the other hand, revenue collections are assumed to lag behind. Thus, inflation is expected to result in widening the gap between the government's expenditure and revenue, enlarging the budget deficit which is financed by the banking system and thereby leading to further increases in money supply and, consequently, prices. This self-generating process of inflation was first suggested by Olivera (1967). A model of the inflationary process was empirically tested by Dutton (1971) and Aghevli and Khan (1977). The self-generating aspect of it was empirically shown to exist for selected developing countries by Aghevli and Khan (1978). The interrelationships between inflation, money supply and the government deficit for the Indian economy is generally assumed to follow the basic framework of the Aghevli-Khan hypothesis.

The budget deficit is defined as the difference between government expenditure (GE) and government revenue (GR), i.e.,

$$BD = GE - GR \qquad \text{---(3.34i)}$$

Government expenditure is the sum of government developmental expenditure (GDE) - which is assumed to be an instrument of policy, non-developmental expenditure (NDE) - which is an endogenous variable, and defence expenditure (DEF) - which is assumed to be an exogenous (uncontrollable) variable.Therefore,

$$GE = GDE + NDE + DEF \qquad \text{---(3.34ii)}$$

Government revenue is the sum of tax revenue (TR), non-tax revenue (NTR) and capital receipts (CR). All three variables are endogenous with respect to the model. Therefore,

$$GR = TR + NTR + CR \qquad \text{---(3.34iii)}$$

Substituting eqs.(3.34ii) and (3.34iii) into eq.(3.34i), we obtain the following identity for defining the budget deficit of the Central government

$$BD = GDE + NDE + DEF - TR - NTR - CR \qquad \text{---(3.34)}$$

Thus, the government deficit itself is endogenized by relating it to the various individual components of government expenditure and revenue which, by themselves, could be either endogenous, control or exogenous variables.

3.4.9(ii) Non-developmental expenditure (NDE):

The non-developmental expenditure of the government is specified to be a function of the NDP at market prices at current prices (YNMN) within a partial-adjustment framework. The estimated equation was:

$$NDE = \underset{(1.32)}{383.101} + \underset{\substack{(3.00)\\0.64}}{0.6607}\ NDE(-1) + \underset{\substack{(1.63)\\0.35}}{0.0294}\ YNMN \qquad ---(3.35)$$

$\bar{R}^2 = 0.9507$ $h = -1.079$ S.E.E. = 669.030

F = 184.10 $\rho = -.083$ Mean = 5148.55

C.V. = 12.9945

The speed of adjustment of actual nominal expenditure to desired nominal expenditure is not very high (0.34). The time-constant of the lag is estimated at 2.41 years. The elasticity of nominal government expenditure with respect to nominal income is 0.35 in the short run and 1.03 in the long run. The latter result is an important one as it indicates that, in the long run, both incomes and expenditures move proportionately. The long-term behaviour of the equation also indicates that a unit increase in nominal income would push up nominal expenditure by 0.09 units in the long run.

3.4.9(iii) Tax revenue (TR):

The tax revenue of the government is specified to be a function of the tax rate (T) and NDP at factor cost at current prices (YNFN). The latter was a proxy for the tax base. The estimated equation was:

$$TR = \underset{(-3.03)}{-2883.86} + \underset{\substack{(2.58)\\0.19}}{30089.1}\ T + \underset{\substack{(11.37)\\0.82}}{0.0798}\ YNFN \qquad ---(3.36)$$

$\bar{R}^2 = 0.9837$ D.W. = 0.865   S.E.E. = 389.061

F = 604.61 $\rho = .514$ Mean = 4232.52

C.V. = 9.1922

The equation explains more than 98 percent of the variation in the tax revenue with both the regressors being very significant. It is estimated that a hike in the tax rate by one percentage point would increase government revenue by more than Rs. 300 crores ($ 240 million), thereby decreasing the budget deficit by an equivalent amount _ceteris paribus_. The rate of inflation, under the circumstances, is subject to two conflicting forces: one, a direct effect brought about by the instantaneous impacting of the tax rate on the wholesale price level, due to the forward shifting of the indirect tax content, which forces the price level upwards; and two, an indirect effect as a result of the reduced budget deficit which brings in its wake a deceleration in the growth rate of money supply which forces the price level downwards. One of the more interesting problems in applied macroeconomic planning is to identify that countervailing force which is expected to prevail.

3.4.9(iv) Non-tax revenue (NTR):

The non-tax revenue of the government is specified to be a function of the NDP at factor cost at current prices (YNFN) within a partial-adjustment framework. The estimated equation was:

$$\underset{\substack{(-1.56)}}{NTR = -127.164} + \underset{\substack{(2.47)\\0.42}}{0.4673}\, NTR(-1) + \underset{\substack{(3.33)\\0.57}}{0.0202}\, YNFN \qquad ---(3.37)$$

$\bar{R}^2 = 0.9815$ $h = -0.752$ S.E.E. = 149.518
$F = 503.93$ $\rho = -.154$ Mean = 1517.20
C.V. = 9.8549

The speed of adjustment of actual revenue to desired revenue is about 0.53. The time-constant of the lag is estimated at 1.31 years. The elasticity of revenue with respect to income is 0.42 in the short run and 1.07 in the long run. Combining this result with the one obtained earlier for eq.(3.35), we observe that, in the long run, both government expenditure as well as revenue move quite proportionately with nominal income. The derivation of the long-run multiplier of income indicates that a unit step-up in it would increase revenue by about 0.04 units. The results therefore show that any increase in nominal income, as a result of inflation, will increase long-run government expenditure at a faster rate than long-run government revenue. Thus, the model broadly confirms the Aghevli-Khan hypothesis bearing out the contention that inflation results in widening the gap between government expenditure and revenue enlarging, in the process, the fiscal deficit leading thereby to further increases in money supply and prices.

3.4.9(v) Capital receipts (CR):

As far as the Indian government is concerned, its capital receipts basically comprise the following items: (i) internal and external market loans, (ii) net small savings and net contributions from state provident funds and (iii) recoveries of loans and advances from state governments. The first term is capable of being proxied by the sum of net domestic borrowing and net foreign aid (DBFA) which is an exogenous variable in our model.The second term is a major component of net domestic savings at current prices. While our model did not specifically consider savings in nominal terms, we could have introduced it easily by defining nominal savings as the product of savings at constant prices (SNR) and the national income deflator (P), both of which are endogenous to the system. As far as the third item is concerned, we assumed that the repayment of loans is an inverse function of the price level, i.e., the capability of state governments to repay loans decreases with every increase in the price level, all else remaining unchanged. Therefore, our equation would have been specified by the following linear functional form:

$$CR = a_0 + a_1 DBFA + a_2(SNR.P) + a_3 P \qquad \text{---(3.38i)}$$

where $a_1 > 0$, $a_2 > 0$ and $a_3 < 0$. However, it was decided to take advantage of the fact that capital receipts, if they are positively correlated with savings in nominal terms and negatively correlated with the price level, should be positively correlated with savings in real terms. Therefore, our equation was specified by the following linear functional form instead:

$$CR = b_0 + b_1 DBFA + b_2 SNR \qquad \text{---(3.38ii)}$$

where $b_1 > 0$ and $b_2 > 0$.

The estimated equation was:

$$CR = \underset{(-2.97)}{-626.373} + \underset{\substack{(5.53)\\0.30}}{0.2385}\,DBFA + \underset{\substack{(12.92)\\0.74}}{0.6605}\,SNR \qquad \text{---(3.38)}$$

$\bar{R}^2$ = 0.9736	D.W. = 1.937	S.E.E. = 378.139
F = 369.80	ρ = -.069	Mean = 3695.29
		C.V. = 10.2330

The estimated equation performs very well and explains more than 97 percent of the variation in capital receipts with both the coefficients being extremely significant. The elasticity of capital receipts with respect to real savings is estimated at 0.74. The role of savings in increasing capital receipts, and thereby reducing the budget deficit and consequently the rate of inflation, is clearly underscored by the model. The savings rate therefore becomes an important variable and policy measures to raise it by increasing the income share of the non-agricultural sector have to be strongly implemented. In this respect, it needs to be mentioned that thirty five years of macroeconomic planning in India has succeeded in raising this ratio from 0.40 in 1950 to almost 0.60 currently.

3.4.10 The miscellaneous sector

This sector is concerned only with the estimation of the total area, i.e., both irrigated as well as non-irrigated, under foodgrains and non-foodgrains, respectively. Most macroeconometric models of the Indian economy have considered these variables as exogenous. However, in view of the fact that the area under cultivation is a very important determinant of crop production (be it foodgrains or non-foodgrains), this treatment was considered hazardous, especially from the viewpoint of forecasting and control. As such, great care was taken to formulate appropriate behavioural response functions for the determination of these areas.

3.4.10(i) Area under foodgrains (AF):

The area under foodgrains is specified as a function of the relative price of foodgrains as against non-foodgrains (WPF/WPNF), total investment in agriculture by both the public as well as the private sectors (IAR) and rainfall (R) within a partial-adjustment framework. The estimated equation was:

$$
\begin{aligned}
AF = {} & \underset{\substack{(2.37)}}{40.7078} + \underset{\substack{(1.66)\\0.24}}{0.2440}\,AF(-1) + \underset{\substack{(3.17)\\0.46}}{0.0053}\,IAR(-1) \\
& + \underset{\substack{(2.51)\\0.24}}{8.9278}\,(WPF/WPNF) + \underset{\substack{(5.72)\\0.53}}{0.4063}\,R \qquad \text{---(3.39)}
\end{aligned}
$$

$\bar{R}^2 = 0.8276$ $\quad h = -1.648$ $\quad S.E.E. = 1.8454$

$F = 25.00$ $\quad \rho = -.284$ $\quad Mean = 122.286$

$C.V. = 1.5091$

The speed of adjustment of the actual area to the desired area is high (0.76). The time constant of the lag is 7.1 years. The explanatory power of the equation is quite good, considering the volatile nature of the phenomenon being modelled. As expected, it is rainfall that is the dominant variable both in terms of significance as well as elasticity. A 10 percent increase in the wholesale price index of foodgrains relative to the price of non-foodgrains would result in a long-run increase in the area under foodgrains by as much as 1.2 million hectares. Considering that the mean value of the dependent variable is 122.3 million hectares, this would imply an order of magnitude of around 1 percent. The long-run elasticity of acreage with respect to total investment in agriculture is quite high at 0.61. The estimates also reveal that an increase in agricultural investment by Rs. 100 crores ($ 80 million) would bring about an additional 0.5 million hectares under foodgrains' cultivation.

3.4.10(ii) Area under non-foodgrains (ANF):

The area under non-foodgrains is specified, in an identical manner, as a function of the relative price of foodgrains as against non-foodgrains (WPF/WPNF), total investment in agriculture (IAR) and rainfall (R) within a partial-adjustment framework. The estimated equation was:

$$
\begin{aligned}
ANF = {} & \underset{\substack{(3.62)}}{16.8596} + \underset{\substack{(2.17)\\0.40}}{0.3134}\,ANF(-1) + \underset{\substack{(2.92)\\0.50}}{0.0010}\,IAR(-1) \\
& - \underset{\substack{(-1.54)\\-0.19}}{1.0493}\,(WPF/WPNF) + \underset{\substack{(3.78)\\0.44}}{0.0546}\,R \qquad \text{---(3.40)}
\end{aligned}
$$

$\bar{R}^2 = 0.7757$ $\quad h = 0.101$ $\quad S.E.E. = 0.3436$

$F = 18.30$ $\quad \rho = -.028$ $\quad Mean = 31.6429$

$C.V. = 1.0860$

The speed of adjustment of the actual area to the desired area is fairly high (0.69). The time constant of the lag is estimated at 8.6 years. The explanatory power of the equation is good but not as satisfactory as the one obtained for the previous equation. Once again it is discerned that rainfall is the most dominant variable. A 10 percent rise in the price of foodgrains relative to the price of non-foodgrains would result in a long-run decrease in the acreage devoted to the latter by as much as 0.15 million hectares. Considering that the mean value of the dependent variable is 31.6 million hectares, this would imply an order of magnitude of around 0.5 percent. The long-run elasticity of acreage with respect to the total investment in agriculture is quite high at 0.73. It is also estimated that an increase in agricultural investment by Rs. 100 crores ($ 80 million) would bring about an additional 0.1 million hectares under non-foodgrains' cultivation.

3.5 The Model

3.5.1 An overview

The model comprises 40 equations, of which 27 are behavioural equations and the remaining 13 are identities. Of the former, 3 equations can be technically termed as statistical constructs. The model is compartmentalized into 10 modules (or blocks), with feedbacks and feedforwards linking most of them in a fairly simultaneous manner. The central module is the production sector, the output of which is determined once the exogenous variables (i.e., rainfall index and man-days lost), the policy variable (i.e., government developmental expenditure) and the values of the concerned variables from the miscellaneous and investment sectors are provided. The consumption and savings sectors are auxiliary modules and are uniquely determined once the central module has been solved. The savings sector, in conjunction with the trade and output sectors, helps to jointly ascertain the investment sector, given the predetermined values of the policy variables (i.e., public sector investment in agriculture and transport). The accelerator feedback, where income determines investment which in turn influences income is very much apparent here. Given the 'two-gap' version, the trade sector is determined simultaneously along with the investment sector, given the prior solution of the consumption sector and the feedforward effect of the price sector. The price sector, which is the sub-central module forming the all-important link between the real and nominal sectors of the economy, is determined given the solution of the output sector, the value of the policy variable (i.e., the tax rate) and the feedforward effect of the monetary sector. The income sector is determined given the solutions of the output and price sectors. The monetary and government sectors are the satellite modules linked together by means of an overheating feedback which runs from money supply to nominal income via prices through to an endogenized government receipt and expenditure system onto the budget deficit and back to money supply. A secondary version of the overheating feedback runs from prices to real money stock through to velocity and back onto prices. Finally, the miscellaneous sector is determined given the value of the exogenous variable (i.e., rainfall) and the solutions of the price and investment sectors. The model also involves a self-containing feedback which runs from the exogenous tax rate directly onto the price level onwards through nominal income and tax revenue and back onto prices via the budget deficit and money supply. The dynamics of this circuitry can suggest measures to optimally stabilize the instrument without triggering off any undesirable consequences.

Figure 3.1
Flow Chart Of The Macroeconometric Model

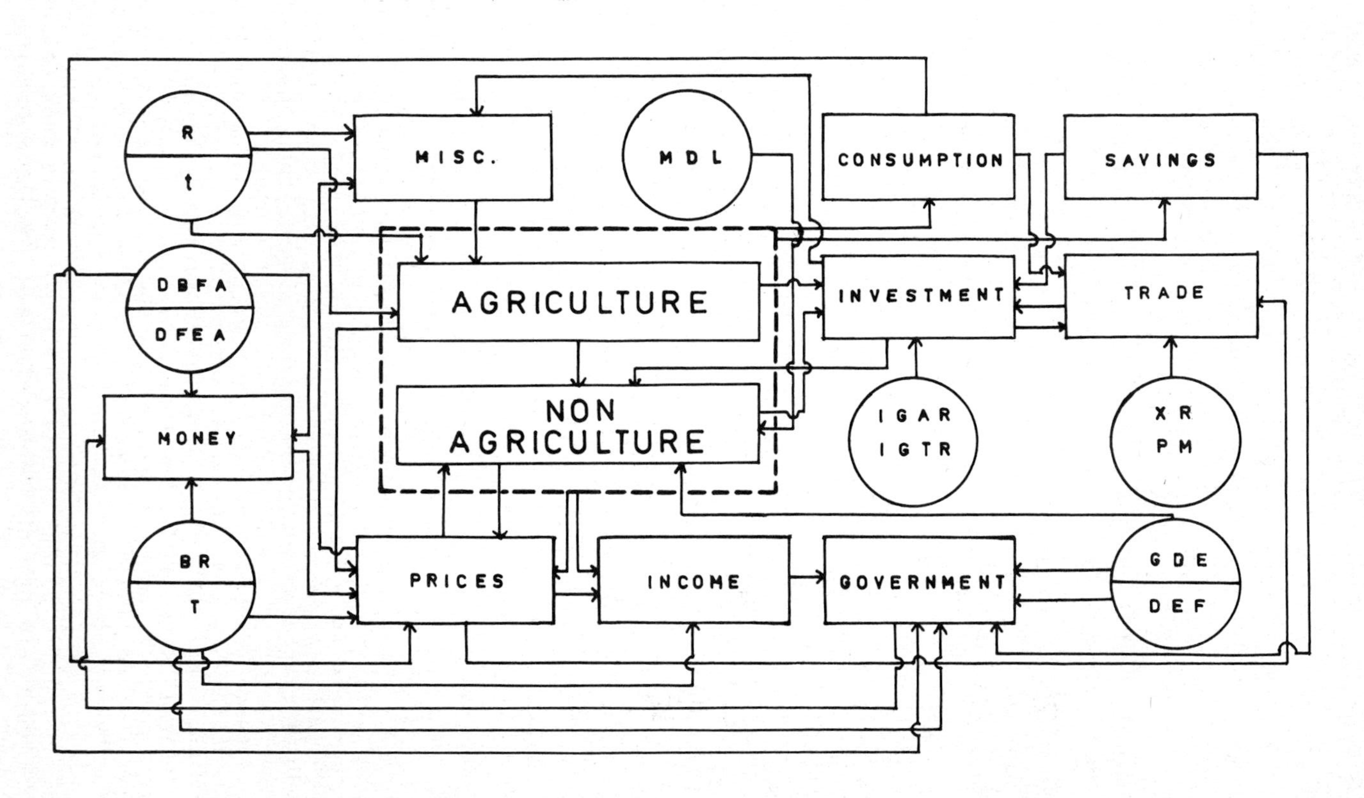

3.5.2 Some limitations

In our model building attempt, emphasis was laid on the crucial role played by supply constraints as well as monetary and fiscal factors in the economy. On the supply side, attention was largely given to the determination of agricultural output. The determinants of manufactured output could perhaps have been analyzed in greater detail than that managed in the present study. While we did consider government spending in our model, we underplayed its role because we paid attention only to the impact of government developmental expenditure on output via the services sector. However, government spending, by way of generating demand in the economy, often provides an impetus to investment in the manufacturing sector which leads to an increase in the output of this sector. Moreover, the other effect of government spending is to raise the level of capital stock in public enterprises as well as in infrastructural activities - both of which lead ultimately to increases in output. These links have, however, not been built into the model.While the impact of fiscal operations on prices via the government deficit has been brought out in the study, the role of credit and money in influencing the level of output has not been investigated. Yet another limitation of the model is that the linkages between the foreign sector and the domestic economy have not been considered in detail especially with regard to foreign exchange constraints, debt servicing problems and capital account inflows. Needless to say, the most serious limitation of the model is the fact that labour market descriptions are totally missing from it. While the complete absence of reliable data on unemployment ruled out any testing for the existence of trade-offs between inflation and unemployment for the Indian economy; the possibility of model misspecification and the need for a robust design structure compelled us to preclude the incorporation of any wage-price dynamics into the system because the assumptions underlying its operational relevance have not yet been fully explored within the context of the Indian labour market.

3.6 Conclusions

When econometric modelling had its genesis at the Cowles Foundation during the mid-forties, it was, indeed, the "best of times". The progenitors of modern econometrics honestly felt that existing economic theory was capable of being modelled accurately enough to encompass most, if not all, significant economic phenomena. This faith was fuelled by an equally strong belief in the ability of classical statistics to handle all the complicated problems posed by economic theory.

Forty years later these primordial hopes for econometric modelling have all but vanished. Inspite of major breakthroughs in knowledge as well as notable theoretical and empirical accomplishments, even the most diehard protagonists of econometric modelling have been forced to concede that econometric models have fallen very short of expectations and it is,indeed, the "worst of times".

However, in the opinion of many econometricians (see Karchere and Kuh 1982), this has been a direct consequence of unbounded aspirations rather than mere unfounded optimism.Thus, in most instances, the very complexity of the economic phenomena under scrutiny has preempted their successful synthesis via modelling. In other cases, the assumptions regarding exogeneity as well as causality have been incapable of direct verification and the separation and quantification of theoretically meaningful components, in the absence of any possibility of controlled experiments, through applied regression analysis have been wholly inadequate. Thus, statistical paradigms which implicitly assume experimental control have very often been found unequal to the task. There is, under the circumstances, a very dangerous possibility of being led into what has been aptly termed 'John Wayne econometrics' (see Waelbroeck 1982), in which the valiant econometrician strides through his sample, his trusted OLS colt at his hip, avoiding outliers, ignoring residuals, shooting down observations and riding off into the sunset with the beautiful regression at his side.

It is vital that modellers avoid such a nemesis, which surprises even the most 'wary transgressor', because the best that they can realistically hope to accomplish is a reasonable approximation of economic reality - an approximation that is bound to decay exponentially with time. In such a charged environment, model assessment becomes strategically important and the use of procedures that measure the extent of model adequacy and robustness is unequivocally recommended if model builders are to make warrantable claims regarding the informational contents of their systems.

As the use of models for policy evaluation is quantitatively different from the use of models to expand the frontiers of knowledge *per se*, the guidelines of what constitutes minimum acceptable informational content must adapt themselves to the rapidly evolving economic scenarios and become increasingly stricter.

As the overwhelming emphasis of econometricians has been solely on the building and using of models to the comparitive neglect of their evaluation, it is to be expected that models will continue to be inherently controversial unless and until suitably appropriate criteria are specified that lead to the articulation of explicit scientific standards for testing the reliability of a model and thereby blunting some of the very valid criticisms levelled against the use of models in the policy process. It is to this end that the remainder of the study will be devoted.

CHAPTER 4

STRUCTURAL ANALYSIS OF THE MODEL

4.1 Introduction

We had ended the last chapter by mentioning that evaluating the reliability of an econometric model is as important as its estimation. Such an evaluation can take place at three distinct levels and to perform the proper assessment of a model we have to initially identify the purpose(s) behind its original construction. The purposes can range from structural analysis to forecasting to control. These correspond, respectively, to the descriptive, predictive and prescriptive uses of econometrics. This chapter shall be concerned only with the first aspect of it and it is with this notion in mind that we launch into the following preliminaries.

4.2 The Nature Of Structural Analysis

One of the main reasons for performing an econometric study is that of using the estimated model for structural analysis. By this is meant an investigation of the underlying relationships in the system under consideration in order to comprehend and explain relevant economic phenomena. Structural analysis therefore involves the quantitative estimation of all the essential causalities, static as well as dynamic, implied by the model.

The basic step in structural analysis is the estimation of the coefficients of the reduced form of the system. In addition to the estimation of these coefficients themselves, it is also concerned with the interpretation of certain coefficients or combinations of coefficients. Three important ways of interpreting the coefficients are via the multipliers, elasticities and simulations. These will be the subjects of discussion in the remaining sections of this chapter.

4.3 The Reduced Form Of The Model

4.3.1 Linearized approximation of the nonlinear dynamic system

The mathematical model constructed in the last chapter was supposed to represent the annual macroeconomic activity in the Indian economy and is in a form known in the literature as the 'structural form'. As noted earlier, many of the definitions involving value, quantity and price contain nonlinearities. These identities *inter alia* impart a nonlinear content into the structural form of the model. Such a nonlinear structural form of simultaneous equations is not readily amenable to linear control and filtering techniques and therefore, in order to be able to apply the algorithms described in Appendices A and B, suitable transformations on the nonlinear structural form need to be carried out.

While we can sometimes deal with nonlinear dynamic systems directly, very often we must resort to some approximations. If we treat nonlinear systems as perturbations on linear dynamic systems, then we can generally say more about the local behaviour of such nonlinear systems. Thus, in order to apply the results available in linear systems and control theory, we may construct approximations to nonlinear dynamic models that are locally valid, i.e., in some small neighbourhoods about the reference point or path.

Let us consider such a transformation on a nonlinear dynamic system. Let $x(t)$ be the path corresponding to $u(t)$ for the system governed by

$$x(t+1) = f(x(t),u(t),t), \quad t(0) < t < T. \qquad \text{---(4.1)}$$

This implies that $x(t)$ is the reference time-path associated with the reference instrument time-path $u(t)$. Let $y(t)$ be the path corresponding to $u(t)+du(t)$, so that

$$y(t+1) = f(y(t),u(t)+du(t),t), \quad t(0) < t < T; \quad y(t(0)) = x(t(0)). \qquad \text{---(4.2)}$$

Define $z(t)$ to be the difference $y(t)-x(t)$. Then, it is governed by

$$\begin{aligned} z(t+1) &= y(t+1) - x(t+1) \\ &= f(x(t)+z(t),u(t)+du(t),t) - f(x(t),u(t),t), \qquad \text{---(4.3)} \\ &\qquad t(0) < t < T; \quad z(t(0)) = 0. \end{aligned}$$

Define the right-hand side as $h(z(t),du(t),t)$, suppressing the dependence of $\underline{x}$ and $\underline{u}$ since they are known functions of time. Then we have

$$z(t+1) = h(z(t),du(t),t), \quad t(0) < t < T \qquad \text{---(4.4)}$$

with

$$h(0,0,t) = 0, \quad t(0) < t < T.$$

The initial condition $z(t(0))$ need not be zero if $y(t(0))$ is taken to be different from $x(t(0))$. The nonlinear difference equation, eq.(4.4), describes the motion of a control system about the reference path $x(t)$. Typically, in this kind of approximation $du(t)$ is small in some sense, e.g. $\|du(t)\| < \underline{s}$ for all $\underline{t}$ in $[t(0),T]$ for some $T > t(0)$. If $z(t)$ also remains small, then we may approximate eq.(4.4) by linear equations (see Aoki 1976). We thus obtain the associated linear difference equation of eq.(4.4) by expanding $\underline{h}$, retaining linear terms only. The only drawback with this reference path procedure is that we obtain a dynamic control system with time-varying coefficient matrices in its state transition equation. Under the circumstances the control model, with this time-varying feature, implies different transitions of the system state at different times (the system is 'non-stationary') and the Kalman-Bucy theorem on the stability of the filter cannot be applied. To preempt this problem, we adopted the reference point procedure where the underlying assumptions are identical to the ones stated above, the only difference being that we now have only one single reference point rather than a path comprising many such points (see Vishwakarma 1974).

A similar linear approximation of the first-order can be obtained by linearizing the model around this reference value or 'operating point', as it is sometimes called in engineering, in the 40-dimensional space of the endogenous variables. Deviations of second and higher orders can then be neglected as usual. Under this method, however, we end up with a time-invariant transfer function.

Great care needs to be exercised in the selection of the reference values. If the chosen set of values does not satisfy the definitional identities, that is,does not conform to the nonlinearities in the model, the recursive filtering may lead to divergent results (Kau and Kumar 1969).

The reference values selected were the actual historical values taken by all the variables in the base year, i.e., 1970-71, and,as such, all of them satisfied the definitional identities. In all the subsequent calculations reported in this study (except where specifically mentioned otherwise), the results will be related to these reference values, which are provided in Table 4.1.

Table 4.1
Reference Values

Endogenous Variables

No.	Symbol	Value	No.	Symbol	Value
1.	QF	112.9	21.	MR	1634.0
2.	QNF	108.7	22.	WPF	1.0000
3.	QA	111.6	23.	WPNF	1.0000
4.	YAR	16980.0	24.	WP	1.0000
5.	YMR	7117.0	25.	P	1.0000
6.	YTR	1574.0	26.	YNFN	34519.0
7.	YSR	8848.0	27.	YNMN	38047.0
8.	YNAR	17539.0	28.	CC	4371.0
9.	YNFR	34519.0	29.	SCBR	364.0
10.	YGFR	36736.0	30.	TBR	452.0
11.	YPDR	33062.0	31.	M1	7140.0
12.	CPR	29838.0	32.	MS	7140.0
13.	s	0.1323	33.	v	5.1451
14.	SNR	4567.0	34.	BD	285.0
15.	INR	4961.0	35.	NDE	3474.0
16.	KR	96826.0	36.	TR	2451.0
17.	DR	2217.0	37.	NTR	891.0
18.	IAR	901.0	38.	CR	2524.0
19.	ITR	609.0	39.	AF	124.3
20.	FR	394.0	40.	ANF	31.5

Exogenous And Control Variables

No.	Symbol	Value	No.	Symbol	Value
1.	R	96.5	1.	IGAR	292.0
2.	t	10.0	2.	IGTR	428.0
3.	MDL	19.56	3.	GDE	1477.0
4.	XR	1240.0	4.	T	0.1022
5.	PM	1.0000	5.	BR	6.0000
6.	DFEA	-36.0			
7.	DBFA	1681.0			
8.	DEF	1200.0			

In the equations of the model, all the variables were replaced by the sums of their reference values and the deviations from these reference values. In the subsequent simplification, deviations of the second and higher order were ignored. For example, consider eq.(3.26) where we obtain national income at current prices. We have

$$YNFN = P \cdot YNFR$$

Each variable is replaced by the sum of its respective reference value and the deviation from it. We thus have

$$(\hat{YNFN} + dYNFN) = (\hat{P} + dP)(\hat{YNFR} + dYNFR)$$

where ^ designates the reference value and the symbol $\underline{d}$ designates the difference between the value of a variable and its reference value. Substituting their respective reference values from the ones given in Table 4.1, we obtain

$$(34519 + dYNFN) = (1.0000 + dP)(34519 + dYNFR)$$

Simplifying and ignoring all residuals, i.e., deviations of the second order, we end up with the following linearized (in dYNFN) variant of the nonlinear identity:

$$dYNFN = dYNFR + 34519.0\ P$$

The procedure was adopted for all the equations. We note, in passing, that the reference values are the same whether the variable appears in the lagged or unlagged version in an equation. The linearized structural form of the model is as follows:

$$dQF = -2.3764 + 1.4650\ dAF + 0.8031\ dR + 1.9210\ dt$$

$$dQNF = -0.0130 + 11.5406\ dANF + 0.3454\ dR + 1.2239\ dt$$

$$dQA = 0.6812\ dQF + 0.3188\ dQNF$$

$$dYAR = -188.532 + 128.361\ dQA$$

$$dYMR = 204.887 + 0.6836\ dYMR(-1) + 18.5186\ dQNF + 0.7642\ dYTR - 22.8235\ dMDL$$

$$dYTR = 95.1237 + 0.9820\ dYTR(-1) + 0.3307\ dITR$$

$$dYSR = 271.563 + 0.7138\ dYSR(-1) + 0.0909\ dYAR + 0.9368\ dYTR + 0.1987\ dGDE - 21.9170\ dWP$$

$$dYNAR = dYMR + dYTR + dYSR$$

$$dYNFR = dYAR + dYNAR$$

$$dYGFR = dYNFR + dDR$$

$$dYPDR = 103.855 + 0.9608\ dYNFR$$

$$dCPR = -290.531 + 0.1633\ dCPR(-1) + 0.6607\ dYPDR$$

$$ds = 0.008002 + 0.0000091399\ dYNAR - 0.0000008604\ dYNFR$$

$$dSNR = 0.1323\ dYNFR + 34519.0\ ds$$

$$dINR = dSNR + dFR$$

$$dKR = 4961.0 + dKR(-1) + dINR(-1)$$

$$dDR = -154.606 + 0.0213\ dKR$$

$$dIAR = -97.5322 + 0.0420\ dYAR + 2.1759\ dIGAR$$

$$dITR = -7.5062 + 0.0749\ dYTR + 0.9032\ dIGTR$$

$$dFR = dMR - dXR$$

$$dMR = 167.886 + 0.7742\ dMR(-1) + 0.0550\ dCPR + 0.0550\ dINR - 576.443\ dPM + 744.329\ dWP$$

$$dWPF = -0.0259 + 0.6339\ dWPF(-1) - 0.0097\ dQF - 0.0097\ dQNF(-1) + 0.000098\ dCPR$$

$$dWPNF = -0.1016 - 0.0102\ dQNF - 0.0102\ dQNF(-1) + 0.000250\ dYMR + 0.7266\ dWPF$$

$$dWP = -0.0117 + 0.5287\ dWPF + 0.2963\ dWPNF - 0.000086\ dYMR + 0.000039\ dM1 + 3.6484\ dT$$

$$dP = 0.0111 + 0.2349\ dv + 0.00010854\ dM1 - 0.00002213\ dYNFR$$

$$dYNFN = dYNFR + 34519.0\ dP$$

$$dYNMN = 1.1022\ dYNFN + 34519.0\ dT$$

$$dCC = 289.315 + dCC(-1) + 1.0079\ dBD + 0.6126\ dDFEA - 0.2550\ dDBFA$$

$$dSCBR = -219.552 + 0.6102\ dCC - 199.441\ dBR$$

$$dTBR = -24.9188 + 1.1733\ dSCBR$$

$$dM1 = 129.708 + 1.5073\ dCC + 1.5073\ dTBR$$

$$dMS = dM1 - 7140.0\ dWP$$

$$dv = 0.00014006\ dYGFR - 0.0007206\ dMS$$

$$dBD = dGDE + dNDE + dDEF - dTR - dNTR - dCR$$

$$dNDE = 322.955 + 0.6607\ dNDE(-1) + 0.0294\ dYNMN$$

$$dTR = 494.862 + 30089.1\ dT + 0.0798\ dYNFN$$

$$dNTR = 95.4841 + 0.4673\ dNTR(-1) + 0.0202\ dYNFN$$

$$dCR = 267.049 + 0.2385\ dDBFA + 0.6605\ dSNR$$

$$dAF = -0.3519 + 0.2440\ dAF(-1) + 0.0053\ dIAR(-1) + 8.9278\ dWPF - 9.2797\ dWPNF + 0.4063\ dR$$

$$dANF = 0.3523 + 0.3134\ dANF(-1) + 0.0010\ dIAR(-1) - 1.0493\ dWPF + 1.4016\ dWPNF + 0.0546\ dR$$

4.3.2 Reduced form of the linearized small perturbation model

The system of equations presented in the last section represents the structural form of the linearized small perturbation model. These equations express the endogenous variables as functions of other endogenous variables, predetermined variables and disturbances. On the other hand, the reduced form equations describe the endogenous variables in terms of the predetermined variables and disturbances of the structural equations alone.

The statistical justification for including lagged endogenous variables amongst the set of predetermined variables in the reduced form is based on the assumption that the disturbances are independently distributed. This implies that the lagged endogenous variables are independent of the current operation of the equation system. The current endogenous variables are called the jointly dependent variables of the system, while all the other variables are called predetermined. For a lagged endogenous variable we should interpret predetermined in a temporal sense. For a current exogenous variable we interpret it in a causal sense, the variable being determined 'from the outside'. Both these interpretations apply to a lagged exogenous variable.

In the language of dynamic analysis, the reduced form coefficients are called impact multipliers, since they measure the immediate response of the endogenous variables to changes in the predetermined variables. As there exists a definite relationship between the reduced form coefficients and the structural parameters, it is possible to initially estimate the structural parameters and then substitute these estimates into the system of parameters' relationships to obtain the reduced form coefficients.

A quick glance at the linearized version of the model will confirm that all the exogenous variables are current exogenous variables devoid of any lags whatsoever, while endogenous variables appear with a maximum lag of only one period. (Being an annual model, this is a very reasonable order of magnitude). This implies that we can rewrite the linearized small perturbation model in the following vector notation as

$$dx(t) = A_0 dx(t) + A_1 dx(t-1) + B_0 du(t) \qquad ---(4.5)$$

where dx(t-i) are the 40x1 vectors of endogenous variables appearing with lag $\underline{i}$ and du(t) is the 14x1 vector of exogenous variables (including control variables), while the prefix $\underline{d}$ is used to denote small perturbation values of the variables about their reference points. A_0, A_1 and B_0 are appropriately dimensioned time-invariant matrices.

The vector dx is given by:

1	dQF
2	dQNF
3	dQA
4	dYAR
5	dYMR
6	dYTR
7	dYSR
8	dYNAR
9	dYNFR
10	dYGFR
11	dYPDR
12	dCPR
13	ds
14	dSNR
15	dINR
16	dKR
17	dDR
18	dIAR
19	dITR
20	dFR
21	dMR
22	dWPF
23	dWPNF
24	dWP
25	dP
26	dYNFN
27	dYNMN
28	dCC
29	dSCBR
30	dTBR
31	dM1
32	dMS
33	dv
34	dBD
35	dNDE
36	dTR
37	dNTR
38	dCR
39	dAF
40	dANF

The vector of exogenous variables du is given by:

1	dR
2	dt
3	dMDL
4	dXR
5	dPM
6	dDFEA
7	dDBFA
8	dDEF
9	dGDE
10	dIGAR
11	dIGTR
12	dT
13	dBR
14	1

To convert eq.(4.5) into its reduced form, we rewrite it as follows:

$$x(t) = Ax(t-1) + Bu(t) \qquad ---(4.6)$$

where

$$A = (I - A_0)^{-1}A_1 \quad \text{and} \quad B = (I - A_0)^{-1}B_0.$$

Here the prefix d has been omitted both for convenience and because this is usual in most analysis. (This omission of the prefix can sometimes be dangerous since it may make the analyst lose sight of the small-perturbation nature of the equations).

Referring to eq.(4.6) we see that the relationship between the control input vector u(t) and the output vector x(t) can be obtained by applying L or the lag operator, i.e., Lx(t)=x(t-1), to this equation to yield the following:

$$x(t) = (I - AL)^{-1}Bu(t) \qquad ---(4.7)$$

where $(I - AL)^{-1}B$ is the multivariable control input-output transfer function of the system.

This term is of primary importance because it helps to define the mechanisms operating between the input variables and the output variables. Thus the control analyst should give first priority to the identification and estimation of a model which describes the mechanisms implicit in this term. It is in this sense that we can refer to eq.(4.7) as a 'control model' of the system.

In Table 4.2 we provide the numerical values for the matrices A and B which constitute the reduced form of the linearized small perturbation model.

Table 4.2

The Reduced Form Of The Model

A	COL1 COL11 COL21 COL31	COL2 COL12 COL22 COL32	COL3 COL13 COL23 COL33	COL4 COL14 COL24 COL34	COL5 COL15 COL25 COL35	COL6 COL16 COL26 COL36	COL7 COL17 COL27 COL37	COL8 COL18 COL28 COL38	COL9 COL19 COL29 COL39	COL10 COL20 COL30 COL40
ROW1	-0.0306616	0.137026	0	0	-0.00216113	-0.00199811	0.000140492	0	0	0
	0	.0000505865	0	0	7.8177E-10	7.8177E-10	0	0.00867921	-.000672888	0
	3.2313E-23	0.947026	1.41151	0	0	0	0	-4.7436E-07	0	0
	0	0	0	0	-3.1589E-07	0	2.2342E-07	0	0.353208	0.315617
ROW2	3.5939E-17	7.0446E-17	0	0	0	-3.5324E-17	8.5587E-18	0	0	0
	0	-6.1523E-19	0	0	-7.3904E-20	-7.3904E-20	0	0.0115406	-1.1896E-17	0
	-1.5097E-36	-12.1096	16.1753	0	0	0	0	2.0491E-20	0	0
	0	0	0	0	1.9341E-20	0	-1.2311E-20	0	-4.6225E-16	3.61682
ROW3	-0.0208867	0.0933421	0	0	-0.00147216	-0.00136111	.0000957032	0	0	0
	0	.0000344595	0	0	5.3254E-10	5.3254E-10	0	0.00959142	-.000458371	0
	2.2012E-23	-3.21541	6.11821	0	0	0	0	-3.2313E-07	0	0
	0	0	0	0	-2.1518E-07	0	1.5219E-07	0	0.240605	1.36804
ROW4	-2.68103	11.9815	0	0	-0.188969	-0.174714	0.0122846	0	0	0
	0	0.00442326	0	0	6.8357E-08	6.8357E-08	0	1.23116	-0.058837	0
	2.8255E-21	-412.733	785.34	0	0	0	0	-.000041478	0	0
	0	0	0	0	-.000027621	0	.0000195357	0	30.8843	175.603
ROW5	6.6553E-16	1.3046E-15	0	0	0.6836	0.750444	1.5850E-16	0	0	0
	0	-1.1393E-17	0	0	-1.3686E-18	-1.3686E-18	0	0.213716	0.252721	0
	-2.7957E-35	-224.252	299.544	0	0	0	0	3.7947E-19	0	0
	0	0	0	0	3.5817E-19	0	-2.2799E-19	0	-8.5603E-15	66.9785
ROW6	0	0	0	0	0	0.982	0	0	0	0
	0	0	0	0	0	0	0	0	0.3307	0
	0	0	0	0	0	0	0	0	0	0
	0	0	0	0	0	0	0	0	0	0
ROW7	-0.13751	1.1461	0	0	-0.0170989	0.904551	0.71507	0	0	0
	0	0.000226868	0	0	0.000003979	0.000003979	0	0.113107	0.304618	0
	1.6447E-19	-45.0386	72.1636	0	0	0	0	-0.00241438	0	0
	0	0	0	0	-0.00160778	0	0.00113715	0	2.83686	16.1359
ROW8	-0.13751	1.1461	0	0	0.666501	2.637	0.71507	0	0	0
	0	0.000226868	0	0	0.000003979	0.000003979	0	0.326823	0.888039	0
	1.6447E-19	-269.29	371.708	0	0	0	0	-0.00241438	0	0
	0	0	0	0	-0.00160778	0	0.00113715	0	2.83686	83.1144
ROW9	-2.81854	13.1276	0	0	0.477533	2.46228	0.727355	0	0	0
	0	0.00465013	0	0	.0000040474	.0000040474	0	1.55799	0.829202	0
	1.6729E-19	-682.024	1157.05	0	0	0	0	-0.00245586	0	0
	0	0	0	0	-0.0016354	0	0.00115669	0	33.7212	258.718
ROW10	-2.81854	13.1276	0	0	0.477533	2.46228	0.727355	0	0	0
	0	0.00465013	0	0	0.021304	0.021304	0	1.55799	0.829202	0
	1.6729E-19	-682.024	1157.05	0	0	0	0	-0.00245586	0	0
	0	0	0	0	-0.0016354	0	0.00115669	0	33.7212	258.718

A	COL1 COL11 COL21 COL31	COL2 COL12 COL22 COL32	COL3 COL13 COL23 COL33	COL4 COL14 COL24 COL34	COL5 COL15 COL25 COL35	COL6 COL16 COL26 COL36	COL7 COL17 COL27 COL37	COL8 COL18 COL28 COL38	COL9 COL19 COL29 COL39	COL10 COL20 COL30 COL40
ROW11	-2.70806	12.613	0	0	0.458813	2.36576	0.698843	0	0	0
	0	0.00446784	0	0	.0000038887	.0000038887	0	1.49691	0.796698	0
	1.6074E-19	-655.288	1111.69	0	0	0	0	-0.00235959	0	0
	0	0	0	0	-0.0015713	0	0.00111135	0	32.3993	248.576
ROW12	-1.78921	8.33339	0	0	0.303138	1.56306	0.461725	0	0	0
	0	0.166252	0	0	.0000025693	.0000025693	0	0.989011	0.526378	0
	1.0620E-19	-432.949	734.494	0	0	0	0	-0.00155898	0	0
	0	0	0	0	-0.00103815	0	0.000734266	0	21.4062	164.234
ROW13	.0000011682	-8.1977E-07	0	0	.0000056809	.0000219833	.0000059099	0	0	0
	0	-1.9274E-09	0	0	3.2885E-11	3.2885E-11	0	.0000016466	.0000074031	0
	1.3593E-24	-0.00187447	0.00240185	0	0	0	0	-1.9954E-08	0	0
	0	0	0	0	-1.3288E-08	0	9.3982E-09	0	-3.0851E-06	0.000537057
ROW14	-0.332566	1.70848	0	0	0.259276	1.0846	0.300231	0	0	0
	0	0.00054868	0	0	.0000016706	.0000016706	0	0.262962	0.365253	0
	0	-154.937	235.987	0	0	0	0	-0.00101371	0	0
	0	0	0	0	-.000675048	0	0.000477448	0	4.35482	52.767
ROW15	-4.27252	0.24524	0	0	0.289192	1.2209	0.339058	0	0	0
	0	0.0165532	0	0	-.000140856	-.000140856	0	0.292919	0.411152	0
	0.819259	81.1397	264.574	0	0	0	0	0.0854687	0	0
	0	0	0	0	0.0569152	0	-0.040255	0	4.79497	59.1591
ROW16	0	0	0	0	0	0	0	0	0	0
	0	0	0	0	1	1	0	0	0	0
	0	0	0	0	0	0	0	0	0	0
	0	0	0	0	0	0	0	0	0	0
ROW17	0	0	0	0	0	0	0	0	0	0
	0	0	0	0	0.0213	0.0213	0	0	0	0
	0	0	0	0	0	0	0	0	0	0
	0	0	0	0	0	0	0	0	0	0
ROW18	-0.112603	0.503222	0	0	-0.00793668	-0.00733798	0.000515951	0	0	0
	0	0.000185777	0	0	2.8710E-09	2.8710E-09	0	0.0517089	-0.00247115	0
	1.1867E-22	-17.3348	32.9843	0	0	0	0	-1.7421E-06	0	0
	0	0	0	0	-1.1601E-06	0	8.2050E-07	0	1.29714	7.37533
ROW19	0	0	0	0	0	0.0749	0	0	0	0
	0	0	0	0	0	0	0	0	0	0
	0	0	0	0	0	0	0	0	0	0
	0	0	0	0	0	0	0	0	0	0
ROW20	-3.93995	-1.46324	0	0	0.0299162	0.136297	0.0388271	0	0	0
	0	0.0160045	0	0	-.000142527	-.000142527	0	0.0299567	0.0458997	0
	0.819259	236.076	28.5871	0	0	0	0	0.0864824	0	0
	0	0	0	0	0.0575903	0	-0.0407325	0	0.44015	6.39213

A	COL1 COL11 COL21 COL31	COL2 COL12 COL22 COL32	COL3 COL13 COL23 COL33	COL4 COL14 COL24 COL34	COL5 COL15 COL25 COL35	COL6 COL16 COL26 COL36	COL7 COL17 COL27 COL37	COL8 COL18 COL28 COL38	COL9 COL19 COL29 COL39	COL10 COL20 COL30 COL40
ROW21	-3.93995	-1.46324	0	0	0.0299162	0.136297	0.0388271	0	0	0
	0	0.0160045	0	0	-.000142527	-.000142527	0	0.0299567	0.0458957	0
	0.819259	236.076	28.5871	0	0	0	0	0.0864824	0	0
	0	0	0	0	0.0575903	0	-0.0407325	0	0.44015	6.39213
ROW22	-0.00957793	-.000512479	0	0	.0000506705	0.000172561	.0000438863	0	0	0
	0	0.000015802	0	0	2.4421E-10	2.4421E-10	0	.000127347	.0000581121	0
	1.0094E-23	0.582285	0.0582887	0	0	0	0	-1.4818E-07	0	0
	0	0	0	0	-9.8675E-08	0	6.9791E-08	0	-0.00132831	0.0130335
ROW23	-0.00695932	-0.0105724	0	0	0.000207717	0.000312994	.0000318878	0	0	0
	0	.0000114817	0	0	1.7744E-10	1.7744E-10	0	-.000055032	0.000105404	0
	7.3342E-24	0.490543	-0.0477495	0	0	0	0	-1.0767E-07	0	0
	0	0	0	0	-7.1697E-08	0	5.0710E-08	0	-.000965147	-0.0106769
ROW24	-0.00484538	-0.00259975	0	0	-3.5763E-06	-.000022598	-7.0076E-06	0	0	0
	0	.0000079941	0	0	-1.8127E-07	-1.8127E-07	0	-.000054478	-7.6101E-06	0
	-7.4924E-21	0.343163	-0.0354166	0	0	0	0	0.000109988	0	0
	0	0	0	0	.0000732432	0	-.000051803	0	-0.00134472	-0.00791921
ROW25	-0.0094375	-0.00425225	0	0	.0000523981	0.000220375	0.000061119	0	0	0
	0	.0000155703	0	0	7.6428E-07	7.6428E-07	0	-7.7575E-06	.0000742138	0
	2.6251E-21	0.608772	0.0106496	0	0	0	0	-.000038537	0	0
	0	0	0	0	-.000025662	0	.0000181506	0	-.000706941	0.00238126
ROW26	-328.591	-133.656	0	0	2.28626	10.0694	2.83712	0	0	0
	0	0.542122	0	0	0.0263863	0.0263863	0	1.2902	3.39059	0
	9.0784E-17	20332.2	1524.66	0	0	0	0	-1.33271	0	0
	0	0	0	0	-0.887478	0	0.627696	0	9.31828	340.916
ROW27	-362.174	-147.315	0	0	2.51992	11.0985	3.12707	0	0	0
	0	0.597527	0	0	0.029083	0.029083	0	1.42206	3.73755	0
	1.0006E-16	22410.1	1680.48	0	0	0	0	-1.46891	0	0
	0	0	0	0	-0.978178	0	0.691846	0	10.2706	375.758
ROW28	22.6081	7.96851	0	0	-0.328366	-1.40806	-0.393161	0	0	0
	0	-0.0372997	0	0	-0.00179879	-0.00179879	0	-0.262959	-0.474181	0
	-7.4351E-17	-1282.07	-260.975	0	0	0	0	1.09147	0	0
	0	0	0	0	0.726832	0	-0.514074	0	-3.53393	-58.3543
ROW29	13.7955	4.86238	0	0	-0.200369	-0.859199	-0.239907	0	0	0
	0	-0.0227602	0	0	-0.00109762	-0.00109762	0	-0.160458	-0.289345	0
	-4.5369E-17	-782.322	-159.247	0	0	0	0	0.666016	0	0
	0	0	0	0	0.443513	0	-0.313688	0	-2.1564	-35.6078
ROW30	16.1862	5.70503	0	0	-0.235093	-1.0081	-0.281482	0	0	0
	0	-0.0267046	0	0	-0.00128784	-0.00128784	0	-0.188265	-0.339489	0
	-5.3231E-17	-917.898	-186.844	0	0	0	0	0.781436	0	0
	0	0	0	0	0.520374	0	-0.36805	0	-2.53011	-41.7787

A	COL1 COL11 COL21 COL31	COL2 COL12 COL22 COL32	COL3 COL13 COL23 COL33	COL4 COL14 COL24 COL34	COL5 COL15 COL25 COL35	COL6 COL16 COL26 COL36	COL7 COL17 COL27 COL37	COL8 COL18 COL28 COL38	COL9 COL19 COL29 COL39	COL10 COL20 COL30 COL40
ROW31	58.4747	20.6101	0	0	-0.849302	-3.64188	-1.01689	0	0	0
	0	-0.0964736	0	0	-0.00465249	-0.00465249	0	-0.68013	-1.22644	0
	-1.9230E-16	-3316.02	-674.997	0	0	0	0	2.82303	0	0
	0	0	0	0	1.87991	0	-1.32963	0	-9.14032	-150.93
ROW32	93.0707	39.1723	0	0	-0.823767	-3.48053	-0.966855	0	0	0
	0	-0.153551	0	0	-0.00335825	-0.00335825	0	-0.291159	-1.17211	0
	-1.3881E-16	-5766.2	-422.123	0	0	0	0	2.03772	0	0
	0	0	0	0	1.35696	0	-0.959749	0	0.460988	-94.3873
ROW33	-0.0674615	-0.0263889	0	0	0.00066049	0.00285293	0.000798589	0	0	0
	0	0.0001113	0	0	.0000054038	.0000054038	0	0.000428021	0.000960759	0
	1.0005E-19	4.0596	0.466238	0	0	0	0	-0.00146872	0	0
	0	0	0	0	-.000978052	0	0.000691757	0	0.0043908	0.104252
ROW34	22.4309	7.90605	0	0	-0.325792	-1.39702	-0.390079	0	0	0
	0	-0.0370073	0	0	-0.00178469	-0.00178469	0	-0.260898	-0.470464	0
	-7.3768E-17	-1272.03	-258.929	0	0	0	0	0.0907546	0	0
	0	0	0	0	0.721135	0	-0.510045	0	-3.50623	-57.897
ROW35	-10.6479	-4.33107	0	0	0.0740856	0.326296	0.091936	0	0	0
	0	0.0175673	0	0	0.00085504	0.00085504	0	0.0418087	0.109884	0
	2.9418E-18	658.858	49.4061	0	0	0	0	-0.0431861	0	0
	0	0	0	0	0.631942	0	0.0203403	0	0.301956	11.0473
ROW36	-26.2216	-10.6657	0	0	0.182444	0.803538	0.226402	0	0	0
	0	0.0432613	0	0	0.00210563	0.00210563	0	0.102958	0.270601	0
	7.2446E-18	1622.51	121.668	0	0	0	0	-0.10635	0	0
	0	0	0	0	-0.0708208	0	0.0500901	0	0.743598	27.2051
ROW37	-6.63755	-2.69985	0	0	0.0461825	0.203402	0.0573098	0	0	0
	0	0.0109509	0	0	0.000533004	0.000533004	0	0.0260621	0.068498	0
	1.8338E-18	410.71	30.7981	0	0	0	0	-0.0269208	0	0
	0	0	0	0	-0.0179271	0	0.479979	0	0.188229	6.88651
ROW38	-0.21966	1.12845	0	0	0.171252	0.71638	0.198303	0	0	0
	0	0.000362403	0	0	.0000011035	.0000011035	0	0.173686	0.241249	0
	6.7632E-17	-102.336	155.869	0	0	0	0	-.000669554	0	0
	0	0	0	0	-.000445869	0	0.000315354	0	2.87636	34.8526
ROW39	-0.0209294	0.0935331	0	0	-0.00147518	-0.0013639	0.000095899	0	0	0
	0	.0000345301	0	0	5.3363E-10	5.3363E-10	0	0.00592437	-.000459309	0
	2.2057E-23	0.646434	0.963491	0	0	0	0	-3.2380E-07	0	0
	0	0	0	0	-2.1562E-07	0	1.5251E-07	0	0.241097	0.215438
ROW40	0	0	0	0	0	0	0	0	0	0
	0	0	0	0	0	0	0	0.001	0	0
	0	-1.0493	1.4016	0	0	0	0	0	0	0
	0	0	0	0	0	0	0	0	0	0.3134

B	COL1	COL2	COL3	COL4	COL5	COL6	COL7
	COL8	COL9	COL10	COL11	COL12	COL13	COL14
ROW1	1.46682	2.00495	0.0721543	-4.1738E-23	-2.4059E-20	-2.9059E-07	2.3499E-07
	-4.7811E-07	-6.7124E-18	0	.0000386306	0.00308132	.0000646887	-2.11759
ROW2	0.975517	1.2239	0	1.9500E-36	1.1241E-33	1.2553E-20	-1.2905E-20
	3.2201E-20	1.7550E-17	0	2.4147E-18	-1.1819E-16	-5.8383E-19	4.05275
ROW3	1.31019	1.75595	0.0491515	-2.8432E-23	-1.6389E-20	-1.9795E-07	1.6008E-07
	-3.2569E-07	1.0225E-18	0	.0000263151	0.00209899	.0000440659	-0.150484
ROW4	168.178	225.396	6.30913	-3.6495E-21	-2.1037E-18	-.000025409	.0000205475
	-.000041806	-3.4983E-15	0	0.00337784	0.269429	0.00565634	-207.848
ROW5	18.0652	22.6649	-22.8235	3.6111E-35	2.0816E-32	2.3246E-19	-2.3899E-19
	5.9631E-19	3.2501E-16	0	4.4717E-17	-2.1886E-15	-1.0812E-17	352.632
ROW6	0	0	0	0	0	0	0
	0	0	0	0	0	0	95.1237
ROW7	15.4495	20.7056	0.570883	-2.1244E-19	-1.2246E-16	-0.00147905	0.00119605
	-0.00243345	-3.2501E-16	0	0.19662	15.6831	0.329249	344.558
ROW8	33.5147	43.3705	-22.2526	-2.1244E-19	-1.2246E-16	-0.00147905	0.00119605
	-0.00243345	0	0	0.19662	15.6831	0.329249	792.314
ROW9	201.693	268.766	-15.9435	-2.1608E-19	-1.2456E-16	-0.00150446	0.00121659
	-0.00247526	-3.4983E-15	0	0.199998	15.9526	0.334506	584.465
ROW10	201.693	268.766	-15.9435	-2.1608E-19	-1.2456E-16	-0.00150446	0.00121659
	-0.00247526	-3.4983E-15	0	0.199998	15.9526	0.334906	535.529
ROW11	193.786	258.231	-15.3185	-2.0761E-19	-1.1968E-16	-0.00144548	0.0011689
	-0.00237823	-3.3612E-15	0	0.192158	15.3272	0.321777	665.409
ROW12	128.035	170.613	-10.1209	-1.3717E-19	-7.9071E-17	-.000955031	0.000772294
	-0.0015713	-2.2207E-15	0	0.126959	10.1267	0.212598	149.105
ROW13	0.000132785	0.000165156	-.000189669	-1.7557E-24	-1.0121E-21	-1.2224E-08	9.8850E-09
	-2.0112E-08	-7.3722E-21	0	0.000001625	0.000129617	.0000027212	0.0147388
ROW14	31.2675	41.2588	-8.6565	0	0	-.000620997	0.000502175
	-0.00102172	-5.6860E-16	0	0.0825535	6.58477	0.13824	586.093
ROW15	34.7134	45.7872	-9.65532	-1.0582	-609.993	0.0523581	-0.0423398
	0.0861439	-8.8490E-16	0	0.180527	-555.181	-11.6554	706.751
ROW16	0	0	0	0	0	0	0
	0	0	0	0	0	0	4961
ROW17	0	0	0	0	0	0	0
	0	0	0	0	0	0	-48.9367

B	COL1 COL8	COL2 COL9	COL3 COL10	COL4 COL11	COL5 COL12	COL6 COL13	COL7 COL14
ROW18	7.06347	9.46662	0.264984	-1.5328E-22	-8.8357E-20	-1.0672E-06	8.6299E-07
	-1.7558E-06	2.1759	0	0.000141869	0.011316	0.000237566	-106.262
ROW19	0	0	0	0	0	0	0
	0	0	0.9032	0	0	0	-7.5062
ROW20	3.44586	4.52842	-0.99882	-1.0582	-609.993	0.0529791	-0.042842
	0.0871656	-3.1630E-16	0	0.0979738	-561.765	-11.7936	120.657
ROW21	3.44586	4.52842	-0.99882	-0.0582011	-609.993	0.0529791	-0.042842
	0.0871656	-3.1630E-16	0	0.0979738	-561.765	-11.7936	120.657
ROW22	-0.00168079	-0.00272795	-0.00169175	-1.3038E-23	-7.5156E-21	-9.0774E-08	7.3405E-08
	-1.4935E-07	-9.3114E-20	0	.0000120672	0.000962528	.0000202072	0.00925291
ROW23	-0.00665524	-0.00879968	-0.0069351	-9.4733E-24	-5.4608E-21	-6.5957E-08	5.3336E-08
	-1.0852E-07	4.0417E-19	0	.0000087681	0.000699373	.0000146825	-0.048057
ROW24	-0.00739628	-0.00990635	0.000119402	9.6776E-21	5.5786E-18	.0000673787	-.000054486
	0.000110857	3.1991E-19	0	0.000108906	-0.714452	-0.0149991	-0.126692
ROW25	-0.00212318	-0.00299332	0.00174943	-3.3908E-21	-1.9546E-18	-.000023608	.0000190906
	.0000388841	-1.5036E-19	0	.000021828	5.93144	0.00525528	-0.0200504
ROW26	128.402	165.44	-76.3319	-1.1726E-16	-6.7595E-14	-0.816419	0.660204
	-1.34324	-1.3568E-14	0	0.553472	204763	181.742	-107.655
ROW27	141.525	182.348	-84.133	-1.2925E-16	-7.4503E-14	-0.899857	0.727677
	-1.48052	-1.4954E-14	0	0.610037	260209	200.316	-118.657
ROW28	-29.5633	-38.738	10.9632	9.6036E-17	5.5359E-14	0.668635	-0.540698
	1.10009	1.3029E-15	0	0.99065	-43258.7	-12.474	-632.186
ROW29	-18.0395	-23.6379	6.68976	5.8601E-17	3.3780E-14	0.408001	-0.329934
	0.671277	7.9503E-16	0	0.604495	-26396.5	-207.053	-605.312
ROW30	-21.1658	-27.7344	7.8491	6.8757E-17	3.9634E-14	0.478708	-0.387111
	0.78761	9.3281E-16	0	0.709254	-30971	-242.935	-735.131
ROW31	-76.464	-100.194	28.3558	2.4839E-16	1.4318E-13	1.72939	-1.39849
	2.84534	3.3699E-15	0	2.56227	-111886	-384.978	-1931.25
ROW32	-23.6545	-29.4624	27.5033	1.7929E-16	1.0335E-13	1.24831	-1.00945
	2.05382	1.0857E-15	0	1.78467	-106785	-277.884	-1026.67
ROW33	0.0452945	0.058874	-0.0220519	-1.2923E-19	-7.4493E-17	-.000899741	0.000727583
	-0.00148033	-2.7603E-18	0	-0.00125802	76.9516	0.20029	0.814821
ROW34	-29.3316	-38.4344	10.8773	9.5283E-17	5.4925E-14	0.0555963	-0.283458
	1.09147	1.2927E-15	0	0.982886	-42919.6	-12.3762	-914.278

B	COL1 COL8	COL2 COL9	COL3 COL10	COL4 COL11	COL5 COL12	COL6 COL13	COL7 COL14
ROW35	4.16084 -0.0435272	5.36102 -4.3966E-16	-2.47351 0	-3.7998E-18 -0.0179351	-2.1904E-15 7650.15	-0.0264558 5.88929	0.0213937 319.466
ROW36	10.2465 -0.10719	13.2021 -1.0827E-15	-6.09129 0	-9.3575E-18 -0.0441671	-5.3941E-15 46429.2	-0.0651502 14.503	0.0526843 486.271
ROW37	2.59373 -0.0271334	3.34188 -2.7407E-16	-1.5419 0	-2.3687E-18 -0.0111801	-1.3654E-15 4136.22	-0.0164917 3.67119	0.0133361 93.3095
ROW38	20.6522 -.000674843	27.2514 -3.7556E-16	-5.71762 0	-8.7357E-17 0.0545266	-5.0356E-14 4.34924	-.000410169 0.0913072	0.238832 654.164
ROW39	0.453053 -3.2635E-07	0.0573038 -4.5818E-18	0.0492521 0	-2.8490E-23 0.000026369	-1.6423E-20 0.00210329	-1.9836E-07 .0000441561	1.6040E-07 0.176662
ROW40	0.0546 0	0 0	0 0	0 0	0 0	0 0	0 0.3523

4.4 Dynamic Policy Multipliers

4.4.1 The final form of the model

Policy multipliers of an economic model convey extremely useful information on the magnitude and timing of influences of the instruments on the endogenous variables. A special type of multipliers are the well-known impact multipliers which were computed in the last section. However, equally important are the long run multipliers which indicate the permanent effects on the endogenous variables as a result of a unit impulse (or a once-and-for-all change) in the policy variables.

There are as many long run multipliers as there are pairs of targets and instruments. In a multidimensional dynamic system, we have a multiplier matrix where each element of the matrix is the dynamic policy multiplier of a target and instrument pair. When there are preferred or assigned pairings of targets and instruments, the diagonal elements of the multiplier matrix show the effects of the instruments on the intended targets, while the off-diagonal elements represent spillover effects or the unintended effects of instruments on non-assigned targets, if the pairs are numbered by the same subscript.

In order to obtain this matrix, we can rewrite our dynamic econometric system given by eq.(4.6) in the following lag operator notation which is essentially equivalent to the control model version of eq.(4.7):

$$A(L)x(t) = B(L)u(t) \qquad \text{---(4.8)}$$

where A(L) and B(L) are matrices of polynomials in the lag operator $\underline{L}$ given by

$$A(L) = I - AL \quad \text{and}$$
$$B(L) = B.$$

It is the presence of $\underline{I}$ in the equation defining the polynomial A(L) that generates 'instantaneous coupling' (see Salmon and Wallis 1982) between the endogenous variables. This coupling is removed in the reduced form which therefore expresses each endogenous variable as an explicit function of predetermined variables only.

Thus an alternative solution of the reduced form is obtained by multiplying through by the inverse of A(L), under appropriate stability conditions, to result in the final form approximation of the equation system. This is given by

$$x(t) = A(L)^{-1}B(L)u(t) \qquad \text{---(4.9)}$$

in which each endogenous variable is expressed as an infinite distributed-lag function of the exogenous variables. The matrix $A(L)^{-1}B(L)$ is known in systems literature as the transfer function matrix where each of its elements is a polynomial in $\underline{L}$. The coefficients in the expansion of $A(L)^{-1}B(L)$ provide the dynamic multipliers of the system, describing the response of x(i,t) to a unit shock in u(j,t-k), where x(i,t) represents the value of the ith endogenous variable in time t. In empirical work the infinite distributed-lag is generally approximated by a ratio of finite-degree polynomials in L, but we can also obtain an explicit expression by writing $A(L)^{-1}=a(L)/|A(L)|$, where a(L) is the adjoint matrix of A(L) and |A(L)| is the determinant. Thus the final form may be written as

$$x(t) = \frac{a(L)}{|A(L)|} B(L)u(t) \qquad ---(4.10)$$

giving a set of multi-input 'transfer function' equations. Multiplying through by |A(L)|, we obtain a further representation known as the final equations (Tinbergen 1939, Goldberger 1959):

$$|A(L)|\, x(t) = a(L)B(L)u(t) \qquad ---(4.11)$$

Since |A(L)| is a scalar polynomial, each final equation relates a given endogenous variable to its own past values and to the exogenous variables, current and past, but to no other endogenous variable, current and past. In effect, the dynamic interrelations with the other endogenous variables have been solved out. The interesting property of this representation is that the autoregressive operator |A(L)| is common to all the endogenous variables unless, as noted by Goldberger (1959), the model is decomposable for this results in the cancellation of common factors across some of the final equations. Thus, the characteristic dynamic behaviour of the endogenous variables can, in principle, be studied by considering the common characteristic polynomial |A(L)|.

4.4.2 Stability conditions

In the earlier section we had mentioned, in passing, that there should exist appropriate stability conditions on A(L) if the final form of the model is to be obtained. This basically implies that the transfer function matrix $A(L)^{-1}B(L)$ should be stable in the sense that an impulse should not produce a divergent impulse response. A rational transfer function for discrete systems is considered to be stable if and only if all its poles lie strictly inside the unit circle (see Aoki 1983).

To demonstrate this equivalence, let us rewrite eq.(4.8) in its expanded final form representation given by

$$x(t) = T(L)u(t) = \sum_{i=0}^{\infty} T_i u(t-i) \qquad ---(4.12)$$

where T(L) is a matrix whose elements are rational polynomials in the lag operator L and is given by

$$T(L) = (I - AL)^{-1}B \qquad ---(4.13a)$$

$$= \left(\sum_{i=0}^{\infty} A^i L^i\right)B \qquad ---(4.13b)$$

so that

$$T_0=B,\ T_1=AB,\ T_2=A^2B,\ \ldots,\ T_k=A^kB,\ \ldots \qquad ---(4.14)$$

The final form matrices T_i in eq.(4.14), expressed in terms of the matrices A and B, are called Markov matrices in the control literature. In econometric literature, these matrices are referred to as matrix multipliers. Thus T_0 (=B) is designated as the impact matrix multiplier, since its elements measure the immediate effect of a unit variation in any control variable on any endogenous variable. Similarly, the matrix T_i is termed as the interim, delay or dynamic matrix multiplier with lag $\underline{i}$. Finally, assuming that A is stable, the steady-state static matrix multiplier representing the equilibrium effects on the endogenous variables as a result of a sustained unit change in the control variables is given by $(I - A)^{-1}B$. An inspection of the matrices T_i in eq.(4.14) reveals that it is the state transition matrix A that determines their dynamic evolution.

Thus, an important, yet oft neglected, property of the system governed by the time-invariant state transition equation given by eq.(4.6) is whether or not the solution of the difference equation tends to grow indefinitely as $\underline{t} \to \infty$. This implies that we should be able to ascertain whether the linear system is dynamically stable in the sense that as time tends to infinity there is no danger that the system will tend to move outside a finite range. Such a kind of stability is considered essential if the model is expected to provide a realistic representation of an economic system. In order to study this question, we assume that we are dealing with an autonomous system, i.e., without an input u(t). Thus, we reduce our attention to the time-invariant linear discrete system given by

$$x(t) = Ax(t-1) \qquad \text{---(4.15)}$$

If such a system does indeed settle down towards the zero state, i.e., the solution $x(t) = 0$ as $\underline{t} \to \infty$, then it is considered to be stable.

It has been shown (see Athans and Falb 1966) that such a system is asymptotically, as well as exponentially, stable if and only if all the characteristic values of A have modulii strictly less than unity. Asymptotic stability implies that the terminal state eventually approaches the zero state, no matter what the initial state was originally. Exponential stability implies that the terminal state converges exponentially to the zero state irrespective of the initial state.

In the terminology of discrete-time systems, if the characteristic values (also known as characteristic roots or eigen values) have modulii less (greater) than unity they are supposed to be inside (outside) the unit circle. With systems that are not asymptotically or exponentially stable, it is convenient to refer to those eigen values that lie strictly inside the unit circle as the stable poles of the system, and to all the remaining ones as the unstable poles.

4.4.3 The eigen values of the system

The solution of the eigen value problem was obtained by recourse to the LR transformation method (Rutishauser 1958).

The purpose of this algorithm is to compute all the eigen values (real or complex) of a given nxn matrix A with real or complex elements. This method uses a similarity transformation involving the triangular factors of a matrix to produce another matrix with great diagonal dominance. If this is applied iteratively, the diagonal elements normally converge to the eigen values of the original matrix. When the matrix has certain patterns of zero elements, these are retained by the transformation thereby improving the efficiency of the algorithm.Thus, the transformations produced from A a sequence of similar matrices, $A^{(k)}$, which tended to a form from which the eigen values were easily extracted. This form was block triangular with each diagonal block corresponding to eigen values of equal magnitude. As A was real in our case, so also was $A^{(k)}$ and the limiting form had along its diagonal, blocks of 1x1 and 2x2 principal submatrices. The 2x2 blocks corresponded to the complex conjugate pairs of eigen values.

The existence of complex roots will contribute to the oscillatory behaviour of the model which implies that the solution will be a sinusoidal function of time. To determine the period of oscillation for such complex roots we employ the following relationship:

$$T = 2\pi / \tan^{-1}(b/a) \qquad \text{---(4.16)}$$

where T: time period of oscillation
a: real part of the root
b: imaginary part of the root

As A was a 40x40 matrix, we obtained 40 eigen values of which only 15 were non-zero. These are listed, in Table 4.3, in descending order of absolute magnitude along with their modulus (also known as damping factors) as well as the periodic modes of behaviour of the system contingent upon its complex conjugate characteristic values.

Table 4.3
Eigen Values Of The Model

No.	Eigen Value		Modulus	T
	Real part(a)	Imaginary part(b)	$\sqrt{a^2+b^2}$	
1.	0.9998	0.0	0.9998	-
2.	0.9964	0.0	0.9964	-
3.	0.9820	0.0	0.9820	-
4.	0.8193	0.0	0.8193	-
5.	0.7153	0.0	0.7153	-
6.	0.7036	0.0	0.7036	-
7,8.	0.6330	$\pm$0.1410	0.6485	28.67
9.	0.4779	0.0	0.4779	-
10.	0.3134	0.0	0.3134	-
11.	0.1661	0.0	0.1661	-
12,13.	-0.0968	$\pm$0.0249	0.0999	24.96
14,15.	0.0105	$\mp$0.0222	0.0246	5.56

It is thus noticed that the system possesses 9 real non-zero roots and 3 pairs of complex conjugate roots. The largest root of the system is found to be real with magnitude less than one. This implies that the system is stable. The damping, however, is not very strong. The 6 eigen values which are complex contribute to the oscillatory characteristics of the model and thereby impart a number of periodic modes of behaviour into the system. The largest time-period of oscillation is as big as 28.7 years with a damping factor of about 0.65. This introduces a certain amount of cyclical components in the response of the model. The other two periodic components have time periods of 25.0 and 5.6 years. These, however, have extremely strong damping factors of 0.10 and 0.02, respectively, and therefore do not contribute too much to the cyclical behaviour of the model.

4.4.4 The long run multipliers of the model

The assurance that all the characteristic values of A have modulii strictly less than unity implies that the estimated model can be considered to be asymptotically (as well as exponentially) stable. Therefore we have the result that (see Nikaido 1972)

$$\lim_{t \to \infty} A^t \to 0 \qquad ---(4.17)$$

Therefore the final form of the system as generated by eq.(4.12) does converge and it is possible to obtain its steady state solution. It is to be noted that the mean path of models that explain economic growth endogenously (e.g. the Harrod-Domar model) have no steady-state solutions and in such models we cannot obtain the long run multipliers of the system.

From eqs.(4.13) and (4.14) we obtain the expression for the long run multiplier of the system which is given by

$$(I + A + A^2 + \dots)B \qquad ---(4.18)$$

The power series in eq.(4.18), called a Neumann expansion, converges if A is stable or, equivalently, if all the characteristic roots of A have modulii strictly less than unity. This condition is a necessary and sufficient one for the final form of the system to converge to a steady-state, as well as to ensure that (I-A) is nonsingular. Incorporating the result of eq.(4.17), we obtain the limiting form of eq.(4.18) which is given by

$$F = (I - A)^{-1}B \qquad ---(4.19)$$

In Table 4.4, we provide the numerical values of the matrix F, which constitutes the long run multipliers of the model.

Table 4.4

The Long Run Multipliers Of The Model

F	COL1	COL2	COL3	COL4	COL5	COL6	COL7
	COL8	COL9	COL10	COL11	COL12	COL13	COL14
ROW1	1.71497	2.01625	0.167653	-.000018453	-0.0471092	-2.4370E-17	1.9751E-17
	-4.0095E-17	0.0275381	0.097916	0.000551481	1.2644E-12	-2.0935E-17	27.1214
ROW2	1.24682	1.23249	-0.361241	9.5113E-07	0.00242814	-6.0814E-17	4.9685E-17
	-1.0004E-16	0.034579	-0.525969	-.000028425	3.1803E-12	7.6223E-17	-158.467
ROW3	1.56572	1.76639	-0.00095858	-.000012267	-0.0313167	-3.5581E-17	2.8813E-17
	-5.8537E-17	0.0297827	-0.100979	0.000366607	1.8705E-12	-1.0974E-16	-32.0441
ROW4	200.977	226.735	-0.123044	-0.00157462	-4.01984	-5.6425E-15	4.5680E-15
	-9.2835E-15	3.82294	-12.9617	0.0470581	2.9575E-10	-1.4164E-14	-4301.74
ROW5	72.9751	72.1362	-93.2781	.0000556689	0.142117	-3.7112E-15	2.9979E-15
	-6.1053E-15	2.02388	-137.355	-0.00166368	1.8954E-10	2.2797E-14	-41681.3
ROW6	0	0	0	0	0	0	0
	0	0	-44.1231	0	0	0	-13685.3
ROW7	67.1367	75.5963	-1.36549	-0.0240782	-61.4691	-1.9024E-15	1.5400E-15
	-3.1289E-15	1.28609	-153.318	0.719585	1.0001E-10	-6.0645E-15	-46559.4
ROW8	140.112	147.733	-94.6436	-0.0240226	-61.327	-5.5422E-15	4.4816E-15
	-9.1163E-15	3.30997	-334.796	0.717922	2.8824E-10	4.1238E-15	-101926
ROW9	341.089	374.468	-94.7666	-0.0255972	-65.3468	-1.1182E-14	9.0430E-15
	-1.8397E-14	7.13291	-347.758	0.76498	5.8401E-10	-1.0324E-14	-106228
ROW10	-1442.84	-1322.03	2074.04	3.68349	9403.55	24.0906	-19.4811
	39.6358	-46.5171	6874.09	25.6069	-1246733	-4.0615E-12	2088271
ROW11	327.719	359.788	-91.0518	-0.0245938	-62.7852	-8.8865E-15	7.1863E-15
	-1.4619E-14	6.8533	-334.125	0.734993	4.6490E-10	-9.8798E-15	-101960
ROW12	258.783	284.107	-71.899	-0.0194205	-49.5784	-7.0154E-15	5.6740E-15
	-1.1544E-14	5.41171	-263.842	0.580387	3.6711E-10	-7.8722E-15	-80859.8
ROW13	0.000987135	0.00102807	-.000783496	-1.9754E-07	-.000504298	-3.0097E-20	2.4342E-20
	-4.9517E-20	.0000241156	-0.00276079	.0000059035	1.5675E-15	4.3955E-20	-0.832195
ROW14	79.201	85.03	-39.5831	-0.0102054	-26.0533	-2.6051E-15	2.1313E-15
	-4.2778E-15	1.77613	-141.308	0.304991	1.3541E-10	-1.2336E-15	-42780.5
ROW15	0	0	0	0	0	0	0
	0	0	0	0	0	0	-4961
ROW16	-83752.6	-79647.8	101822	174.136	444549	1131.01	-914.604
	1860.84	-2518.78	339054	1166.29	-58532042	-1.9349E-10	103035380
ROW17	-1783.93	-1696.5	2168.81	3.70909	9468.9	24.0906	-19.4811
	39.6358	-53.65	7221.84	24.8419	-1246733	-4.0623E-12	2194499

F	COL1 / COL8	COL2 / COL9	COL3 / COL10	COL4 / COL11	COL5 / COL12	COL6 / COL13	COL7 / COL14
ROW18	8.44105	9.52287	-0.00516786	-.000066134	-0.168833	-2.3700E-16	1.9187E-16
	-3.8993E-16	2.33646	-0.544392	0.00197644	1.2422E-11	-5.9510E-16	-278.205
ROW19	0	0	0	0	0	0	0
	0	0	-2.40162	0	0	0	-1032.53
ROW20	-79.201	-85.03	39.5831	0.0102054	26.0533	2.6475E-15	-2.1666E-15
	4.3526E-15	-1.77613	141.308	-0.304991	-1.3647E-10	-5.7035E-15	37819.5
ROW21	-79.201	-85.03	39.5831	1.01021	26.0533	2.6669E-15	-2.1820E-15
	4.3833E-15	-1.77613	141.308	-0.304991	-1.3679E-10	-1.0983E-14	37819.5
ROW22	-0.0216051	-0.0307915	-0.0281305	-4.2207E-06	-0.0107751	-8.5226E-19	6.6575E-19
	-1.4019E-18	-.000010631	-0.0758156	0.000126138	3.5884E-14	2.7183E-17	-23.153
ROW23	-0.0228895	-0.0294817	-0.0363898	-3.0723E-06	-0.00784319	5.6872E-19	-4.7675E-19
	9.3578E-19	-.000207167	-0.0786966	.0000918159	-3.1179E-14	4.3950E-18	-24.1122
ROW24	-0.0431485	-0.046788	0.0173207	0.000307891	0.786013	1.3665E-18	-1.1102E-18
	2.2428E-18	-0.00093869	0.0623631	-.000135408	-7.2189E-14	6.9856E-18	17.5889
ROW25	-0.0780976	-0.0840851	0.0368386	.0000095315	0.0243329	0.000792583	-.000640929
	0.00130402	-0.00173643	0.131784	0.00101917	-35.3365	-1.2781E-16	40.3127
ROW26	-2354.76	-2528.07	1176.86	0.303421	774.602	27.3592	-22.1242
	45.0136	-52.807	4201.29	35.9457	-1219781	-4.3204E-12	1285328
ROW27	-2595.42	-2786.43	1297.14	0.334431	853.766	30.1553	-24.3853
	49.614	-58.2038	4630.67	39.6194	-1309923	-4.7947E-12	1416688
ROW28	-185.066	-154.348	346.517	3.08351	7871.86	1.3319E-14	-1.0603E-14
	2.1671E-14	-6.91605	1129.68	-2.27463	-36168.8	136.37	331243
ROW29	-112.927	-94.1834	211.445	1.88156	4803.41	8.0891E-15	-6.4362E-15
	1.3132E-14	-4.22017	689.328	-1.38798	-22070.2	-116.228	201905
ROW30	-132.497	-110.505	248.088	2.20763	5635.84	9.7065E-15	-7.7749E-15
	1.5967E-14	-4.95153	808.789	-1.62852	-25895	-136.37	236870
ROW31	-478.663	-399.214	896.248	7.97533	20360.2	3.6971E-14	-2.9615E-14
	6.0747E-14	-17.888	2921.85	-5.88322	-93548.7	-2.1454E-13	856446
ROW32	-170.583	-65.1479	772.578	5.77699	14748	2.6755E-14	-2.1321E-14
	4.4025E-14	-11.1858	2476.58	-4.91641	-93548.7	-3.7557E-14	730861
ROW33	-0.0791626	-0.138218	-0.26623	-0.00364699	-9.31037	0.00337413	-0.00272852
	0.0055514	0.00154527	-0.821836	0.00712927	-107.206	-4.8289E-16	-234.175
ROW34	-5.6960E-14	-5.7057E-14	9.9434E-14	9.0452E-16	2.3092E-12	-0.607798	0.253601
	4.4082E-16	-1.9906E-15	3.2034E-13	-2.0484E-16	-1.7744E-11	8.2611E-14	-287.047

F	COL1	COL2	COL3	COL4	COL5	COL6	COL7
ROW35	-224.89	-241.442	112.396	0.0289781	73.9779	2.61292	-2.11296
ROW36	-187.91	-201.74	93.9137	0.024213	61.8132	2.18326	-1.76551
ROW37	-89.2927	-95.8643	44.6267	0.0115057	29.3729	1.03746	-0.838951
ROW38	52.3123	56.1623	-26.1446	-0.00674066	-17.2082	-1.7163E-15	0.2385
ROW39	0.622434	0.0650167	0.114439	-.000012596	-0.0321564	-1.6634E-17	1.3481E-17
ROW40	0.0781084	0.000743918	-0.0313018	8.2416E-08	0.000210399	1.9938E-18	-1.6067E-18

F	COL8	COL9	COL10	COL11	COL12	COL13	COL14
ROW35	4.299	-5.0433	401.243	3.43298	-113504	-4.0109E-13	123706
ROW36	3.59208	-4.214	335.263	2.86847	-67249.4	-3.6408E-13	103064
ROW37	1.70692	-2.00244	159.313	1.36306	-46254.1	-1.5687E-13	48918.9
ROW38	-2.8239E-15	1.17313	-93.3339	0.201447	8.9483E-11	1.9339E-16	-27989.5
ROW39	-2.7368E-17	0.0187973	0.0668369	0.000376438	8.6303E-13	-1.4330E-17	20.135
ROW40	3.2804E-18	0.00299629	-0.0455756	-.000002463	-1.0006E-13	-1.0080E-17	-13.7301

4.5 Multiplier Analysis

Three specific types of multiplier analysis are discussed in this section in order to indicate the nature of this approach to structural analysis.

4.5.1 Static multiplier analysis

The single most important aspect of multiplier analysis is a validation of certain key impact multipliers of the model. Such a vindication usually forms the cornerstone for policy prescription. Our analysis will thus be based upon certain relevant portions of the panel of impact multipliers provided in Table 4.2. These results which comprise the effects of the key exogenous and control variables on ten of the most important endogenous variables of the system are reproduced in Tables 4.5a and 4.5b, respectively.

Table 4.5a
Impact Multipliers Of The Model

(with respect to certain key exogenous variables)

Endogenous variable affected	Exogenous change in R	MDL	DBFA	DEF
1. YNFR	201.7	-15.94	.0012	-.0024
2. CPR	128.0	-10.12	.0008	-.0016
3. ε	.00013	-.00019	.0	.0
4. SNR	31.27	-8.656	.0005	-.0010
5. INR	34.71	-9.655	-.0423	.0861
6. MR	3.446	-.9988	-.0428	.0872
7. WP	-.0074	.0001	-.0001	.0001
8. YNMN	141.5	-84.13	.7277	-1.481
9. M1	-76.46	28.36	-1.398	2.845
10. BD	-29.33	10.88	-.2835	1.091

The results seem to confirm all our prior assumptions regarding the functioning of the Indian economy. The first column indicates that additional rainfall to the extent of one extra index point would increase NDP by almost Rs. 200 crores ($ 160 million) of which consumption would absorb almost 65 percent. Savings and investment would increase by almost Rs. 30 crores apiece. The budget deficit would decrease by an almost equivalent amount thereby bringing about a fall in the supply of money and consequently the price level. All these clearly underscore the extreme reliance of the Indian economy upon weather conditions. The second column reveals an exactly opposite consequence *vis-à-vis* the effects of man-days lost upon the economy. An *additional* one million man-days lost would serve to decrease NDP by almost Rs. 16 crores ($ 12.8 million). The budget deficit would swell up by Rs. 11 crores and consequently money supply and prices would go up. The third column indicates that an increase in domestic borrowing (including foreign aid) by one crore rupees would decrease the budget deficit by Rs. 0.3 crores and money supply by Rs. 1.4 crores. Both these contractions, however, have a negligible impact upon the price level. Contrary to a popular misconception, defence expenditures do not have any widespread deleterious effects on the economy. The fourth column indicates that its effect in as far as reducing real national income and increasing prices are concerned is quite weak. It is however noted that increasing defence spending by an additional one crore rupees would serve to increase imports by almost Rs. 0.09 crores (at constant prices). At current prices this would amount to almost Rs. 0.32 crores. This implies that about 10 - 30 percent of Indian defence needs are imported. As a consequence of this increase in spending, the budget deficit as well as money supply would register increases of Rs. 1.1 crores and Rs. 2.8 crores, respectively.

Table 4.5b
Impact Multipliers Of The Model
(with respect to certain key control variables)

Endogenous variable affected	Policy change in GDE	T/100	BR
1. YNFR	.2000	.1595	.3349
2. CPR	.1270	.1013	.2126
3. s	.0	.0	.0
4. SNR	.0826	.0658	.1382
5. INR	.1805	-5.552	-11.66
6. MR	.0980	-5.618	-11.79
7. WP	.0001	-.0071	-.0150
8. YNMN	.6100	2602.09	200.3
9. M1	2.5622	-1118.86	-384.978
10. BD	.9829	-429.20	-12.38

As before, all the estimated multipliers are found to be in accordance with our assumptions. The first column indicates that an increase in government developmental expenditure to the tune of one crore rupees would increase NDP by only Rs. 0.2 crores, of which Rs. 0.13 crores would be absorbed by consumption. The budget deficit and money supply do increase, but the subsequent impact on the price level is diluted because of the marginal increase in real output. While the impact of this instrument on the savings rate has been indicated as zero in the table, it actually does have a very small effect whose order of magnitude is 1.625×10^{-6}. This can perhaps be best understood in terms of its reciprocal. To increase the savings rate by one percentage point (via the upward effect of the controller on non-agricultural income from which the propensity to save is higher) would require an increase in government developmental expenditure to the extent of

$$\frac{.01}{(1.625)(10^{-6})}$$

or almost Rs. 6150 crores ($ 5 billion) which provides a rough idea of the enormity of the problem facing Indian planners on the income redistribution front. All these factors serve to undermine the power of this instrument. The second column indicating the effect of the tax rate on the economy is an interesting one. The immediate impact of an increase in the tax rate by one percentage point is to reduce the budget deficit by almost Rs. 430 crores ($ 350 million). This causes a reduction in money supply by almost Rs. 1100 crores. The overall impact on the wholesale price level is to reduce it, thereby answering the interesting question posed in the last chapter as to which of the two countervailing forces would prevail. NDP increases by about Rs. 0.16 crores, thereby indicating that the full potential of this instrument has not yet been reached and for the time being, at least, we can increase the tax rate without suffering any loss of real output. This helps to qualify this instrument as a fairly potent one. The third column indicates the effect of the bank rate on the economy. This is the only instrument of monetary policy that the model possesses and, as such, its role in helping to steer the economy needs to be carefully examined. Although a few Indian economists have indicated that the bank rate is ineffective for controlling the economy due to the pervasive effect of a large non-monetized sector, all the estimated multipliers seem to run counter to this implication. An increase in the bank rate by one percentage point helps to decrease the price level by 1.5 percentage points. This is because of its negative impact on money supply which is reduced by almost Rs. 385 crores. The fall in prices reduces non-developmental expenditure and as a consequence the budget deficit decreases by about Rs. 12 crores. NDP and savings increase by Rs. 0.30 crores and Rs. 0.14 crores, respectively. Investment and imports decline by almost identical margins of Rs. 12 crores. All these factors clearly underscore the effectiveness of monetary policy as a powerful instrument.

4.5.2 Dynamic multiplier analysis

A second type of multiplier analysis is a comparitive study of the impact and long run multipliers of the system. This type of an analysis is useful because it helps to identify which of the effects (of the exogenous variables upon the endogenous variables) are of a transient, as against a sustained, nature. The relevant comparitive extracts from Tables 4.2 and 4.4 are provided in Table 4.6a (with respect to the two uncontrollable exogenous variables) and Table 4.6b (with respect to a fiscal controller).

The results in Table 4.6a are quite revealing because they indicate the nature and magnitude of the dynamic responses of the economy when shocked by exogenous impulses. It is seen that most of the multipliers exhibit buildups to reach terminal limiting values comparable in sign but much greater in magnitude. Thus, while a once-and-for-all shock of a unit increase in the index number of rainfall has an initial impact multiplier of 201.7 vis-à-vis real national income, this multiplier builds up over time to reach a cumulative long run value of almost 341.1. Comparable results are obtained for all the other endogenous variables with the exception of investment, budget deficit, imports and national income at current prices. The first two have long run multipliers which trail off to zero. The former is indicative of a capital-stock adjustment mecha-

Table 4.6a
Impact And Long Run Multipliers Of The Model
(with respect to rainfall and man-days lost)

Endogenous variable affected	Exogenous change in R: Impact	R: Long run	MDL: Impact	MDL: Long run
1. YNFR	201.7	341.1	-15.94	-94.77
2. CPR	128.0	258.8	-10.12	-71.90
3. s	.0001	.0010	-.0002	-.0008
4. SNR	31.27	79.20	-8.656	-39.58
5. INR	34.71	.0	-9.655	.0
6. MR	3.446	-79.20	-.9988	39.58
7. WP	-.0074	-.0431	.0001	.0173
8. YNMN	141.5	-2595.0	-84.13	1297.0
9. M1	-76.46	-478.7	28.35	896.2
10. BD	-29.33	.0	10.88	.0

nism and the latter suggests a built-in stabilizer property of budgets. The long-term fall in imports points to a self-containing feedback (see Waelbroeck 1978) which highlights the natural interdependence between the level of imports and the level of investment (including inventories). In the case of national income at current prices, the fact that the initial multiplier of 141.5 exhibited a reverse buildup to reach a cumulative long run value of -2595.0 is a reflection on the fall in prices over the period. Exactly the same analysis (in reverse) is provided by the comparitive study of the impact and long run multipliers of man-days lost. It is noted that an impulse of an extra one million man-days lost has the immediate effect of reducing NDP by nearly Rs. 16 crores. The long run impact, which typifies a negative multiplier buildup, cumulates to as much as Rs. 95 crores ($ 76 million), a large portion of which is absorbed in the form of reduced consumption. The long term rise in the rate of inflation (to the extent of nearly 2 percentage points) is a direct consequence of the fall in real output. All these are alarming results and serve to highlight the critical need for industrial peace, the absence of which is currently depriving the Indian economy of a vast share of present as well as future potential output.

The results in Table 4.6b are equally revealing because they generally indicate that government developmental expenditure has a fairly sustained impact upon the economy. (It needs to be mentioned that the long run multipliers of the tax rate as well as the bank rate were practically zero throughout thereby indicating that the responses they originally triggered in the endogenous variables were entirely of a transient nature devoid of any sustained characteristics). Most of the multipliers do show signs of a buildup albeit of a weak nature. As far as real income is concerned, the initial impact multiplier increases from 0.2 to a cumulative long run value of nearly 0.8. Similar responses are exhibited in the case of two other

Table 4.6b
Impact And Long Run Multipliers Of The Model
(with respect to government expenditure)

Endogenous variable affected	Policy change in GDE: Impact	Policy change in GDE: Long Run
1. YNFR	.2000	.7650
2. CPR	.1270	.5804
3. s	.0	.0
4. SNR	.0826	.3050
5. INR	.1805	.0
6. MR	.0980	-.3050
7. WP	.0001	-.0001
8. YNMN	.6100	39.62
9. M1	2.562	-5.883
10. BD	.9829	.0

variables, i.e., consumption and savings. It needs to be noted that while government developmental expenditure is measured in current prices; income, consumption and savings are all at constant prices. Thus, in terms of current prices, the long run multipliers vis-à-vis these three variables would have a considerably higher order of magnitude. Investment and the budget deficit have long run multipliers which trail off to zero for reasons indicated earlier. The latter prompts a long term fall in money supply as well as prices.

4.5.3 Comparitive static multiplier analysis

The third type of multiplier analysis is to determine the relative efficacy of monetary versus fiscal policy via a comparitive static analysis. Within our model we have two comparable instruments, i.e., the tax rate and the bank rate, the former representing a fiscal tool and the latter a monetary one. As both their multipliers could be computed with reference to percentage point changes in these controllers, the relative efficiency of the two instruments can easily be gauged. Thus the results can suggest which of the two should be more powerfully activated in order to optimally steer the economy.

The multiplier indicating the effect of the tax rate on NDP, dYNFR/dT, was approximately 0.1595 over the sample period. The multiplier indicating the effect of the bank rate on NDP, dYNFR/dBR, was approximately 0.3349 over the same period. A measure of the relative effectiveness of fiscal versus monetary policy is the ratio of the multipliers, which may then be interpreted as the marginal rate of substitution of monetary policy for fiscal policy. We therefore define

$$MRS^{YNFR}_{BR,T} = \frac{dYNFR/dT}{dYNFR/dBR} = \frac{dBR}{dT} \qquad \text{---(4.20)}$$

This marginal rate of substitution was estimated at 0.47.

Thus, it is established that monetary policy (in terms of increasing the bank rate by one percentage point) is more effective than fiscal policy (in terms of increasing the tax rate by an equivalent factor) for stepping-up national income. This is quite an interesting finding because it sidelines the Keynesian contention adopted by many Indian economists that fiscal policy is more effective than monetary policy under conditions of high unemployment which is an endemic feature of the Indian economy. We provide a comparitive analysis of the marginal rates of substitution between monetary and fiscal policy for the same set of ten endogenous variables in Table 4.7.

Table 4.7
The Marginal Rates Of Substitution
Between Monetary And Fiscal Policy

Endogenous variable affected	Policy change in T and BR: $\frac{d(.)}{dT}$	$\frac{d(.)}{dBR}$	$MRS^{(.)}_{BR,T}$
1. YNFR	0.1595	0.3349	0.4763
2. CPR	0.1013	0.2126	0.4765
3. s	.00000130	.00000272	0.4779
4. SNR	0.0658	0.1382	0.4761
5. INR	-5.552	-11.66	0.4762
6. MR	-5.618	-11.79	0.4765
7. WP	-.0071	-.0150	0.4733
8. YNMN	2602.	200.3	12.99
9. M1	-1119.	-384.9	2.91
10. BD	-429.2	-12.38	34.67

The results are extremely interesting. They clearly indicate a demarcation line between the sectors on the basis of the marginal rates of substitution between monetary and fiscal policy. Out of the 10 sectors in the model, monetary policy is seen to be superior in 6 of them, these being the output sector, the consumption sector, the savings sector, the investment sector, the trade sector and the price sector. On the other hand, fiscal policy is much more effective in the income sector, the monetary sector and the government sector. In short, it is seen that monetary policy is more effective in the real sector of the economy while fiscal policy is more effective in the financial sector of the economy. This strange reversal of roles wherein monetary (fiscal) policy has a comparitive advantage over fiscal (monetary) policy in the real (financial) sector in a true Mundellian fashion (Mundell 1968) could be the reason for the hitherto poor performance of the Indian economy seemingly brought about by a wrong assignment problem and an incorrect tracking solution.

4.6 Elasticities

It is often more convenient to express the comparitive static results of the last section in the form of elasticities.

An advantage of this representation is that it provides a dimensionless measure of the sensitivity of the endogenous variable to changes in the exogenous variable. The elasticity of a variable, thus, does not depend upon the units in which it is measured.

The elasticity of x(i) with respect to u(j) at some equilibrium levels of the concerned variables is given by:

$$e(i,j) = \frac{u(j)}{x(i)} \cdot \frac{\partial x(i)}{\partial u(j)}, \quad i=1,2,..,n; \; j=1,2,..,s. \qquad \text{---(4.21)}$$

This can be interpreted as the proportionate change in the endogenous variable x(i) given a unit proportionate change in the exogenous variable u(j), all other parameters and variables being held constant. As the fraction $\partial x(i)/\partial u(j)$ in eq.(4.21) is the impact multiplier of u(j) on x(i), the elasticity of x(i) with respect to u(j) is nothing more than this impact multiplier weighted by the fraction u(j)/x(i) which can be considered as the ratio of their respective equilibrium (or reference year) values. As far as our analysis was concerned, we assumed these to be reference year values.

Thus, an alternative measure of the relative effectiveness of policy would be the ratio of elasticities. This would be a more accurate approximation because of the lack of units involved. Therefore, if we are comparing the relative efficacy of the tax rate with the bank rate in as far as their impact on NDP is concerned, then the marginal rate of substitution of monetary for fiscal policy in terms of elasticities would be given by the following relationship:

$$MRS_{e\;BR,T}^{YNFR} = \frac{\frac{dYNFR}{dT} \cdot \frac{T(0)}{YNFR(0)}}{\frac{dYNFR}{dBR} \cdot \frac{BR(0)}{YNFR(0)}} = \frac{dBR}{dT} \cdot \frac{T(0)}{BR(0)} \qquad \text{---(4.22)}$$

where YNFR(0), T(0) and BR(0) stand for the reference year values of NDP, the tax rate and the bank rate, respectively. Therefore, the ratio of elasticities is obtained by weighting the marginal rates of substitution, as provided in Table 4.7, by a constant fraction given by T(0)/BR(0), which is the ratio of the tax rate to the bank rate in the reference year. From Table 4.1 we obtain this ratio which is seen to be approximately 1.7033 (= 10.22/6.00). Therefore we have the following marginal rates of substitution of monetary for fiscal policy in terms of elasticities which are provided in Table 4.8.

The results in Table 4.8 indicate that the revised marginal rates of substitution of monetary policy for fiscal policy in terms of elasticities are biased relatively in favour of fiscal policy. This was to be expected because the ratio of the tax rate to the bank rate in the reference year, i.e., the weighting ratio, was greater than unity. However, and this is very important, the results do not contradict the essential conclusions arrived at in the last section where we had shown that monetary (fiscal) policy tracks the real (financial) sector far more effectively than its counterpart. This seems to clearly suggest that the instrument-target assignment problem

Table 4.8
The Marginal Rates Of Substitution
Between Monetary And Fiscal Policy
(in terms of elasticities)

Endogenous variable	$MRS^{(\cdot)}_{e\ BR,T}$	Endogenous variable	$MRS^{(\cdot)}_{e\ BR,T}$
1. YNFR	0.8113	6. MR	0.8116
2. CPR	0.8116	7. WP	0.8062
3. s	0.8140	8. YNMN	22.13
4. SNR	0.8109	9. M1	4.957
5. INR	0.8111	10. BD	59.05

should be so framed such that each policy variable is allowed to track macroeconomic targets only in that sector in which it has a comparitive advantage. A wrong allignment could very well result in a loss of control effort and an ensuing sub-optimal strategy.

4.7 Simulation

4.7.1 The method

Simulation, as we will use the word, is simply the mathematical solution of a simultaneous set of difference equations. This set of equations is referred to as the simulation model. Given a model whose parameters have been estimated, given initial values for the endogenous variables and given time series data on the exogenous variables, the model can be 'solved' for each period. The simultaneous solution of all equations of the model which yields the time paths for each of the endogenous variables is known as the 'simulation solution'.

Simulations of a model can be performed for four basic reasons: model testing and evaluation, historical policy analysis, forecasting and policy prescription. From the viewpoint of structural analysis, it is the first reason that is the most important because simulation is additional proof (after the preliminaries of multiplier analysis have been carried out) on the basis of which the model can be vindicated.

In general, simulation refers therefore to the determination of the interacting behaviour of a system which is supposed to be sufficiently explicit so that it can be programmed for numerical study using a computer. The system's numerical behaviour is then determined (simulated) under different assumptions and each simulation run is considered an experiment performed on the model, determining values of the endogenous variables for alternative scenarios regarding the exogenous variables and the stochastic disturbance terms. The latter can either be specified *à priori* at given levels (e.g., their expected values of zero) or chosen via random drawings from a distribution with known and appropriate characteristics. If the disturbance terms are preset at zero, the simulation is referred to as a deterministic simulation. In all other cases it is known as a stochastic simulation.

Simulations are usually classified as either dynamic or static (also known as one-period) simulations. A dynamic simulation is one where historical values are provided for the endogenous variables as starting values only. Thereafter the lagged values of the endogenous variables are determined by the simulation solution of the earlier periods. This technique promulgates cumulative errors which is a major potential cause for divergence in all such simulations. A successful dynamic simulation is therefore considered a more stringent test of a model's reliability because a static simulation, by 'reinitializing' the values of the endogenous variables at each stage using historical records, eliminates the possibility of cumulative errors thereby imparting a pseudo-stability to the simulated values - a feature which, if taken at its face value, can cause a model builder to seriously overestimate the reliability of his system.

This section deals with the dynamic deterministic simulation of the estimated macroeconometric model. This simulation, which was carried out using TROLL, has basically two aspects: normalization and block structuring. In normalization, each nonlinear equation is linearized so that only one endogenous variable appears on the left-hand side. In block structuring, the model is divided into blocks such that each block contains only simultaneous equations. The blocks are ordered in recursive fashion so that the solution of each block depends on prior solutions of one or more previous blocks. There are several algorithms available for solving the simulation problem. Newton's method uses partial derivatives and solves linear systems of equations in one step. Nonlinear systems are approximated by a linear function based on a first-order Taylor-series expansion of the equations and their solution requires iterative use of the algorithm. Newton's method is generally the most efficient and it is particularly good if the model involves significant nonlinearities. The two relaxation algorithms, which are the Gauss-Seidel and the Jacobi methods, use different iterative methods to effect a solution. Their basic drawback is that they require equations to be normalized (i.e., each equation to have a different left-hand-side variable without a coefficient) unlike Newton's method. This feature very often results in divergence that arises because of faulty normalization. As such, we avoided these relaxation methods and carried out the simulation using Newton's method, albeit with a slight modification.

This was in the form of specifying the model in its linearized small perturbation version itself. Our motivation for pursuing such a linear analysis rested on the basic presumption that by doing so we were taking advantage of the considerable analytic understanding of linear structures. This was clearly evident in the wealth of information we gleaned in the earlier section on multiplier analysis which would have been impossible had we retained the original nonlinear version of the model. This is because policy multipliers are not uniquely defined and eigen values do not exist in the same sense, since they are all aspects of linear algebra. It is surprising that this fact has been so extensively disregarded, particularly in comparisons of policy multiplier paths.

The obvious drawback with this approach is the possibility of distorting critical model nonlinearities in the linearization stage. While it is the belief of a few econometricians that models are often only mildly nonlinear (see Kuh and Neese 1982), many others are more prone to believe that this will vary from model to model and very often linearization can be a fatal flaw (see Chow and Corsi 1982). Moreover, since the very act of linearization has different effects on different models it is practically impossible to gain unambiguous information through a comparison of alternative (even mildly) nonlinear models.

In any event, we proceeded with the linearization because of its overwhelming advantages and accordingly we were able to convert the original form of the model which was given by:

$$x(t) = f(x(t), x(t-1), u(t)) \qquad \text{---(4.23)}$$

where

$x(t)$ is a vector of $\underline{n}$ endogenous variables,
$\underline{f}$ is a vector of $\underline{n}$ (nonlinear) equations and
$\underline{u}(t)$ is a vector of $\underline{m+s}$ exogenous variables,

into its form provided in eq.(4.5), called in the literature as the linearized structural form deviations model, which was given by:

$$D.dx(t) = E.dx(t-1) + F.du(t) \qquad \text{---(4.24)}$$

where, as mentioned earlier, $dx(t)$ and $du(t)$ are vectors of endogenous and exogenous variables (all measured in terms of deviations from their reference values), respectively; and D ($= I-A_0$), E ($= A_1$) and F ($= B_0$) are partial derivative matrices (which are held constant for the purpose of the study, despite their possible time-varying nature, by specifying a reference point rather than a reference path).

Given that the D matrix is invertible, we obtained the reduced form deviations model as:

$$dx(t) = A\,dx(t-1) + B\,du(t) \qquad \text{---(4.25)}$$

```
dx(t)  =  A dx(t-1)  +  B du(t)
  :           :               :
  :___(*)_____:               :
        :                     :
        :_________(**)________:
```

As expressed diagrammatically beneath eq.(4.25), the reduced form deviations model is composed of two basic components. The system's homogeneous dynamics (*) are governed by the A matrix. Multipliers (**) are generated by exogenous variable changes and their coefficients B, together with the homogeneous dynamics. It is the combined responses of these two phenomena that entail the so-called simulation solution. It should be noted that such a solution will, under the reduced form deviations framework, be provided in deviations form.

4.7.2 The results

As most of the equations of the model were estimated using annual data over the period 1961-82, we could easily have carried out a full-model full-period historical simulation of the system in order to assess its dynamic interactive behaviour. However, in view of the fact that the Indian economy remained fairly stable during the period 1961-73, it was felt that such a full-period simulation of the model would have imparted a certain amount of 'inertial stability' to the overall simulation solution thereby imbuing the model with a false aura of robustness. As such, in the interests of rigour, it was felt that the simulation should be carried out over the decade 1973-82 which has been an extremely trying period for the Indian economy in view of the fact that it was buffeted by innumerable shocks of magnitudes never imagined before during these years. Several events took place practically simultaneously over this period. The Arab oil embargo imposed in 1973 caused import prices to increase so rapidly that they exhibited a runaway annual inflation rate of 14.8 percent during the decade 1973-82 as against their annual inflation rate of a sedate 3.7 percent over the period 1961-73. Man-days lost as a result of industrial disputes which averaged 17.4 million per year over the phase 1961-73 nearly doubled to reach 30.2 million per year during the decade 1973-82. As a result of a staggering increase in inward remittances, the average annual level of net foreign exchange assets was nearly Rs. 3000 crores ($ 2.4 billion) over the period 1973-82 which was a massive ten-fold increase over the meagre average annual level of about Rs. 300 crores evinced during the earlier phase. The sum of net domestic borrowing and foreign aid quadrupled from an average annual level of Rs. 1350 crores during the period 1961-73 to an average annual level of nearly Rs. 5800 crores ($ 4.6 billion) over the subsequent decade. All these factors have led many Indian economists to conclude that the periods 1961-73 and 1973-82 are structurally different and it would be very difficult for any model that is estimated using data over the entire phase 1961-82, without appropriate dummy variables, to successfully replicate the events that occurred after 1973. It is with this contention in mind that we conducted the simulation only over the troubled decade 1973-82 in order to subject the model to the 'acid' test.

The results of the dynamic simulation are displayed graphically on the following pages. An explanatory note on these graphs is in order. In all the cases, the time bounds of the simulation extended from 1972-73 to 1981-82. The historical and simulated levels of all the endogenous variables, which were in terms of deviations from their respective reference values, were plotted against the horizontal axis on which the coordinate 0.0 represented the value taken by the variable in the base year, i.e., 1970-71. All positive and negative deviations were computed with respect to this value. The scale factor (if any) used to compress the axis is provided on the upper right hand corner of the graph. In all the graphs, the unbroken line (——) denoted by KALMAN represented the actual values of the variable, while the broken line (---) denoted by KALSIM represented the simulated values of the variable.

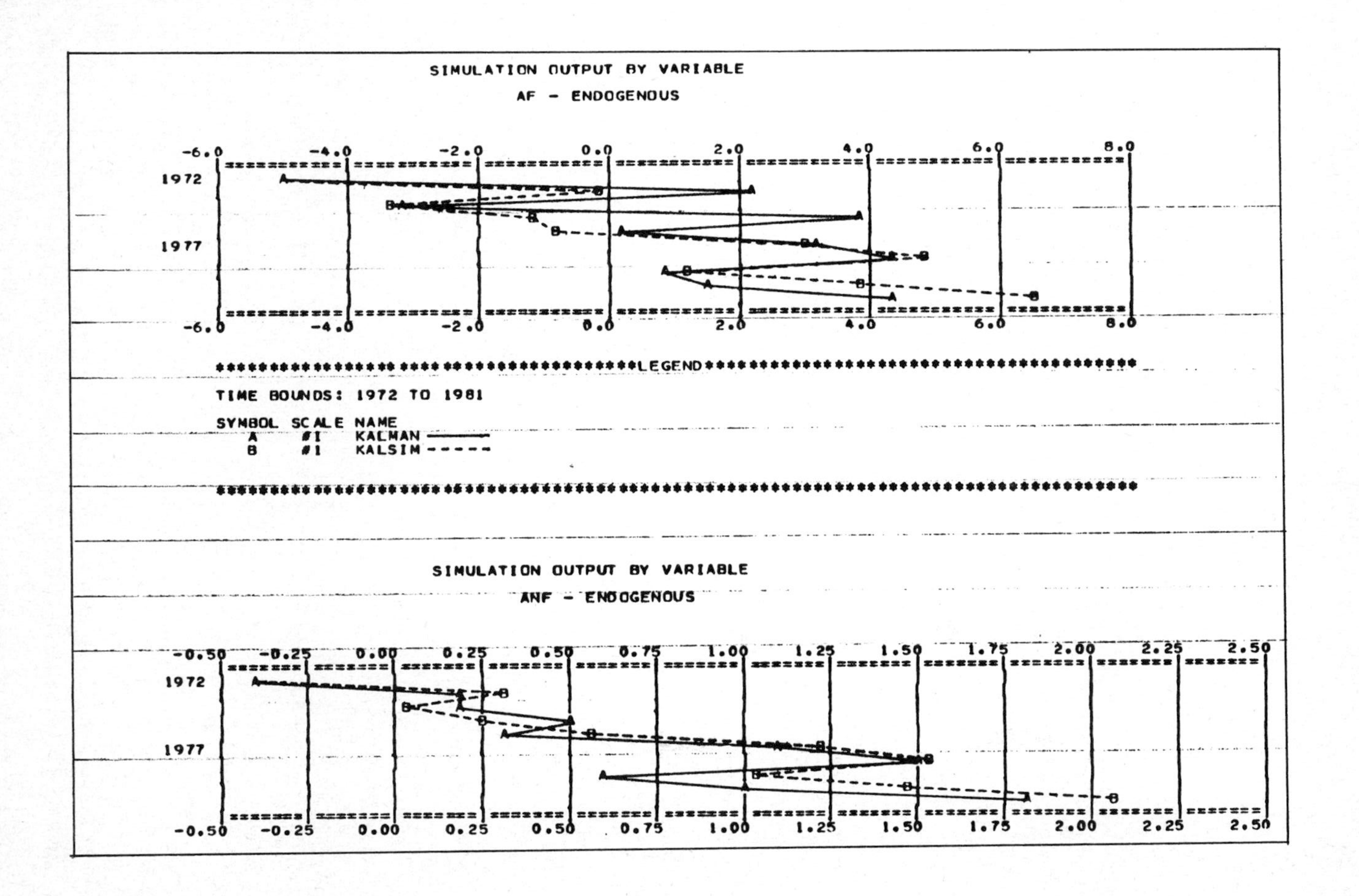
SIMULATION OUTPUT BY VARIABLE
AF - ENDOGENOUS
-6.0
-4.0
-2.0
0.0
2.0
4.0
6.0
8.0
1972
1977
LEGEND
TIME BOUNDS: 1972 TO 1981
SYMBOL SCALE NAME
A #1 KALMAN
B #1 KALSIM
SIMULATION OUTPUT BY VARIABLE
ANF - ENDOGENOUS
-0.50
-0.25
0.00
0.25
0.50
0.75
1.00
1.25
1.50
1.75
2.00
2.25
2.50
1972
1977

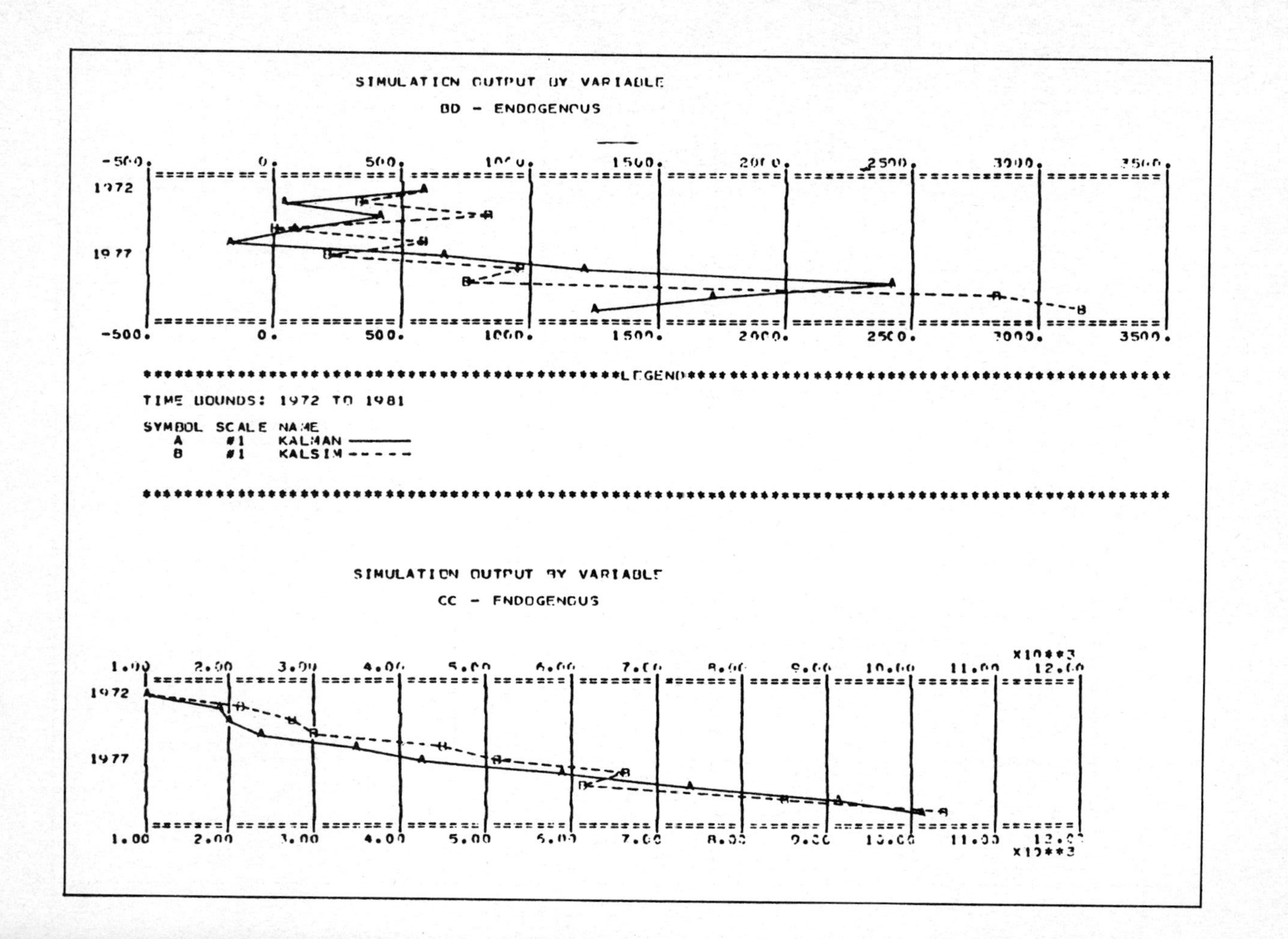
SIMULATION OUTPUT BY VARIABLE
BD - ENDOGENOUS
-500. 0. 500. 1000. 1500. 2000. 2500. 3000. 3500.
1972
1977
LEGEND
TIME BOUNDS: 1972 TO 1981
SYMBOL SCALE NAME
A #1 KALMAN
B #1 KALSIM
SIMULATION OUTPUT BY VARIABLE
CC - ENDOGENOUS
1.00 2.00 3.00 4.00 5.00 6.00 7.00 8.00 9.00 10.00 11.00 12.00
X10**3

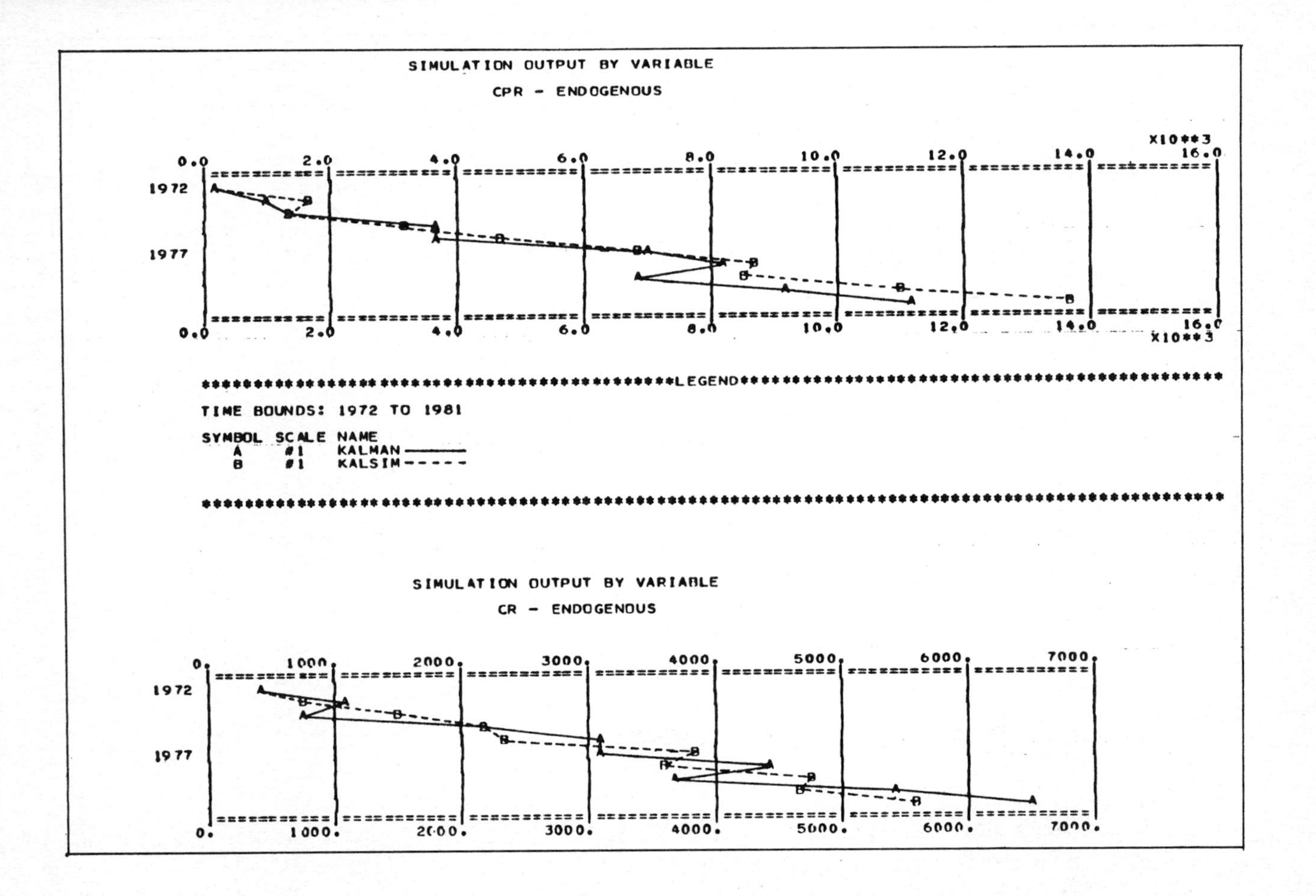

SIMULATION OUTPUT BY VARIABLE
CPR - ENDOGENOUS
X10**3
0.0
2.0
4.0
6.0
8.0
10.0
12.0
14.0
16.0
1972
1977
LEGEND
TIME BOUNDS: 1972 TO 1981
SYMBOL SCALE NAME
A #1 KALMAN
B #1 KALSIM
SIMULATION OUTPUT BY VARIABLE
CR - ENDOGENOUS
0.
1000.
2000.
3000.
4000.
5000.
6000.
7000.

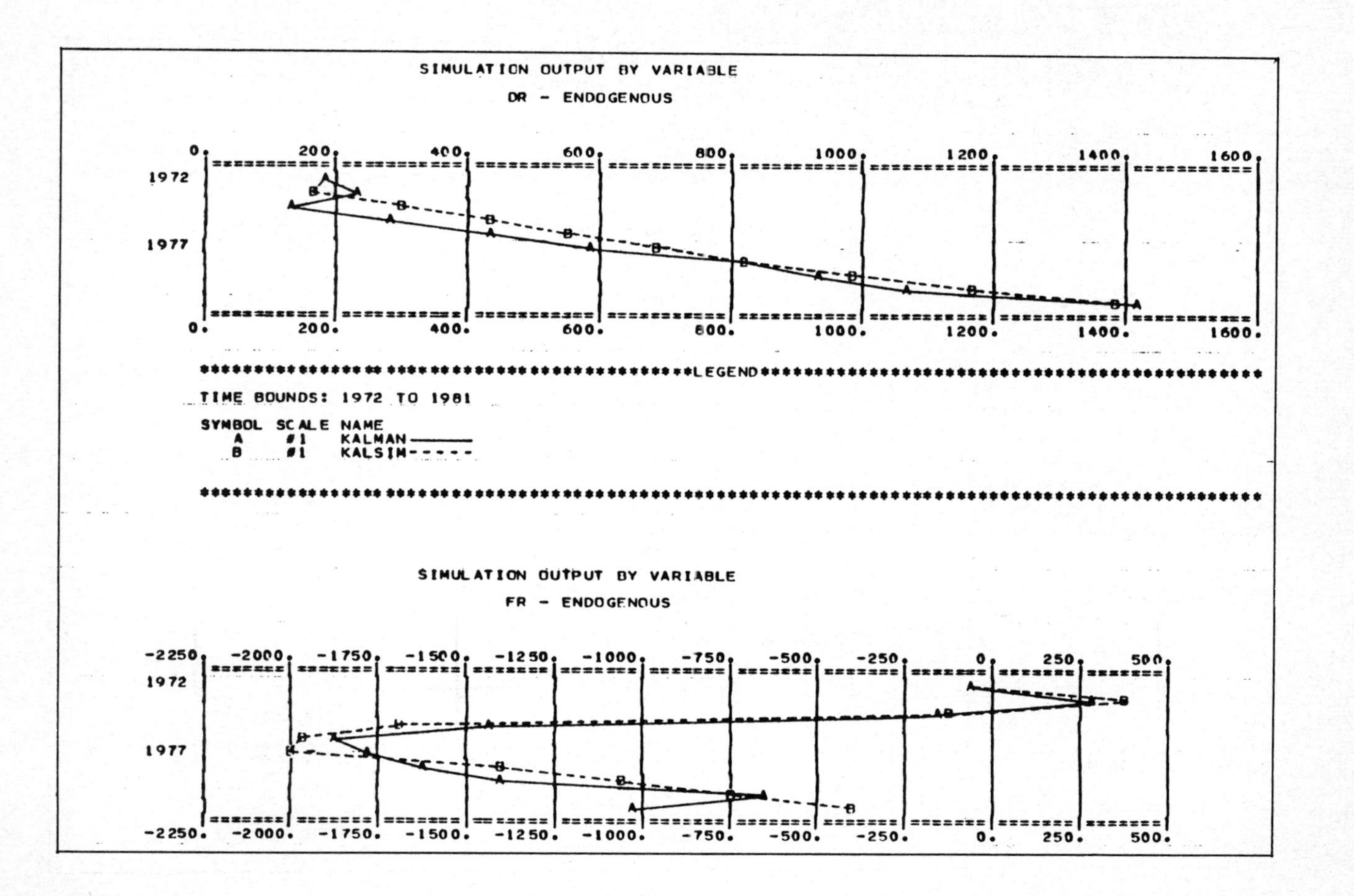

SIMULATION OUTPUT BY VARIABLE
DR - ENDOGENOUS
0. 200. 400. 600. 800. 1000. 1200. 1400. 1600.
1972
1977
LEGEND
TIME BOUNDS: 1972 TO 1981
SYMBOL SCALE NAME
A #1 KALMAN
B #1 KALSIM
SIMULATION OUTPUT BY VARIABLE
FR - ENDOGENOUS
-2250. -2000. -1750. -1500. -1250. -1000. -750. -500. -250. 0. 250. 500.
1972
1977

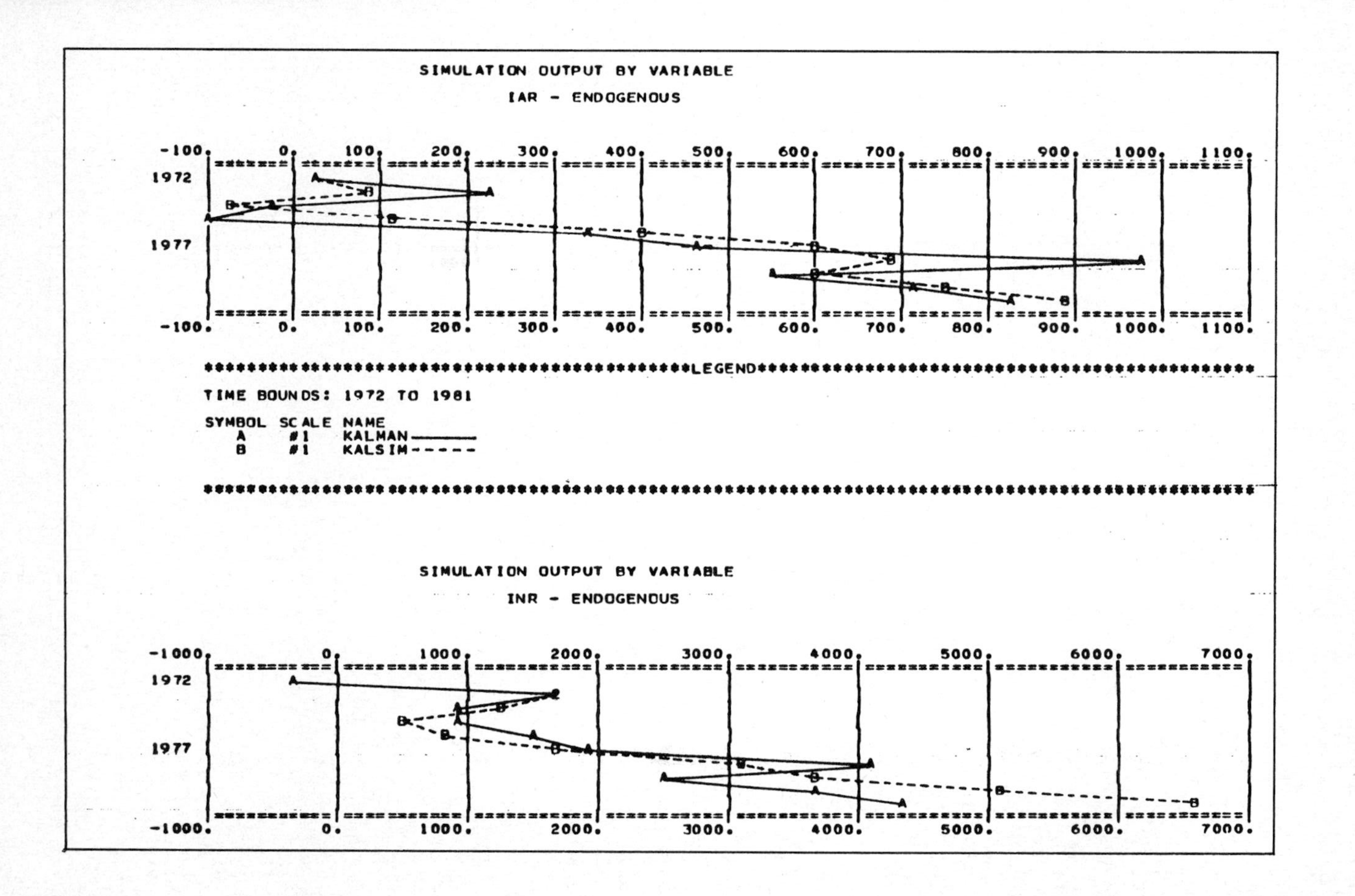
SIMULATION OUTPUT BY VARIABLE
IAR - ENDOGENOUS
-100. 0. 100. 200. 300. 400. 500. 600. 700. 800. 900. 1000. 1100.
1972
1977
LEGEND
TIME BOUNDS: 1972 TO 1981
SYMBOL SCALE NAME
A #1 KALMAN
B #1 KALSIM
SIMULATION OUTPUT BY VARIABLE
INR - ENDOGENOUS
-1000. 0. 1000. 2000. 3000. 4000. 5000. 6000. 7000.
1972
1977

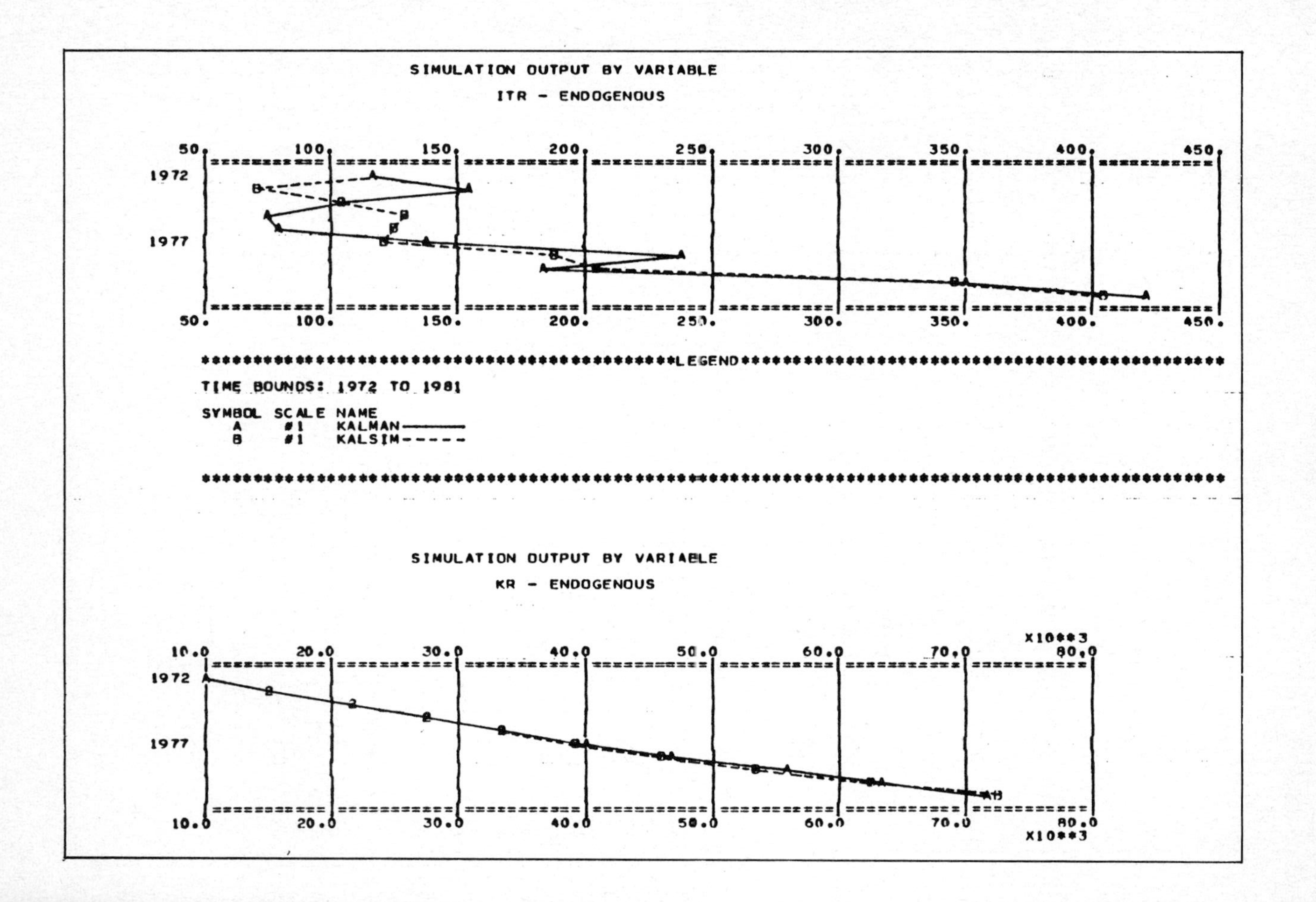
SIMULATION OUTPUT BY VARIABLE
ITR - ENDOGENOUS
50. 100. 150. 200. 250. 300. 350. 400. 450.
1972
1977
LEGEND
TIME BOUNDS: 1972 TO 1981
SYMBOL SCALE NAME
A #1 KALMAN
B #1 KALSIM
SIMULATION OUTPUT BY VARIABLE
KR - ENDOGENOUS
10.0 20.0 30.0 40.0 50.0 60.0 70.0 80.0
X10**3
1972
1977

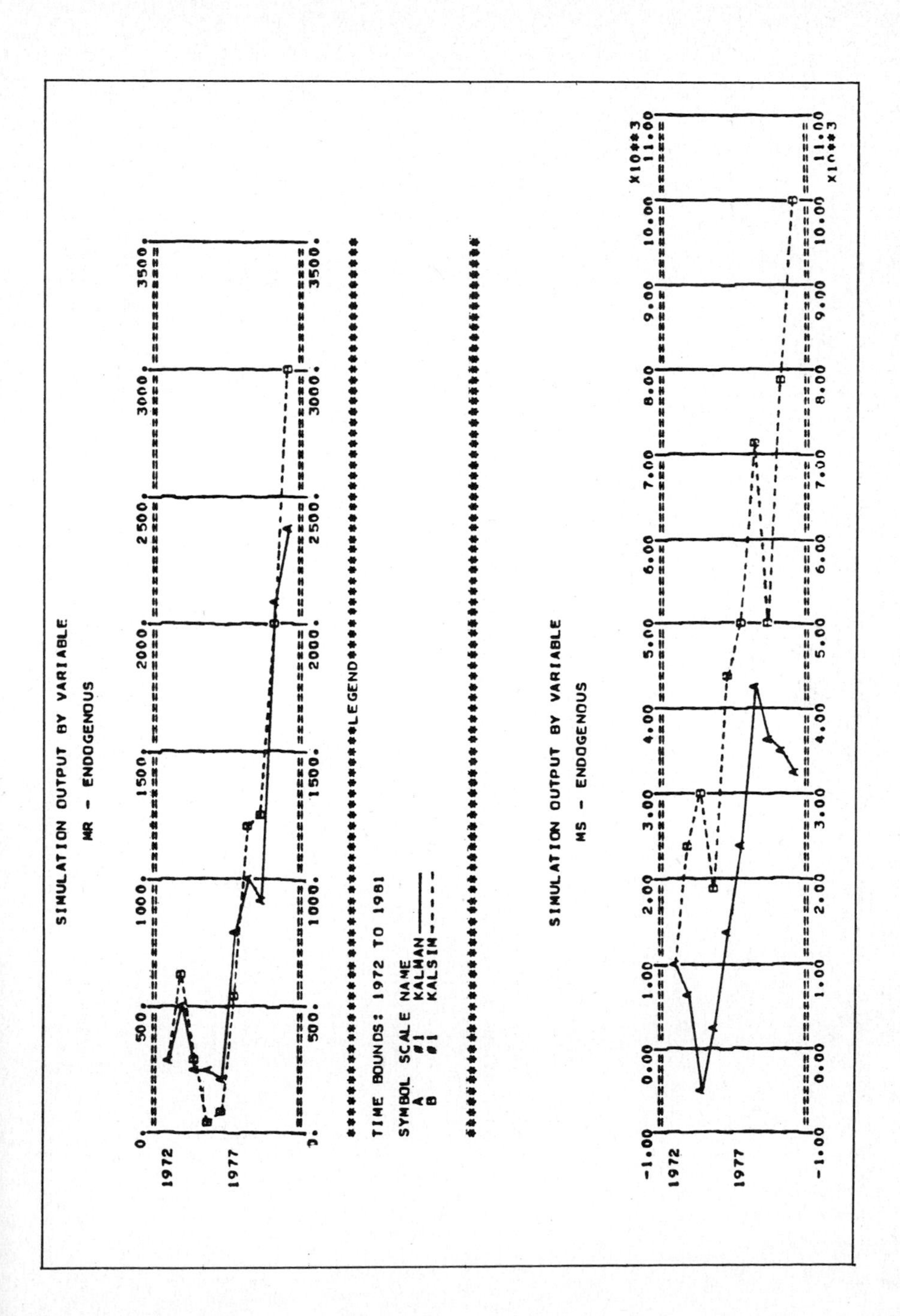
SIMULATION OUTPUT BY VARIABLE
MR - ENDOGENOUS
0 500 1000 1500 2000 2500 3000 3500
1972
1977
LEGEND
TIME BOUNDS: 1972 TO 1981
SYMBOL SCALE NAME
A #1 KALMAN
B #1 KALSIM
SIMULATION OUTPUT BY VARIABLE
MS - ENDOGENOUS
-1.00 0.00 1.00 2.00 3.00 4.00 5.00 6.00 7.00 8.00 9.00 10.00 11.00
X10**3

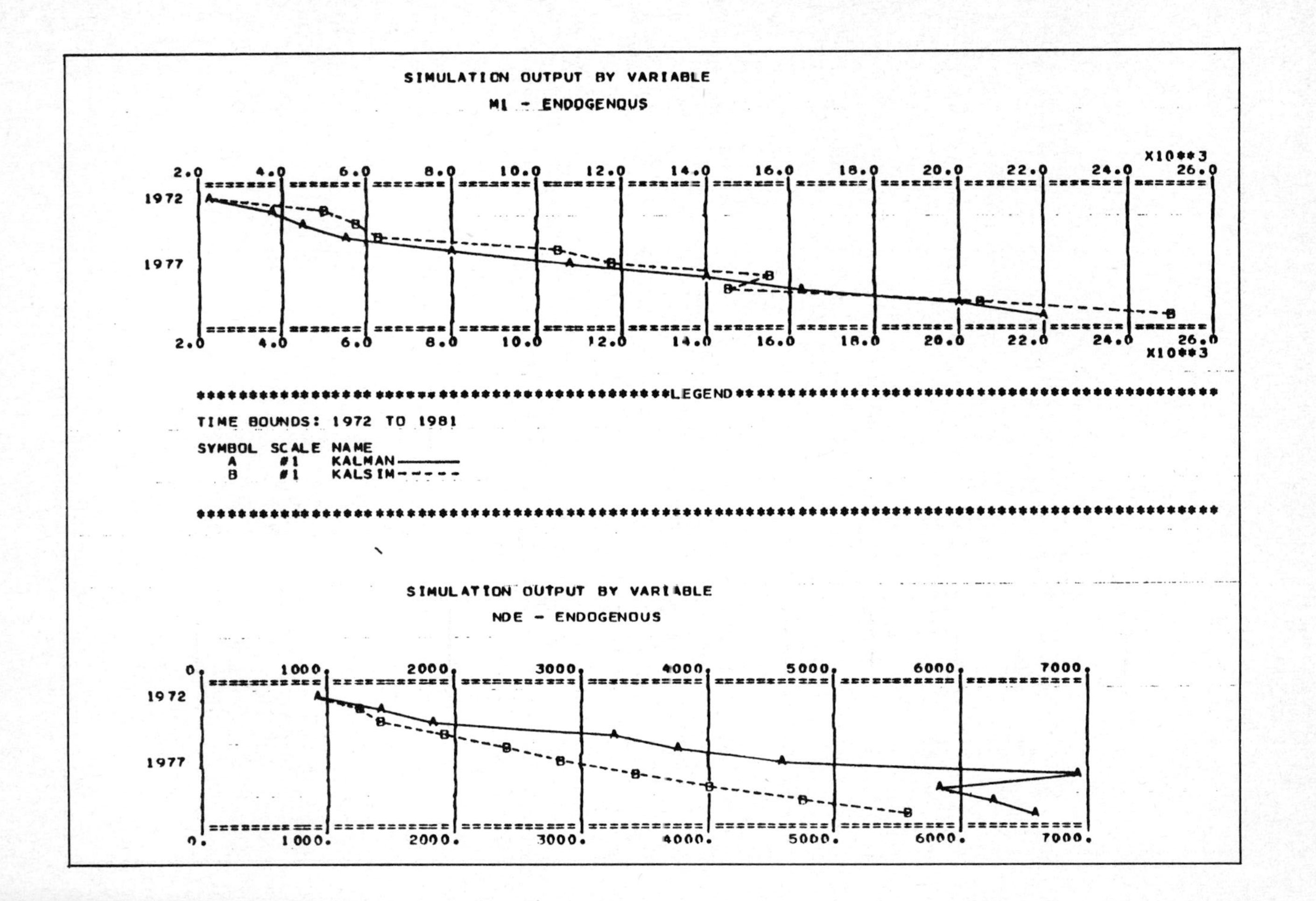
SIMULATION OUTPUT BY VARIABLE
M1 - ENDOGENOUS
2.0 4.0 6.0 8.0 10.0 12.0 14.0 16.0 18.0 20.0 22.0 24.0 26.0
X10**3
1972
1977
LEGEND
TIME BOUNDS: 1972 TO 1981
SYMBOL SCALE NAME
A #1 KALMAN
B #1 KALSIM
SIMULATION OUTPUT BY VARIABLE
NDE - ENDOGENOUS
0. 1000. 2000. 3000. 4000. 5000. 6000. 7000.
1972
1977

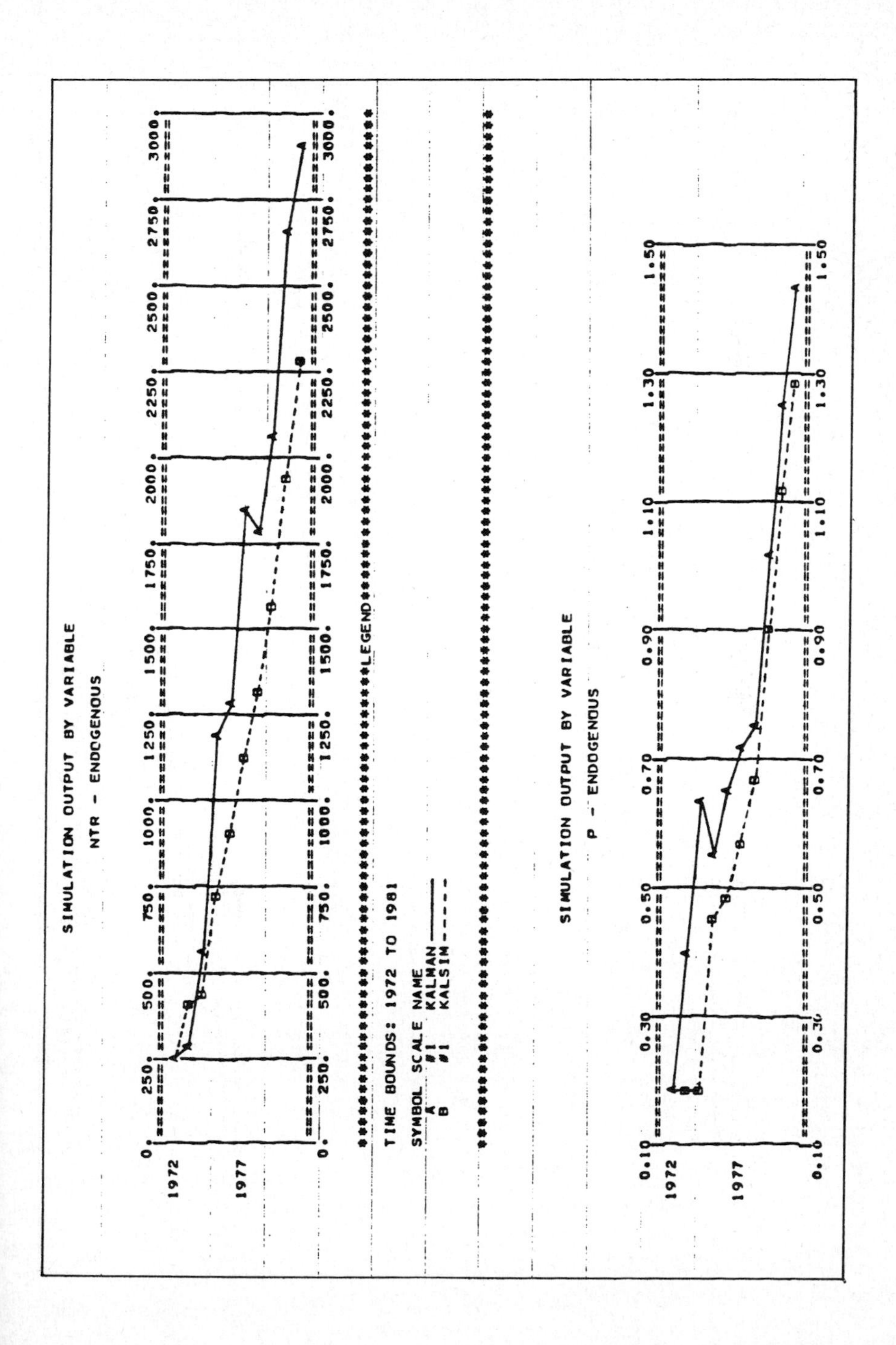
SIMULATION OUTPUT BY VARIABLE
NTR - ENDOGENOUS
0. 250. 500. 750. 1000. 1250. 1500. 1750. 2000. 2250. 2500. 2750. 3000.
1972
1977
LEGEND
TIME BOUNDS: 1972 TO 1981
SYMBOL SCALE NAME
A #1 KALMAN
B #1 KALSIM
SIMULATION OUTPUT BY VARIABLE
P - ENDOGENOUS
0.10 0.30 0.50 0.70 0.90 1.10 1.30 1.50
1972
1977

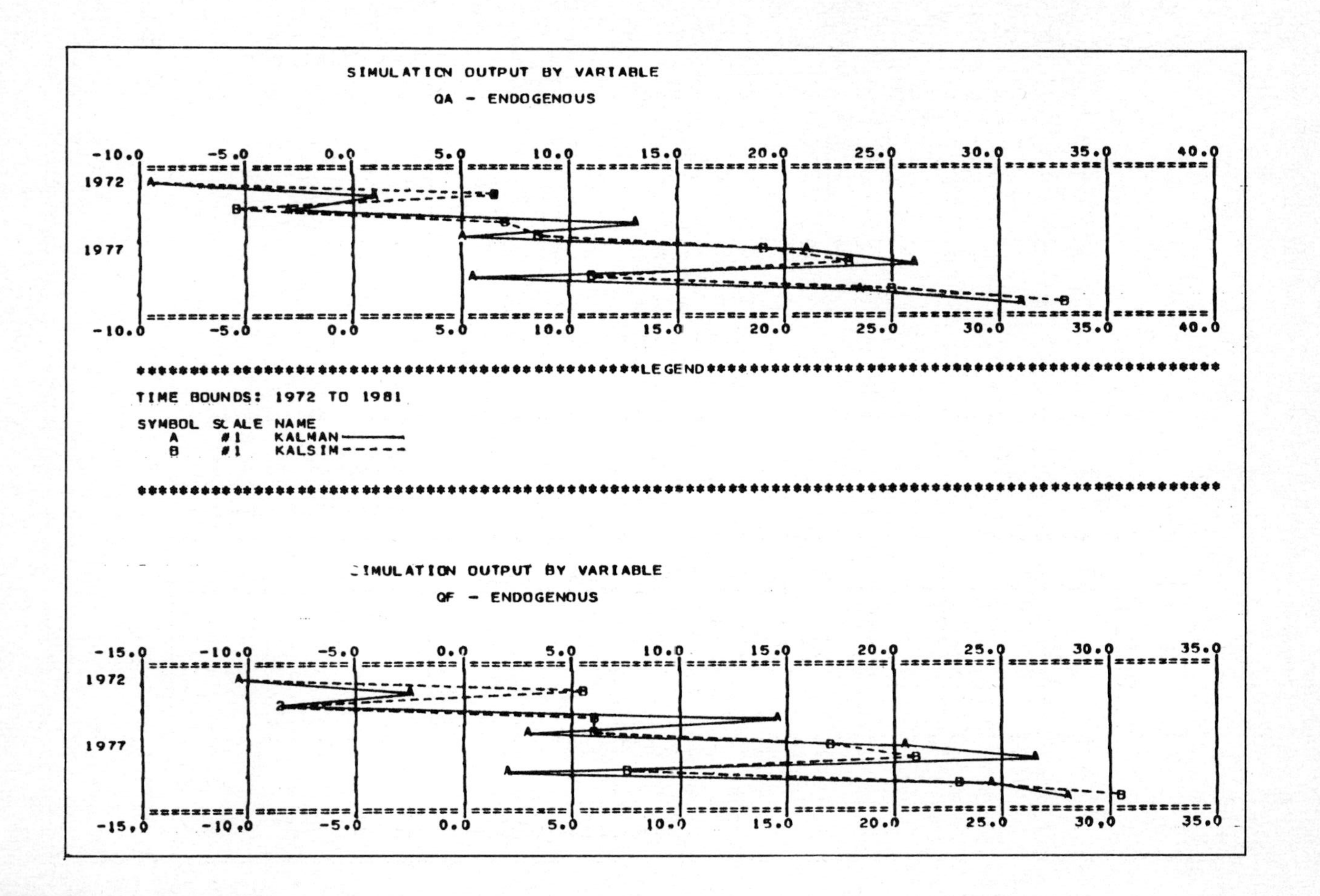
SIMULATION OUTPUT BY VARIABLE
QA - ENDOGENOUS
-10.0
-5.0
0.0
5.0
10.0
15.0
20.0
25.0
30.0
35.0
40.0
1972
1977
LEGEND
TIME BOUNDS: 1972 TO 1981
SYMBOL SCALE NAME
A #1 KALMAN
B #1 KALSIM
SIMULATION OUTPUT BY VARIABLE
QF - ENDOGENOUS
-15.0
-10.0
-5.0
0.0
5.0
10.0
15.0
20.0
25.0
30.0
35.0
1972
1977

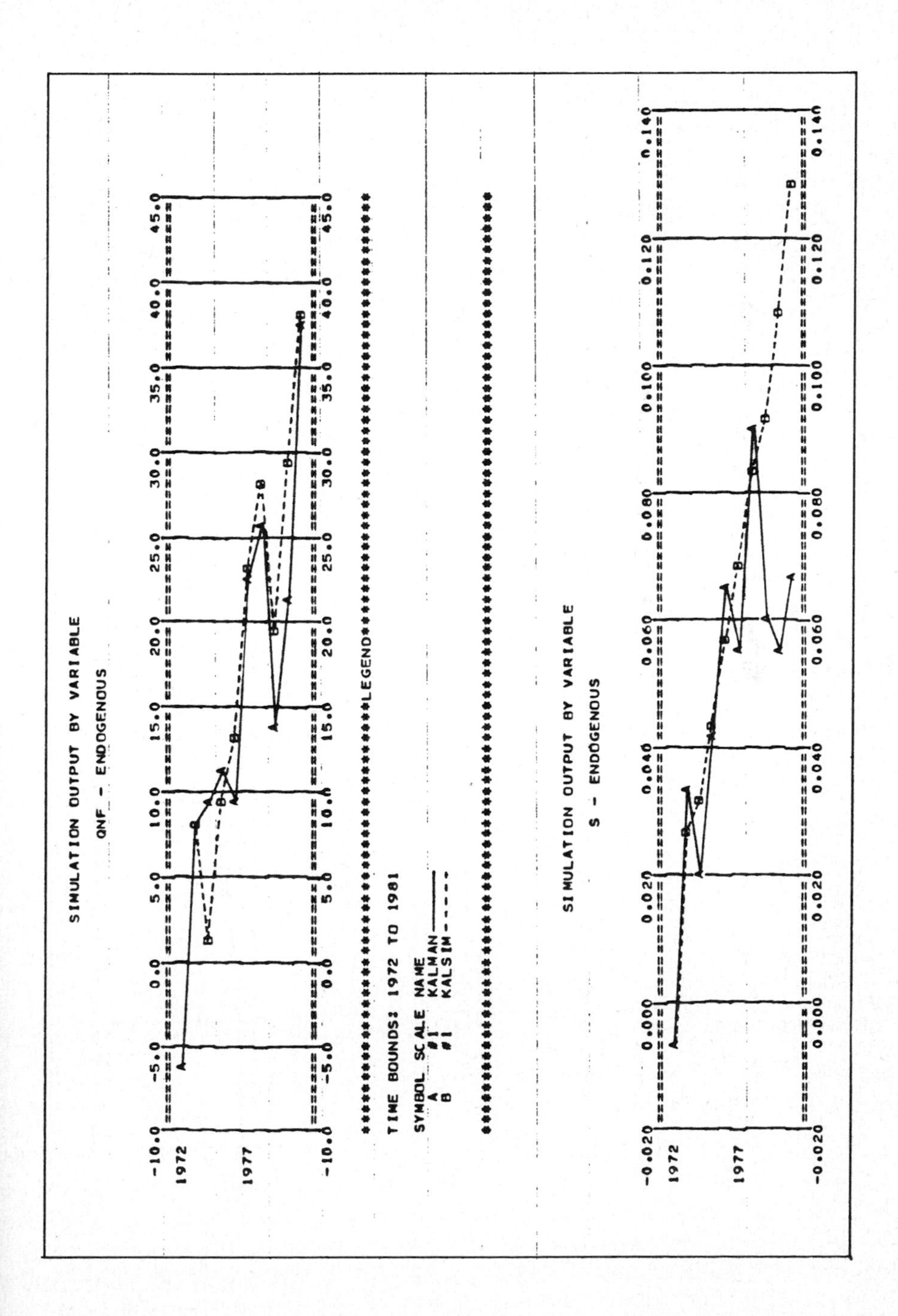
SIMULATION OUTPUT BY VARIABLE
QNF - ENDOGENOUS
-10.0
-5.0
0.0
5.0
10.0
15.0
20.0
25.0
30.0
35.0
40.0
45.0
1972
1977
LEGEND
TIME BOUNDS: 1972 TO 1981
SYMBOL SCALE NAME
A #1 KALMAN
B #1 KALSIM
SIMULATION OUTPUT BY VARIABLE
S - ENDOGENOUS
-0.020
0.000
0.020
0.040
0.060
0.080
0.100
0.120
0.140
1972
1977

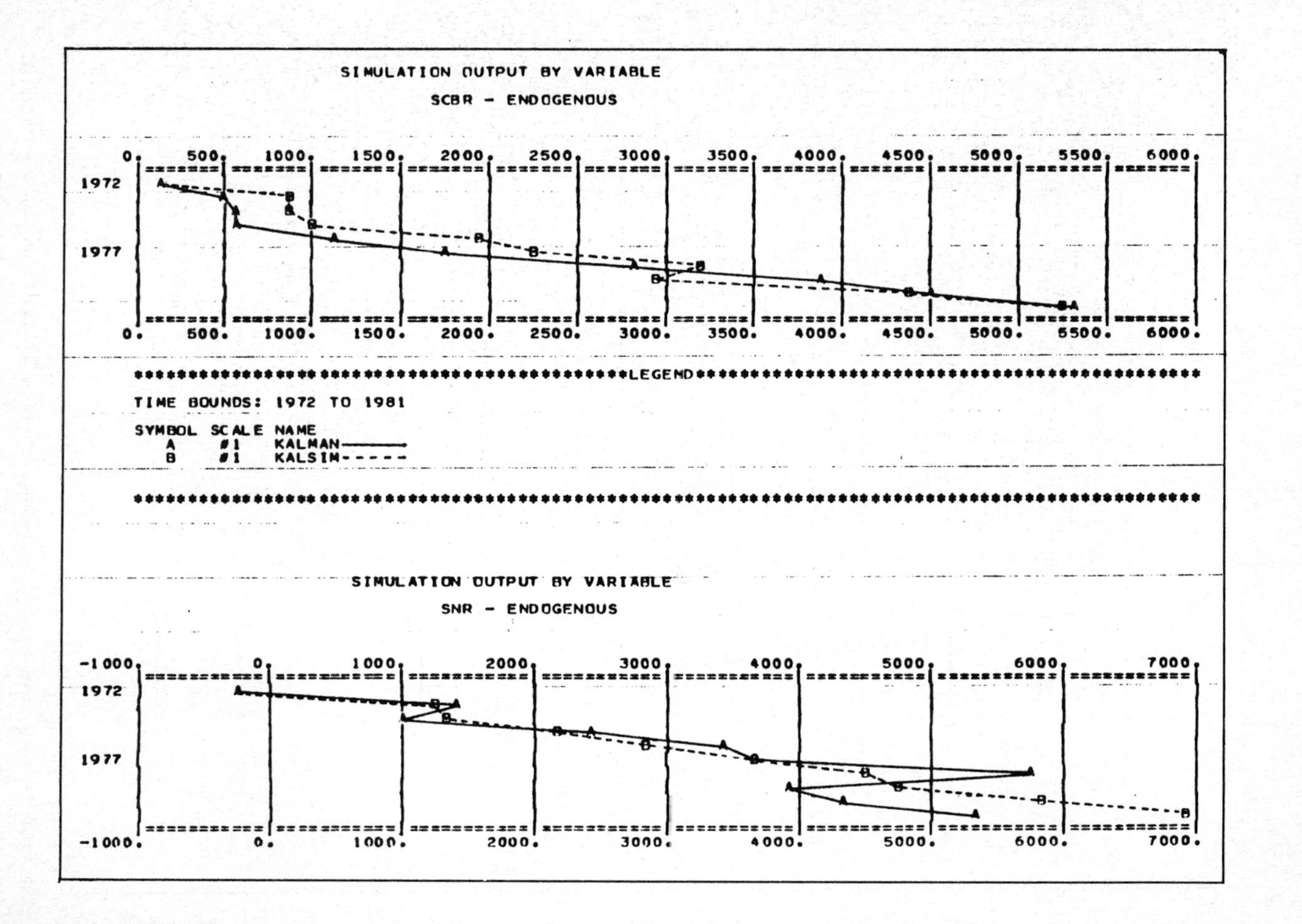
SIMULATION OUTPUT BY VARIABLE
SCBR - ENDOGENOUS
0. 500. 1000. 1500. 2000. 2500. 3000. 3500. 4000. 4500. 5000. 5500. 6000.
1972
1977
LEGEND
TIME BOUNDS: 1972 TO 1981
SYMBOL SCALE NAME
A #1 KALMAN
B #1 KALSIM
SIMULATION OUTPUT BY VARIABLE
SNR - ENDOGENOUS
-1000. 0. 1000. 2000. 3000. 4000. 5000. 6000. 7000.
1972
1977

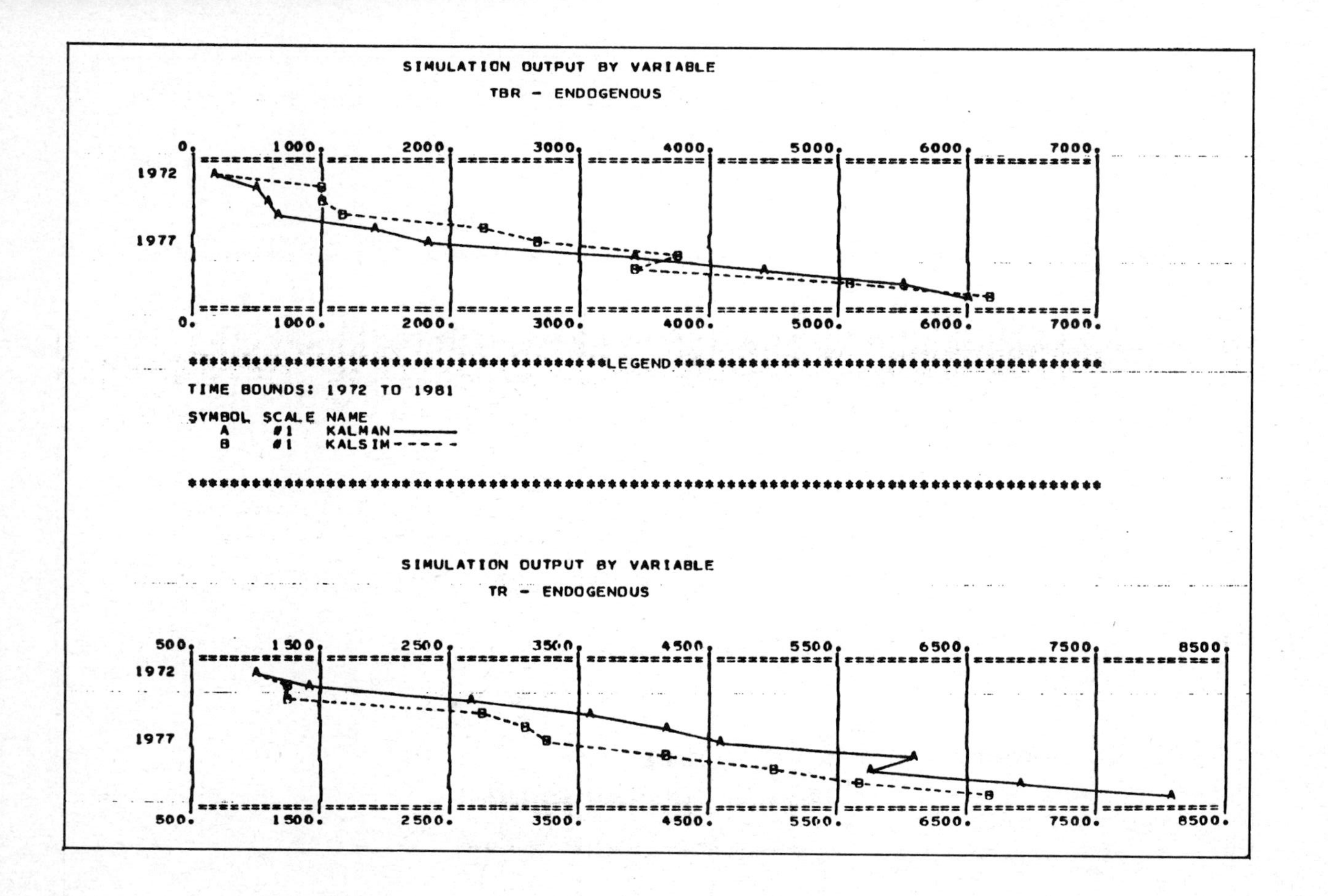
SIMULATION OUTPUT BY VARIABLE
TBR - ENDOGENOUS
0. 1000. 2000. 3000. 4000. 5000. 6000. 7000.
1972
1977
LEGEND
TIME BOUNDS: 1972 TO 1981
SYMBOL SCALE NAME
A #1 KALMAN
B #1 KALSIM
SIMULATION OUTPUT BY VARIABLE
TR - ENDOGENOUS
500. 1500. 2500. 3500. 4500. 5500. 6500. 7500. 8500.
1972
1977

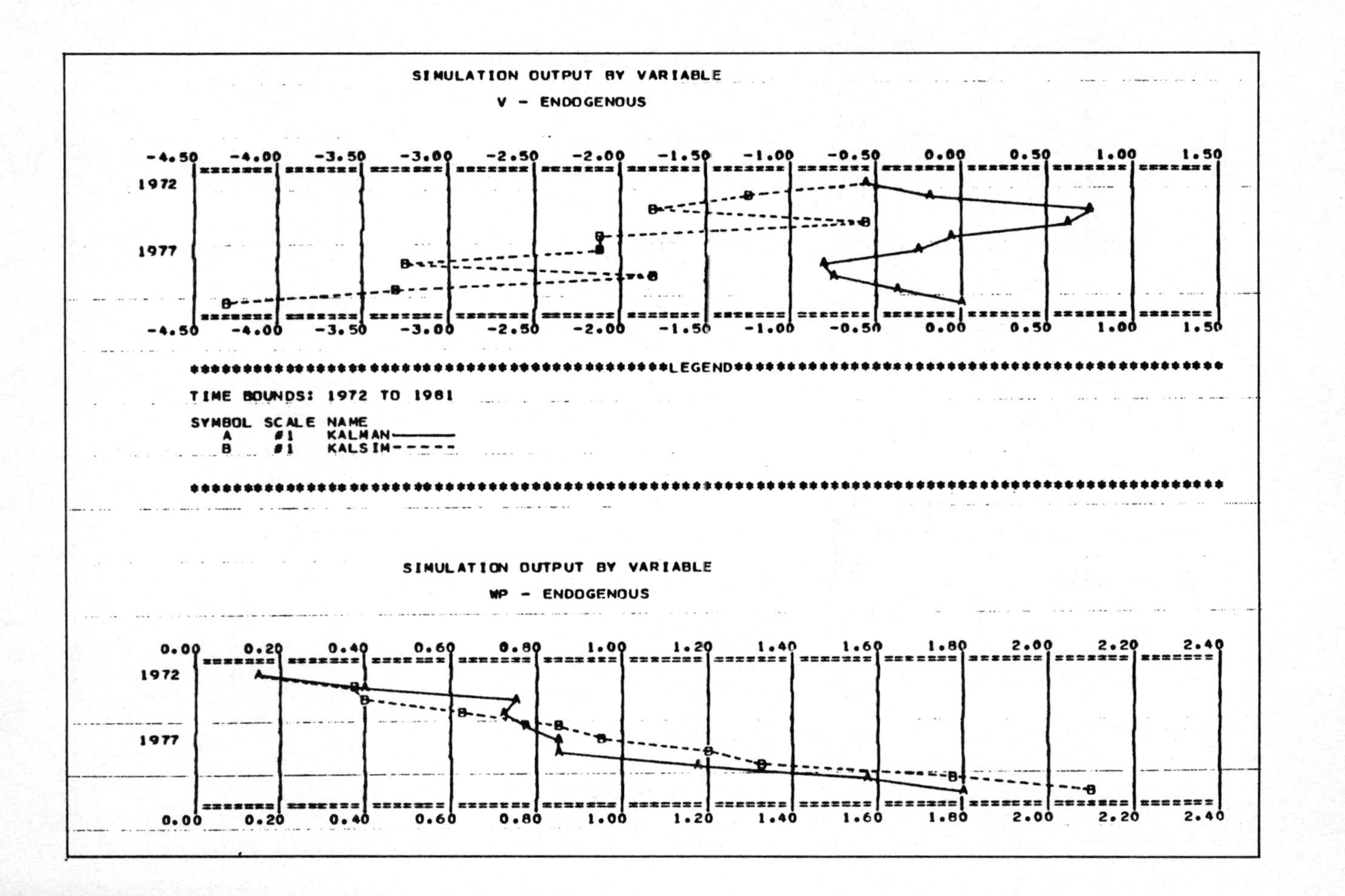
SIMULATION OUTPUT BY VARIABLE
V - ENDOGENOUS
-4.50 -4.00 -3.50 -3.00 -2.50 -2.00 -1.50 -1.00 -0.50 0.00 0.50 1.00 1.50
1972
1977
LEGEND
TIME BOUNDS: 1972 TO 1981
SYMBOL SCALE NAME
A #1 KALMAN
B #1 KALSIM
SIMULATION OUTPUT BY VARIABLE
WP - ENDOGENOUS
0.00 0.20 0.40 0.60 0.80 1.00 1.20 1.40 1.60 1.80 2.00 2.20 2.40
1972
1977

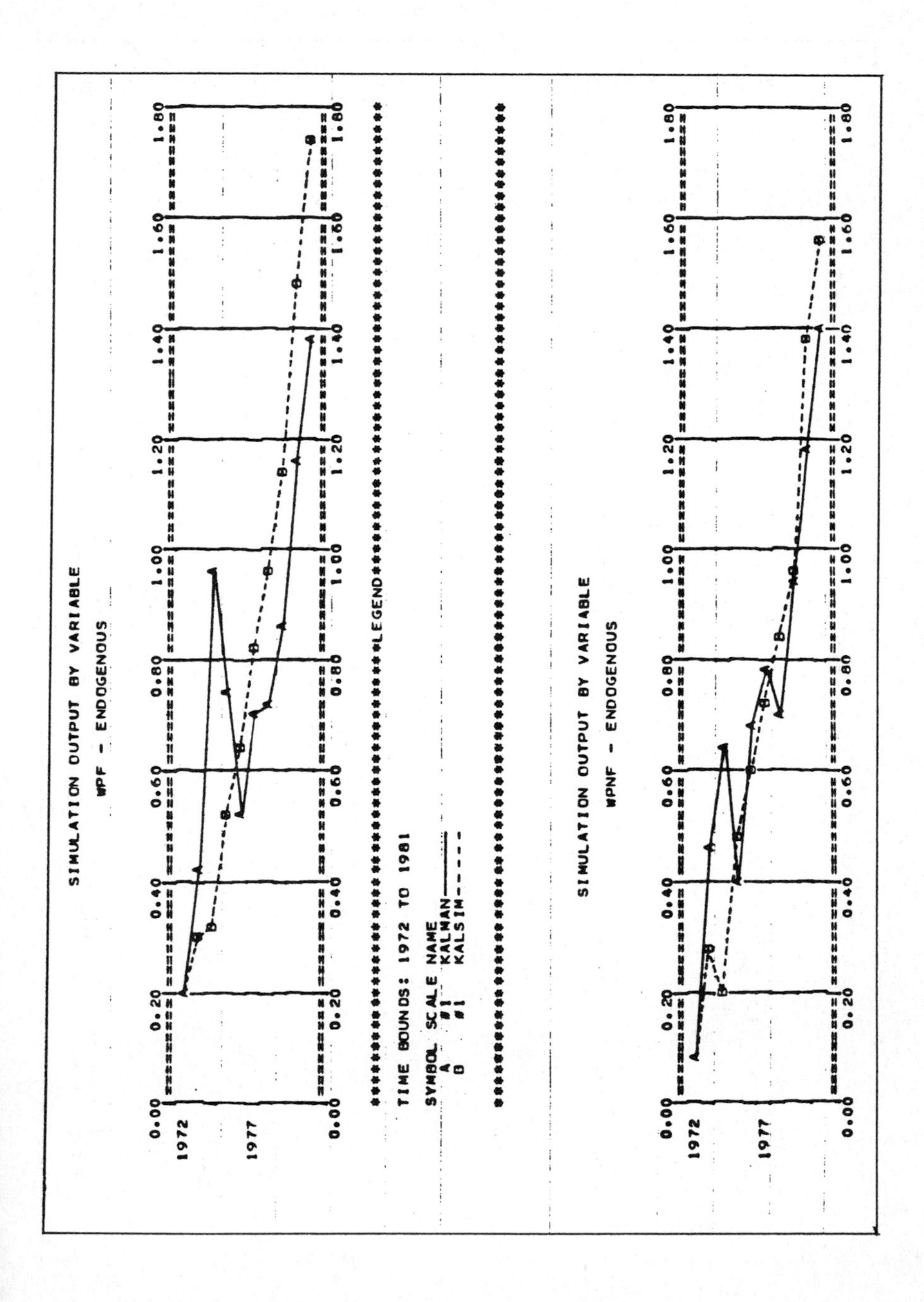
SIMULATION OUTPUT BY VARIABLE
WPF - ENDOGENOUS
0.00
0.20
0.40
0.60
0.80
1.00
1.20
1.40
1.60
1.80
1972
1977
LEGEND
TIME BOUNDS: 1972 TO 1981
SYMBOL SCALE NAME
A #1 KALMAN
B #1 KALSIM
SIMULATION OUTPUT BY VARIABLE
WPNF - ENDOGENOUS

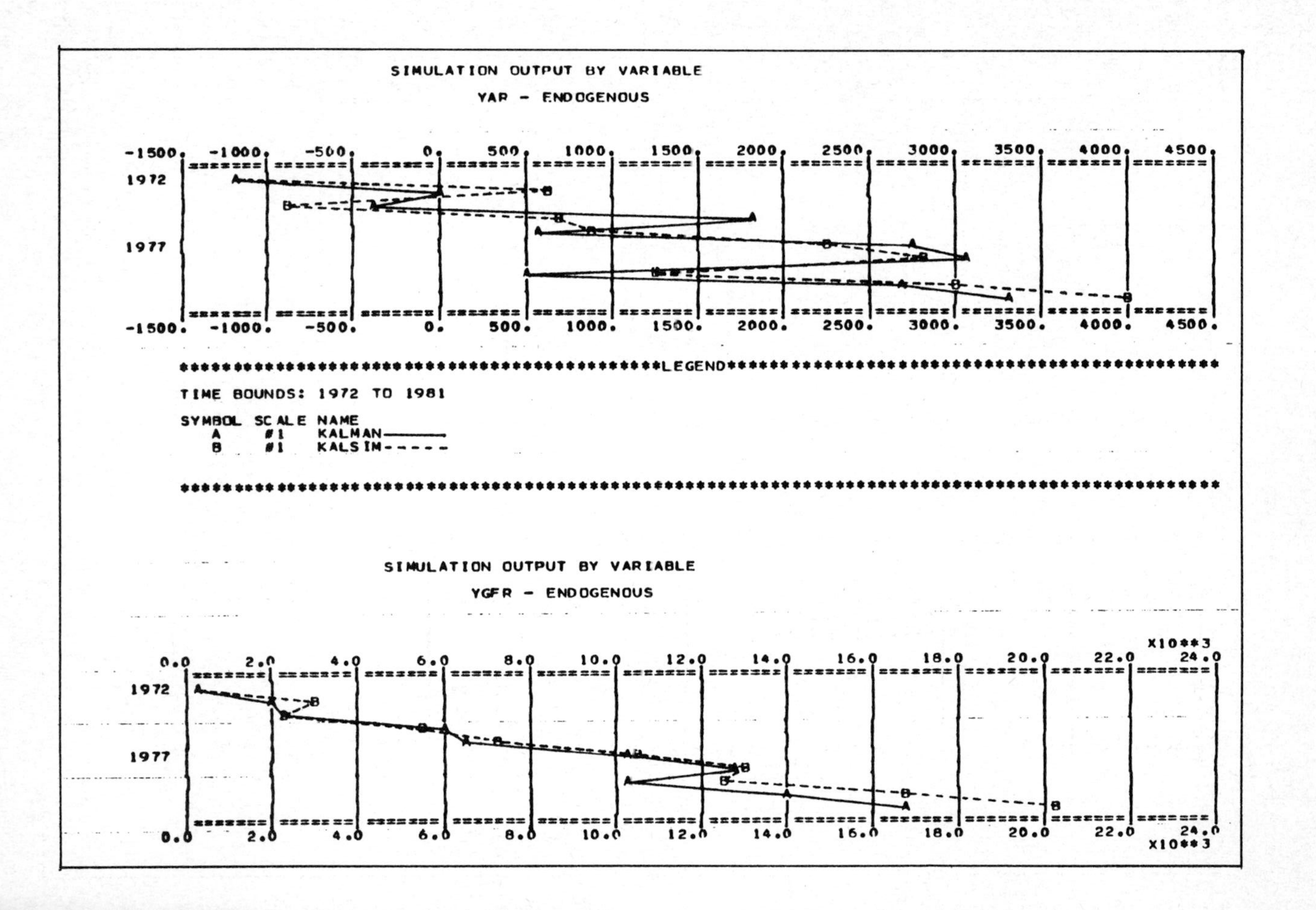
SIMULATION OUTPUT BY VARIABLE
YAR - ENDOGENOUS
-1500. -1000. -500. 0. 500. 1000. 1500. 2000. 2500. 3000. 3500. 4000. 4500.
1972
1977
LEGEND
TIME BOUNDS: 1972 TO 1981
SYMBOL SCALE NAME
A #1 KALMAN
B #1 KALSIM
SIMULATION OUTPUT BY VARIABLE
YGFR - ENDOGENOUS
X10**3
0.0 2.0 4.0 6.0 8.0 10.0 12.0 14.0 16.0 18.0 20.0 22.0 24.0

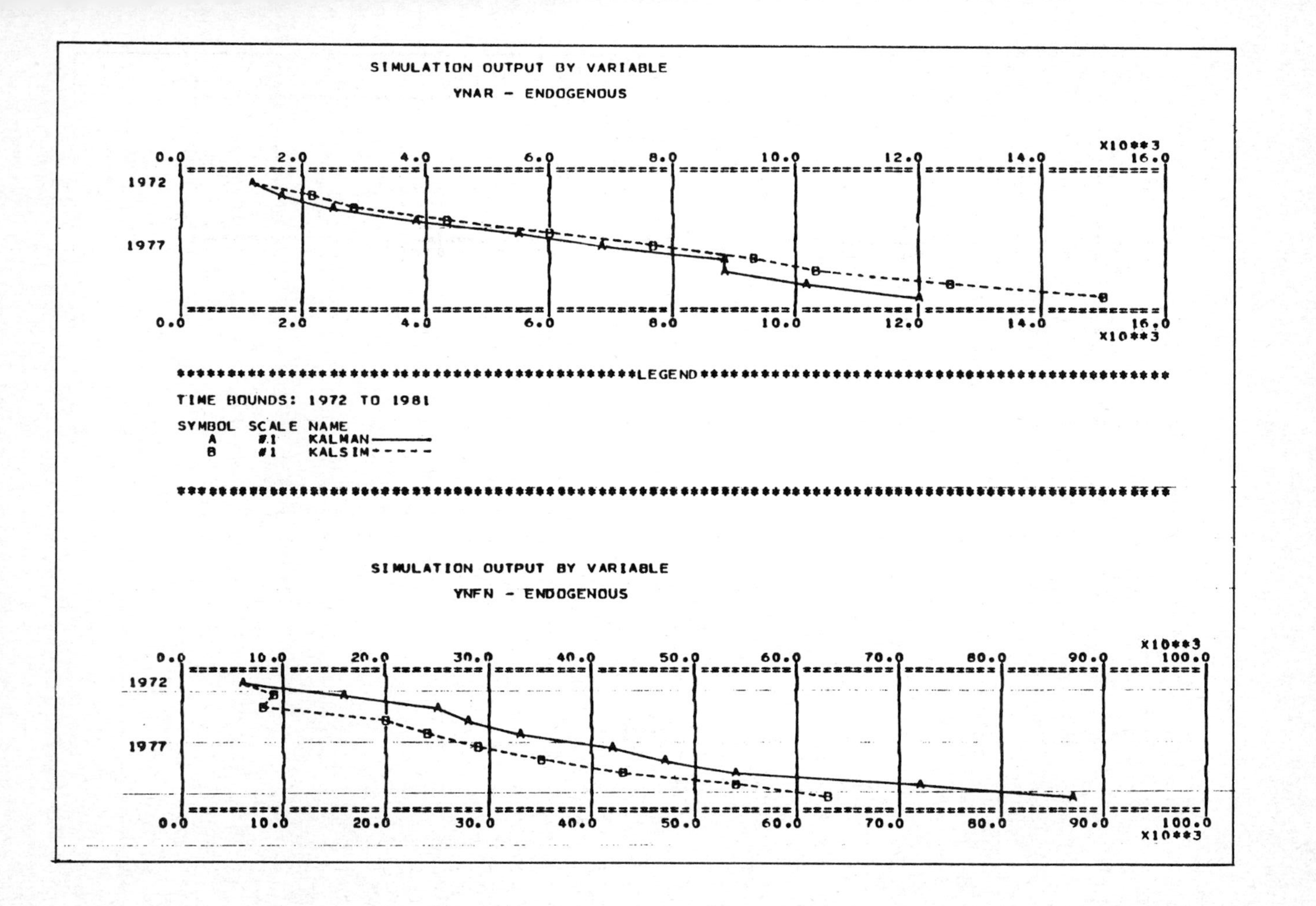

SIMULATION OUTPUT BY VARIABLE
YNAR - ENDOGENOUS
X10**3
0.0 2.0 4.0 6.0 8.0 10.0 12.0 14.0 16.0
1972
1977
0.0 2.0 4.0 6.0 8.0 10.0 12.0 14.0 16.0
X10**3
LEGEND
TIME BOUNDS: 1972 TO 1981
SYMBOL SCALE NAME
A #1 KALMAN
B #1 KALSIM
SIMULATION OUTPUT BY VARIABLE
YNFN - ENDOGENOUS
X10**3
0.0 10.0 20.0 30.0 40.0 50.0 60.0 70.0 80.0 90.0 100.0
1972
1977
0.0 10.0 20.0 30.0 40.0 50.0 60.0 70.0 80.0 90.0 100.0
X10**3

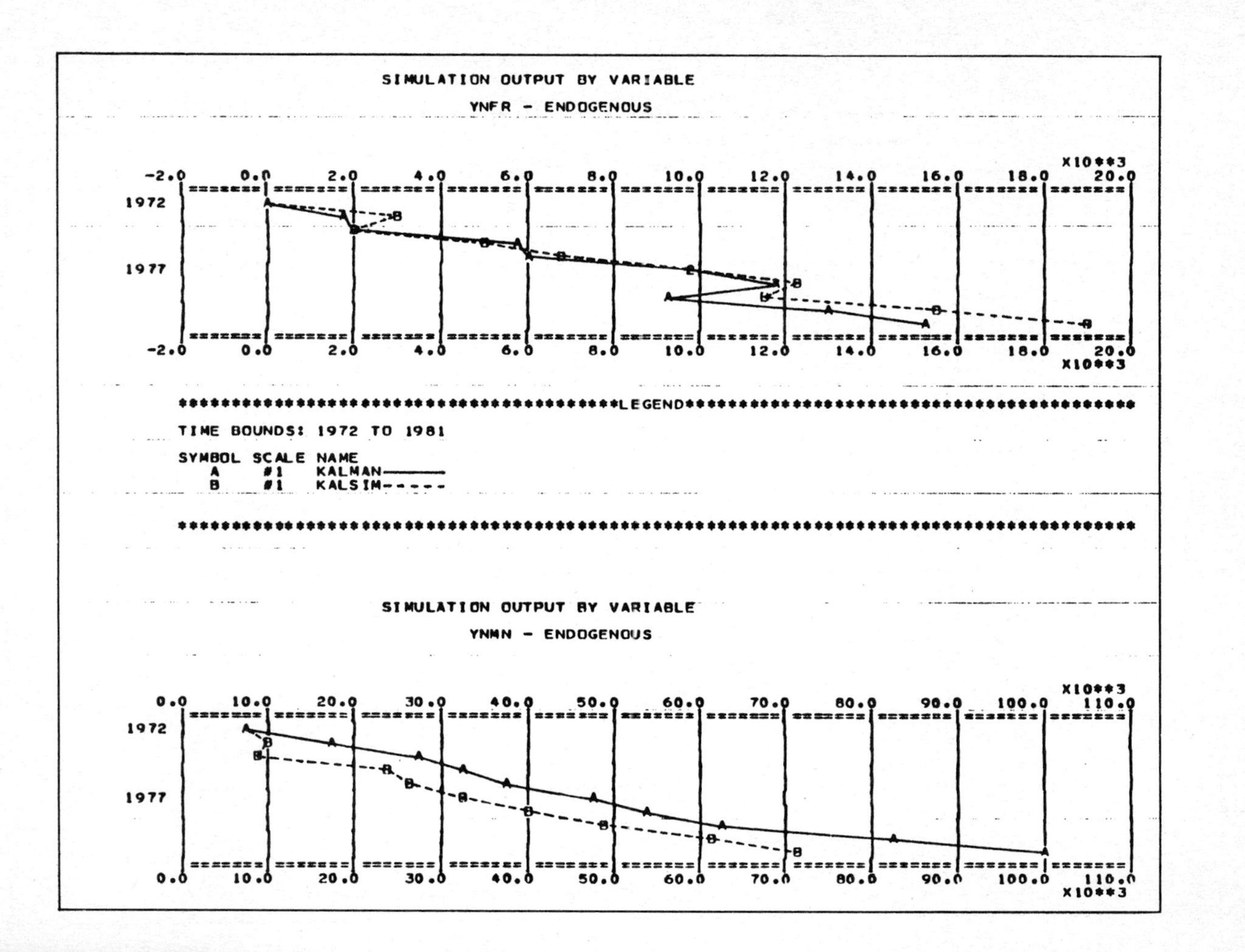
SIMULATION OUTPUT BY VARIABLE
YNFR - ENDOGENOUS
-2.0
0.0
2.0
4.0
6.0
8.0
10.0
12.0
14.0
16.0
18.0
20.0
X10**3
1972
1977
LEGEND
TIME BOUNDS: 1972 TO 1981
SYMBOL SCALE NAME
A #1 KALMAN
B #1 KALSIM
SIMULATION OUTPUT BY VARIABLE
YNMN - ENDOGENOUS
0.0
10.0
20.0
30.0
40.0
50.0
60.0
70.0
80.0
90.0
100.0
110.0
X10**3
1972
1977

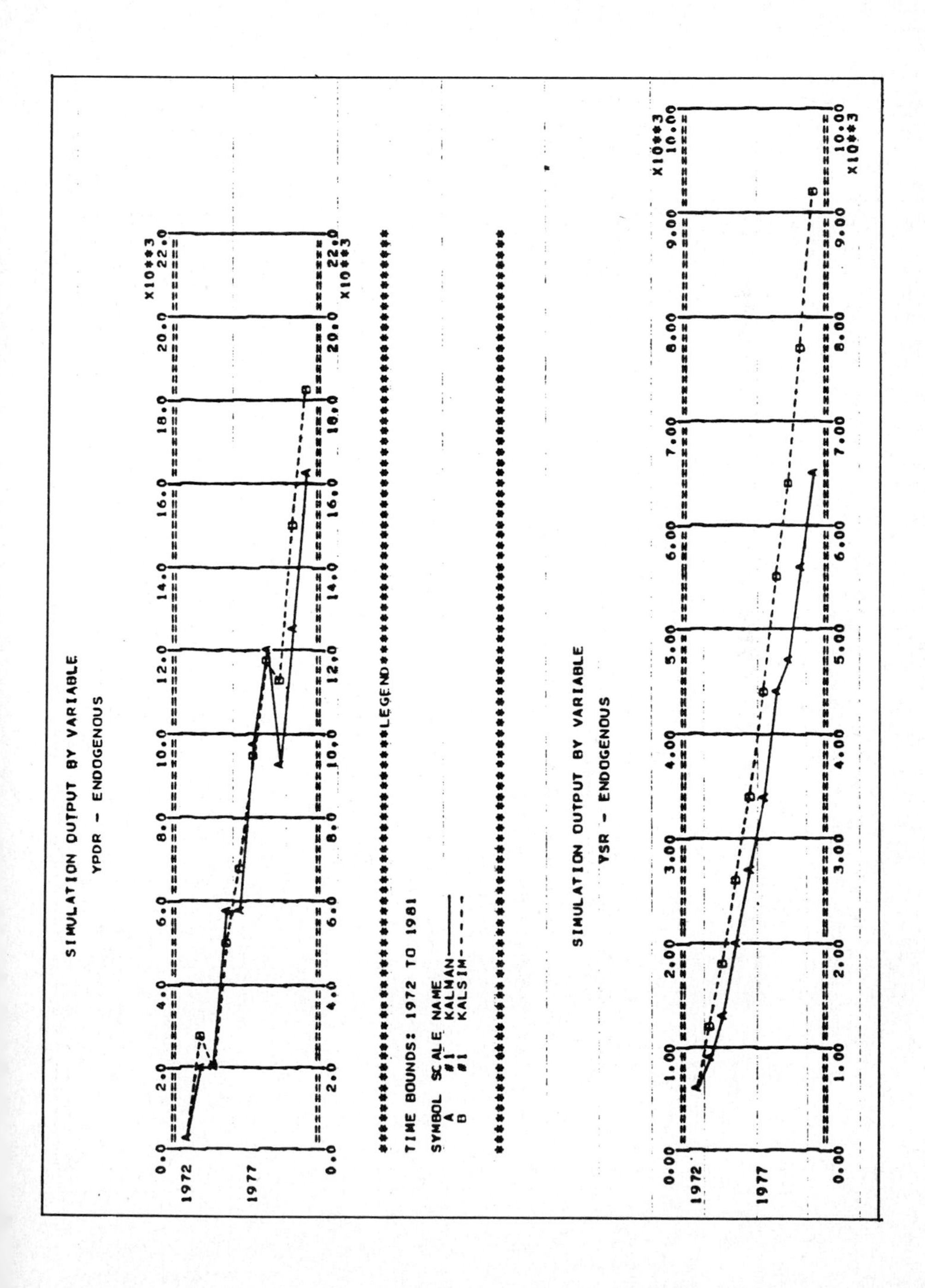
SIMULATION OUTPUT BY VARIABLE
YPDR - ENDOGENOUS
0.0 2.0 4.0 6.0 8.0 10.0 12.0 14.0 16.0 18.0 20.0 22.0
X10**3
1972
1977
LEGEND
TIME BOUNDS: 1972 TO 1981
SYMBOL SCALE NAME
A #1 KALMAN
B #1 KALSIM
SIMULATION OUTPUT BY VARIABLE
YSR - ENDOGENOUS
0.00 1.00 2.00 3.00 4.00 5.00 6.00 7.00 8.00 9.00 10.00
X10**3
1972
1977

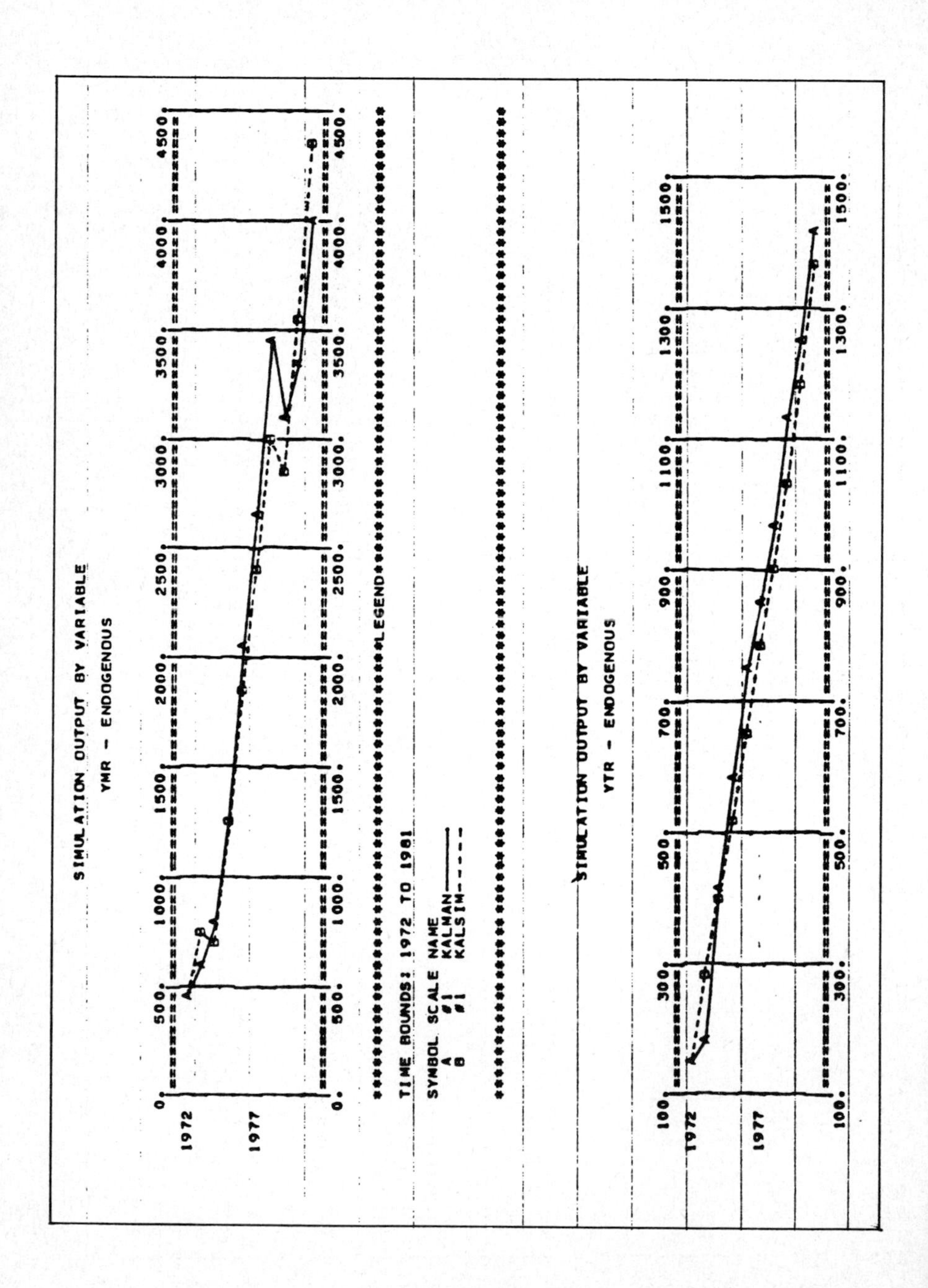
SIMULATION OUTPUT BY VARIABLE
YMR - ENDOGENOUS
0. 500. 1000. 1500. 2000. 2500. 3000. 3500. 4000. 4500.
1972
1977
LEGEND
TIME BOUNDS: 1972 TO 1981
SYMBOL SCALE NAME
A #1 KALMAN
B #1 KALSIM
SIMULATION OUTPUT BY VARIABLE
YTR - ENDOGENOUS
100. 300. 500. 700. 900. 1100. 1300. 1500.
1972
1977

The variables representing the agricultural sector of the economy (QF, QNF, QA and YAR) all perform very well catching most of the turning points in the historical series. This is an important result as this sector is considered very difficult to model in view of its highly volatile nature. As far as the non-agricultural sector was concerned, YMR and YTR perform fairly well. On the other hand, YSR tended to stray away from its historical values, the divergence becoming noticeable towards the end of the simulation. This was primarily because of the use of the autoregressive term in the concerned equation. This divergence, however, did not have any effect on YNFR and YGFR and as a result both these variables did track their historical values and turning points pretty closely. Thus, the performance of the output sector was quite commendable and this was very encouraging, considering its clearly strategic location as the central module of the system.

Both the variables in the consumption sector (YPDR and CPR) tracked quite well. The savings rate (s) did stay close to its historical values initially but the high simulated values of YNAR caused it to diverge away from them towards the close of the experiment. This resulted in even SNR behaving in a like manner.

The effect of this was transmitted to the investment sector. As a result of the 'two-gap' formulation, INR also strayed away from its historical values towards the close of the simulation. This, however, had no bearing on the remaining four variables in the investment sector (KR, DR, IAR and ITR), all of which tracked their historical values very closely.

The trade sector (FR and MR) performed quite well and captured most of their turning points admirably. Considering the fact that it was this sector that took the prime responsibility for bearing the brunt of the exogenous oil price shock of 1973, this is an extremely heartening performance and serves to highlight the robustness of the model.

The performance of the price sector was a bit disappointing. While no cumulative divergence was exhibited, the overall predictions of WPF, WPNF and WP remained consistently above their respective historical values, while in the case of P it remained consistently below. This, combined with the failure of all these variables to capture critical turning points, left a lot to be desired from this sub-central module.

The fact that P failed to track well throughout the simulation period reflected immediately on the performance of the variables in the income sector (YNFN and YNMN) which also remained considerably below their historical levels throughout.

The ripple effects of this were clearly felt in the government sector. Low nominal incomes implied low expenditures (NDE) as well as low revenues (TR and NTR). However, the fall in revenue was more than the fall in expenditure (there being two exogenous expenditure items - GDE and DEF - which were not affected by the fall in nominal income) and consequently BD strayed away considerably from its historical values towards the end of the simulation.

The variables in the monetary sector were partially insulated from any drastic consequences of the above anomaly due to the fact that currency expansion was largely determined by exogenous variables (DBFA and DFEA). As such, the effect of a large BD was minimized and CC tracked its historical values quite closely. This variable being the key to the entire analysis of the monetary sector, the other nominal variables of this sector (SCBR, TBR and M1) followed suit by rendering equally creditable performances. However, the dismal tracking ability of the price sector had its effect on MS and consequently v, both of which performed very poorly.

The variables of the miscellaneous sector (AF and ANF) did track fairly well. Being a function of relative prices, the fact that the simulated values of WPF and WPNF were both consistently lower than their historical values helped not to vitiate their performances.

All in all, the model performed reasonably well considering its size and complexity, as well as the time bounds of the dynamic simulation. One could not help noticing that the simulated values of nearly all the key variables (especially in the real sector of the economy) tracked their historical counterparts fairly well exhibiting no divergent behaviour whatsoever. Most of the weaknesses in its dynamic behaviour resulted from the poor showing of the equations of the price sector, in particular, the equation for the national income deflator. If this variable had tracked better, then the simulated values of the subsequent blocks in the recursive block-structured model would have shown a far closer correlation with their historical values.

While the performance of the model did not rule out the necessity for any fine tuning in its structure via an adjustment of its add or mul factors, - which, in the light of most recent experience with macroeconomic simulation models, have gained considerable importance - it should not be forgotten that these factors reflect the 'expert' opinion on considerations not accounted for by the model. While forecasts (or simulations) with such subjective adjustments have generally proved to be more accurate than those obtained from the purely mechanical application of an econometric model (see Christ 1975, Fromm and Klein 1976), in the case of simulations these factors should be used only as measures of the last resort as they represent a deliberate tampering with the facts and vitiate, to a large extent, the very essence of unbiased modelling (see Pindyck and Rubinfeld 1976).

4.7.3 Analysis of the results

As one would expect the results of a historical simulation to closely match the behaviour of the real world, it is desirable to have some quantitative measure of how closely the individual variables track their actual values. The measure that is most frequently used is the RMS (root-mean-square) error and it is given by:

$$\text{RMS error} = +\sqrt{\frac{1}{T}\sum (P_t - A_t)^2} \qquad \text{---(4.26)}$$

where P_t is the simulated value of the variable in time $\underline{t}$,
A_t is the actual value of the variable in time $\underline{t}$
and T is the length of the simulation period: t=1,..,T.

The RMS error is thus a measure of the deviation of the simulated variable from its actual time path. Needless to say, the magnitude of this error can only be obtained by comparing it with the actual size of the variable in question. Under the circumstances, the measure that is more often invoked is the RMS percent error which is defined as

$$\text{RMS percent error} = +\sqrt{\frac{1}{T}\sum\frac{(P_t - A_t)^2}{(A_t)^2}} \quad \text{---(4.27)}$$

There are several other measures of simulation performance (e.g. the U-statistic) but these are largely concerned with the ability of the simulated variable to accurately capture turning points and, as such, are more often employed to evaluate the *ex-post* forecasting performance of a model. Therefore, we did not use these measures to evaluate our system.

It needs to be noted that it is entirely possible for an equation that has a very good statistical fit to have a very poor simulation fit. Such an occurrence is not only possible, but likely, since the behaviour of the model as a dynamic system may have little, or no, bearing to the way individually estimated equations fit the historical data. It is for this reason that RMS simulation errors are an important criterion for evaluating a multi-equation model.

The RMS error and RMS percent error for each of the 40 endogenous variables of the model, over the simulation period: 1973-82, are presented in Table 4.9 alongwith the mean value of each variable for this decade. It needs to be noted that all these measures do not pertain to the deviations of the endogenous variables from their respective reference values but were computed by re-estimating the historical and simulated values of these variables in their absolute form.

Table 4.9
Results Of The Historical Simulation

Time bounds of the simulation: 1973-82

Variable		Mean	RMS error	RMS percent error
1.	AF	125.5	2.1	1.2
2.	ANF	32.2	0.3	0.7
3.	BD	1107.0	224.	20.2
4.	CC	9120.	757.	7.5
5.	CPR	35054.	1246.	2.7
6.	CR	5584.	759.	12.4
7.	DR	2824.	104.	3.2
8.	FR	-582.	80.	13.8
9.	IAR	1299.	139.	8.7
10.	INR	7083.	1068.	11.9
11.	ITR	794.	41.	4.1
12.	KR	135262.	875.	4.7
13.	MR	2429.	281.	9.2
14.	MS	9103.	3465.	33.9
15.	M1	17830.	1720.	8.6
16.	NDE	7608.	1706.	18.9
17.	NTR	2370.	503.	19.2
18.	P	1.7573	0.2028	9.8
19.	QA	123.0	3.9	2.9
20.	QF	122.6	5.0	3.5
21.	QNF	123.9	4.5	2.8
22.	s	0.1802	0.0302	12.1
23.	SCBR	2478.	191.	7.7
24.	SNR	7665.	916.	9.3
25.	TBR	2909.	224.	7.7
26.	TR	6876.	1216.	16.3
27.	v	4.9846	2.4018	43.8
28.	WP	1.9075	0.2156	9.6
29.	WPF	1.7665	0.3105	15.1
30.	WPNF	1.7251	0.1870	8.5
31.	YAR	18293.	619.	3.1
32.	YGFR	44790.	1771.	2.9
33.	YMR	9281.	243.	2.3
34.	YNAR	23673.	1418.	4.6
35.	YNFN	75334.	13836.	17.1
36.	YNFR	41966.	1753.	3.1
37.	YNMN	84817.	16618.	18.1
38.	YPDR	40596.	1374.	2.6
39.	YSR	12043.	1440.	10.0
40.	YTR	2348.	79.	3.1

The results of the historical simulation have thus shown that some variables have a small RMS simulation error while a few others have comparitively large ones. In this case, model evaluation will involve a consideration of which variables are most critical as well as the reasons as to why the large errors occurred. This will be undertaken in the next chapter when we discuss the concept of 'dimension reduction' from the viewpoint of the determination of robust economic policy.

4.8 Conclusions

This chapter dealt with the structural analysis of the system and was generally directed towards model validation on the basis of its dynamic simulation performance. In the process, the role of multipliers in the determination of policy effectiveness was also indicated. We found the system to be dynamically stable and the impact, as well as the long run, multipliers well in keeping with accepted macroeconomic theory. The concept of the pairing of instruments with targets was established and, contrary to a largely held viewpoint, was found to indicate that monetary policy is superior to fiscal policy in as far as the tracking of targets in the real sector was concerned with the opposite being true in as far as the tracking of financial targets was concerned. This is an extremely interesting result and it is quite possible that a hitherto wrong solution to the assignment problem is the reason for the relatively dismal performance of the Indian economy. The simulation performance of the model over the troubled decade (1973-82) displayed no untoward divergent behaviour and thereby proved that the model was robust enough to handle shocks of severe magnitudes. The next chapter will deal with the reduction of the dimension of the model as a prelude to optimal filtering.

CHAPTER 5

RESOLUTION AND STATE SPACE REPRESENTATION OF THE MODEL

5.1 Introduction

The 'state-variable' or 'state-space' description of a system, which has had a long tradition of use in classical dynamics, quantum mechanics and thermodynamics, has been almost universally adopted in modern control and system theory. Its use has come basically from a motivation to represent any physical system by a number of first-order differential (or difference) equations describing *in toto* the state variables of the system. If, at a given instant in time, the numerical values for each of the variables is known, then the state of the system is completely specified, and if all future inputs to the system are also known, then the state of the system at any future time is also specified.

The present trend amongst control theorists is to resort to the state space description when considering dynamical systems as any other mathematical approach to describe a system (like using frequency-domain transfer-function notation) is considered extremely unwieldy. Zadeh and Desoer (1963), Athans and Falb (1966), Sage (1968), Kalman, Falb and Arbib (1969), Barnett (1976), Aoki (1976,1983), amongst others, have done extensive work in this area.

In this chapter, we shall initially reduce the dimension of the model making use of the results of the last chapter as well as the concept of incidence matrices. After we have resolved and reduced the dimension of the system we shall convert its structural form into a state space representation.

5.2 The Resolution Of The Model

5.2.1 Technical assumptions

A macroeconomic model in structural form can be described by the following system of equations

$$h_i x_t^i = f_t^i(x_t, x_{t-1}, \dots, x_{t-p}, u_t, u_{t-1}, \dots, u_{t-q}) \qquad \text{---(5.1)}$$

$$i = 1,2,\dots,n \quad \text{and} \quad t = 1,2,\dots,T$$

$$x_t^i \text{ given for } i = 1,2,\dots,n \quad \text{and} \quad t = -p+1,-p+2,\dots,0 \qquad \text{--(5.2a)}$$

$$u_t^j \text{ given for } j = 1,2,\dots,m \quad \text{and} \quad t = -q+1,-q+2,\dots,0 \qquad \text{--(5.2b)}$$

where x_t^i stands for the value assumed by the ith endogenous variable in time period t, u_t^j stands for the value assumed by the jth exogenous variable in time period t, p and q are the maximal lags with respect to the endogenous and exogenous variables, respectively, and h_i are given parameters such that

$$h_i = \begin{cases} 1, & \text{if the } ith \text{ equation is solved with respect to } x_t^i, \\ 0, & \text{otherwise.} \end{cases}$$

Note that $h_i=1$ implies $\partial f_t^i/\partial x_t^i=0$. We introduce a diagonal nxn matrix H with ith entry h_i:

$$H = \operatorname{diag}(h_i) \qquad \text{---(5.3)}$$

and an m' vector d_t of available information at the beginning of period t:

$$d_t = (x_{t-1}, x_{t-2}, \ldots, x_{t-p}, u_t, u_{t-1}, \ldots, u_{t-q}) \qquad \text{---(5.4)}$$

where $m'=pn+(q+1)m$ is the vector of predetermined variables.

The simulation of the model (5.1) over the interval 1,T consists therefore, once a sequence $(u_t)_{t=1,2,..,T}$ is given, in the successive solutions for $t=1,2,\ldots,T$ in x_t of the n-dimensional nonlinear system

$$Hx_t - f_t(x_t, d_t) = 0 \qquad \text{---(5.5)}$$

where d_t is given by eq.(5.4).

5.2.2 Loop variables

Let I be a subset of the indices of endogenous variables $\{1,2,..,n\}$ and let S(I) be the subsystem of dimension s=card(I) obtained from the system (5.5) by retaining only the equations of rank $i \in I$ and considering the variables x_t^j for $j \notin I$ as exogenously given. Of course, S(1,2,...,n) coincides with the original system (5.5).

We define the incidence matrix E(I) of the system S(I) as the sxs matrix of elements

$$e_{ij} = \begin{cases} 1, & \text{if the variable } x_t^{p(j)} \text{ occurs in equation } p(i), \\ 0, & \text{otherwise,} \end{cases}$$

where p(i) and p(j) are the ith and jth elements of the ordered set I, respectively.

For example, given the system of equations

$$x_1 = x_2 + x_3 \qquad \text{---(5.6a)}$$

$$x_2 = a_0 + a_1 x_1 \qquad \text{---(5.6b)}$$

$$x_3 = b_1 x_1 - b_2 x_5 \qquad \text{---(5.6c)}$$

$$x_4 = c_1 x_1 \qquad \text{---(5.6d)}$$
$$x_5 = d_1 x_1 - d_2 x_4 \qquad \text{---(5.6e)}$$

where a_0, a_1, b_1, b_2, c_1, d_1 and d_2 are exogenous parameters, we can build incidence matrices as follows:

$$E(1,2,3) = \begin{bmatrix} 1 & 1 & 1 \\ 1 & 1 & 0 \\ 1 & 0 & 1 \end{bmatrix}; \quad E(1,2,5) = \begin{bmatrix} 1 & 1 & 0 \\ 1 & 1 & 0 \\ 1 & 0 & 1 \end{bmatrix}; \qquad \text{---(5.7a)}$$

$$E(3,4,5) = \begin{bmatrix} 1 & 0 & 1 \\ 0 & 1 & 0 \\ 0 & 1 & 1 \end{bmatrix}; \quad E(1,4,5) = \begin{bmatrix} 1 & 0 & 0 \\ 1 & 1 & 0 \\ 1 & 1 & 1 \end{bmatrix}. \qquad \text{---(5.7b)}$$

The intuitive notion of a recursive solution of a system of equations can be formalized in terms of the incidence matrix: system S(I) is recursive if $h_1=1$ for all $\underline{i} \in \underline{I}$ and its incidence matrix E(I) satisfies

$$\begin{cases} e_{ii} = 1, \ \forall \underline{i} \in \underline{I} \\ \text{and} \\ e_{ij} = 0, \ \forall \underline{j} > \underline{i}. \end{cases} \qquad \text{---(5.8)}$$

It is clear from eq.(5.7) that S(1,4,5) is recursive while S(1,2,3), S(1,2,5) and S(3,4,5) are not. It is thus possible, given x_2 and x_3, to directly compute the remaining variables x_1, x_4 and x_5 from eqs. (5.6a), (5.6d) and (5.6e), in this order. Under the circumstances, we say that x_2 and x_3 are the loop variables of system (5.6).

We can extend this result to the general case by using the following definition (see Gabay *et al* 1980):

Definition 5.1 A variable *j* is said to be a loop variable of the system (5.5) if the *jth* equation is not resolved with respect to it (i.e., $h_j=0$) or if the *jth* column of the incidence matrix of system (5.5) has at least one nonzero element above the principal diagonal.

An equivalent characterization is

$$h_j \prod_{i=j+1}^{n} (1 - e_{ij}) = 0. \qquad \text{---(5.9)}$$

Let B be the set of indices of loop variables of cardinality *s*; $\bar{B}$ its complement in $\{1,2,\ldots,n\}$. The following theorem is a direct consequence of the definitions set forth in Gabay *et al* (1980) which stem from the results obtained by Gabay *et at* (1978) and Nepomiastchy *et al* (1978).

Theorem 5.1 In any model, the subsystem $S(\bar{B})$ is recursive.

Remark: Influence of the numbering of variables

System (5.6) can be viewed as an extremely simple macroeconomic model linking together five endogenous variables: Y (national income), C (consumption), I (investment), M (money supply) and R (interest rate), satisfying the equations

$$Y = C + I$$
$$C = a_0 + a_1 Y$$
$$I = b_1 Y - b_2 R$$
$$M = c_1 Y$$
$$R = d_1 Y - d_2 M$$

with the numbering $x_1 = Y$, $x_2 = C$, $x_3 = I$, $x_4 = M$ and $x_5 = R$. Suppose we use another numbering where $x_1 = C$, $x_2 = I$, $x_3 = Y$, $x_4 = R$ and $x_5 = M$, then we can write the model in the form of the system of equations:

$$x_1 = a_0 + a_1 x_3 \qquad \text{---(5.10a)}$$
$$x_2 = b_1 x_3 - b_2 x_4 \qquad \text{---(5.10b)}$$
$$x_3 = x_1 + x_2 \qquad \text{---(5.10c)}$$
$$x_4 = d_1 x_3 - d_2 x_5 \qquad \text{---(5.10d)}$$
$$x_5 = c_1 x_3 \qquad \text{---(5.10e)}$$

which now has three loop variables: $x_3 = Y$, $x_4 = R$ and $x_5 = M$.

However, with the numbering $x_1 = M$, $x_2 = R$, $x_3 = I$, $x_4 = C$ and $x_5 = Y$, the incidence matrix of the resulting system which is given by

$$E(1,2,3,4,5) = \begin{bmatrix} 1 & 0 & 0 & 0 & 1 \\ 1 & 1 & 0 & 0 & 1 \\ 0 & 1 & 1 & 0 & 1 \\ 0 & 0 & 0 & 1 & 1 \\ 0 & 0 & 1 & 1 & 1 \end{bmatrix}$$

indicates the presence of only one loop variable, $x_5 = Y$, and this is obviously the optimal reduction.

The isolation of a small number of loop variables is extremely useful for simplifying the resolution of the system (5.5). In fact, given values for the loop variables, the solutions for the remaining endogenous variables of the subsystem can be directly obtained by recursively applying the formula of the model.

Partitioning of a model therefore basically consists of isolating the loop variables from the other variables of the system. If after several trials corresponding to different numberings of the variables, the number of loop variables $\underline{s}$ is small compared to $\underline{n}$, the incidence matrix of the system is quasilower triangular since most of its nonzero elements lie below the principal diagonal. The model is then said to be in quasitriangular or quasirecursive form.

It is to be noted that the determination of a set of loop variables corresponds to finding arcs of the directed graph associated with the incidence matrix E, the removal of which leaves a tree. Starting with different random orderings of the equations, it is possible to find alternative sets of loop variables and, consequently, to select from them the one yielding the smallest $\underline{s}$ obtained via an optimal reduction. From an econometrician's viewpoint, the knowledge of such sets of coupling variables can be of great interest as it provides valuable insight into the interactions between the variables as a result of the various feedback mechanisms integrated into the system.

5.2.3 Partitioning of the model

The isolation of a set of loop variables, brought about by an appropriate renumbering, induces a partition of the model into four blocks called the prologue, the core, the loop equations block and the epilogue (see Gabay *et al* 1980). They are visualized in Figure 5.1 which shows the configuration of the incidence matrix after renumbering.

The prologue is a recursive subsystem formed by the first $\underline{k} \geqslant 0$ equations where the loop variables do not occur

$$x_t^i = f_t^i(x_t^1,\dots,x_t^{i-1},d_t), \qquad i = 1,2,\dots,k \qquad ---(5.11)$$

Given d_t, the system (5.11) can be solved recursively and independently of the other equations of the model.

After the variables $(x_t^i)_{i=1,..,k}$ defined in the prologue, we can distinguish r-k endogenous variables $x_t^{k+1},\dots,x_t^r$ and the $\underline{s}$ loop variables $x_t^{r+1},\dots,x_t^{r+s}$ which we group in the vector $x_t^b = (x_t^{r+1},\dots,x_t^{r+s})$. Once the prologue variables have been computed and given a value for x_t^b, it is possible to solve the system of equations k+1 to $\underline{r}$ recursively for $x_t^c = (x_t^{k+1},..,x_t^r)$; this subsystem of r-k equations is called the core of the model

$$x_t^i = f_t^i(x_t^1,\dots,x_t^{i-1},x_t^b,d_t), \qquad i = k+1,\dots,r \qquad ---(5.12)$$

The block of the loop equations is constituted by the $\underline{s}$ equations

$$h_i x_t^i = f_t^i(x_t^1,\dots,x_t^{r+s},d_t), \qquad i = r+1,\dots,r+s \qquad ---(5.13)$$

which is not recursive and where $h_i = 0$ or 1.

Once the system (5.11)-(5.13) has been solved, the remaining variables can be computed by a recursive system called the epilogue

$$x_t^i = f_t^i(x_t^1,\dots,x_t^{i-1},d_t), \qquad i = r+s+1,\dots,n \qquad ---(5.14)$$

Figure 5.1
Partitioning Of A Macroeconometric Model

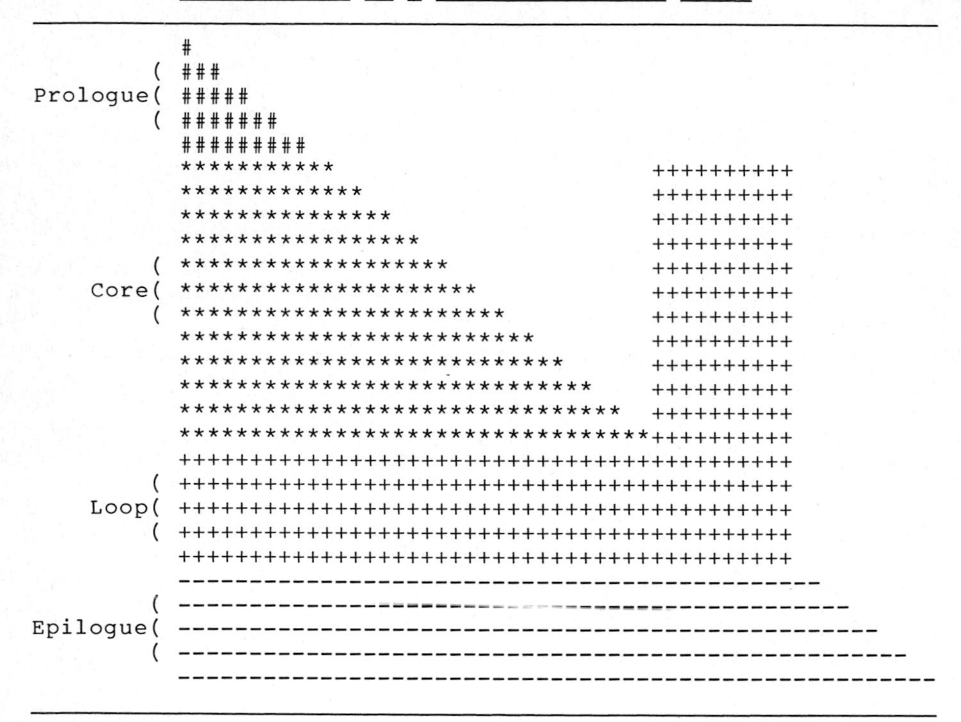

The respective dimensions of these four blocks for our estimated macroeconometric model, after an optimal renumbering, are provided in Table 5.1 and its corresponding incidence matrix is depicted in Figure 5.2.

Table 5.1
Partition Of The Estimated Model

Blocks	Dimension
1. Prologue	2
2. Core	32
3. Loop equations	5
4. Epilogue	1
Total	40

From Figure 5.2 we notice that an optimal resolution of the model yields a global minimum of 5 loop variables. There was an initial recursive subsystem (comprising 2 variables) in which these loop variables did not figure and therefore the renumbering of the variables did help to induce a partitioning

Figure 5.2

Incidence Matrix Of The Estimated Model

```
          ........................................
  1.   KR 1
  2.   DR 11
  3.  YTR   1                                1
  4.   AF    1                                 11
  5.  ANF     1                                11
  6.   QF    1 1
  7.  QNF     1 1
  8.   QA      111
  9.  YAR        11
 10.  YMR   1   1  1
 11.   WP          11                       1  11
 12.  YSR   1     1 11
 13. YNAR   1      1 11
 14. YNFR         1   11
 15. YPDR              11
 16. YGFR  1           1 1
 17.   MS           1     1                 1
 18.    v                111
 19.  CPR               1   1
 20.    s             11     1
 21.  SNR              1     11
 22.  INR                    11              1
 23.   MR           1       1  11
 24.    P              1   1     1          1
 25. YNFN              1         11
 26. YNMN                        11
 27.  NDE                         11
 28.   TR                        1  1
 29.  NTR                        1   1
 30.   CR                    1        1
 31.   BD                          11111
 32.   CC                              11
 33. SCBR                               11
 34.  TBR                                11
 35.   M1                               1 11
 36.  ITR   1                                1
 37.   FR                      1              1
 38.  WPF      1            1                  1
 39. WPNF       1  1                           11
 40.  IAR         1                              1
```

of a unique prologue. In a like manner, the loop variables did not continue right upto the end of the system and there was one variable which could be recursively solved given the solutions of the preceding blocks. This confirmed the existence of an epilogue as well. Therefore, the model resolved itself into 2 prologue variables, a "core" of 32 variables, a loop equations block of 5 variables and 1 epilogue variable.

It needs to be noted that these results were obtained by inspection and not by using any automatic algorithms existing in the literature (see Gabay *et al* 1978, Nepomiastchy *et al* 1978) which are often invoked (especially in the case of large models) to help provide a minimal set (of loop variables), although not necessarily a set of minimal cardinality $\bar{s}$.

From theorem 5.1 we realize that as the subsystem $S(\bar{B})$ is recursive, removing all the 5 loop variables from the model would render the resulting submodel a recursive one. As we wanted such a formulation to facilitate the procedure of optimal filtering *vis-à-vis* the convergence of the Kalman gain, we went ahead with this operation and, in the process, deleted a few specific variables which: (a) on the basis of the simulation results of the last chapter were found to behave erratically and (b) gave rise to superfluous identities. This deletion implied the consequent exogenization of a few other endogenous variables as well. This 'dimension reduction' was carried out with the prime purpose of ending up with a compact subsystem which would render optimal performance during the filtering and control experiments of the subsequent chapters.

5.2.4 Modifications of the model

The following modifications were carried out on the model to ensure the reduction of its dimension:

1. All the 5 loop variables (M1, ITR, FR, WPF and WPNF) were deleted from the system of endogenous variables. Of these, FR was eliminated entirely on grounds of being superfluous. The remaining 4 loop variables were exogenized.
2. The exogenization of M1 implied that all the preceding equations of the monetary block, i.e., CC, SCBR and TBR, were rendered superfluous. As such, all these variables were eliminated. This automatically helped us to dispense with DFEA and BR as exogenous variables.
3. The exogenization of ITR implied that we could eliminate IGTR entirely.
4. The exogenization of WPF and WPNF implied that AF and ANF could both be themselves exogenized because they were now determined entirely by exogenous or predetermined factors.
5. The exogenization of AF and ANF implied that the epilogue variable (IAR) could be dispensed with entirely from the model.
6. The elimination of IAR implied that IGAR could be dropped because it was no longer needed as an exogenous variable.
7. The core variable, v, was eliminated entirely from the model because of its extremely erratic simulation performance. Its continued presence in the system would have jeopardized the determination of robust economic policy.
8. The elimination of v implied, as a corollary, the sequential elimination of the prologue variables, KR and DR, and the core variables, YGFR and MS, because each of them existed only as links in a chain which determined v.
9. CR was exogenized because of its relatively poor simulation performance. This permitted the elimination of DBFA as an exogenous variable.
10. The effects of QF and QNF were directly transmitted onto YAR instead of being channelled through the intermediate variable QA which was eliminated.
11. YNFR was directly considered to be the sum of its indivi-

dual components, i.e., YAR, YMR, YTR and YSR. This allowed us to drop the artificially created variable YNAR.

12. The poor simulation performance of P necessitated its exclusion from the model. Its role, in as far as the determination of YNFN was concerned, was taken over by WP.

It needs to be noted that the elimination of FR from the model implied that we had to replace it by the alternative expression MR-XR in the identity determining INR. Similarly, the elimination of YNAR from the model implied that, in the equation determining s, we had to replace it by the equivalent expression YNFR-YAR.

5.2.5 The resolved structure of the model

The resolution of the model thus lay in the isolation, and subsequent elimination, of a small number of loop and related variables of the system such that the remaining endogenous variables of the dimensionally reduced model could be computed recursively. In our case, the compacted subsystem comprised 20 endogenous variables only. Of these, 14 were determined by reaction equations and the remaining 6 were obtained through identities. The total number of exogenous (including control) variables, under the present format, was 16 (including the intercept term).

The vector dx after the optimal resolution is now given by

1	dYTR
2	dQF
3	dQNF
4	dYAR
5	dYMR
6	dWP
7	dYSR
8	dYNFR
9	dYPDR
10	dCPR
11	ds
12	dSNR
13	dINR
14	dMR
15	dYNFN
16	dYNMN
17	dNDE
18	dTR
19	dNTR
20	dBD

The vector of exogenous variables du is now given by

1	dITR
2	dM1
3	dT
4	dGDE
5	dDEF
6	dWPF
7	dWPNF
8	dXR
9	dAF
10	dANF
11	dR
12	dMDL
13	dCR
14	dPM
15	dt
16	1

Matrix manipulations of the form outlined in the last chapter enabled us to convert the linearized structural form deviations subsystem into its reduced form equivalent given by

$$dx(t) = Adx(t-1) + Bdu(t) \qquad \text{---(5.15)}$$

where A and B are appropriately dimensioned matrices.

In Table 5.2 we provide the numerical values for the matrices A and B which constitute the reduced form of the resolved model.

Table 5.2
The Reduced Form Of The Resolved Model

A	COL1	COL2	COL3	COL4	COL5	COL6	COL7	COL8	COL9	COL10
	COL11	COL12	COL13	COL14	COL15	COL16	COL17	COL18	COL19	COL20
	0.982	0	0	0	0	0	0	0	0	0
	0	0	0	0	0	0	0	0	0	0
	0	0	0	0	0	0	0	0	0	0
	0	0	0	0	0	0	0	0	0	0
	0	0	0	0	0	0	0	0	0	0
	0	0	0	0	0	0	0	0	0	0
	0	0	0	0	1.5179E-16	0	0	0	0	0
	0	0	0	0	0	0	0	0	0	0
	0.750444	0	0	0	0.6836	0	0	0	0	0
	0	0	0	0	0	0	0	0	0	0
	-.000064538	0	0	0	-0.00005879	0	0	0	0	0
	0	0	0	0	0	0	0	0	0	0
	0.921352	0	0	0	0.00128849	0	0.7138	0	0	0
	0	0	0	0	0	0	0	0	0	0
	2.6538	0	0	0	0.684888	0	0.7138	0	0	0
	0	0	0	0	0	0	0	0	0	0
	2.54977	0	0	0	0.658041	0	0.685819	0	0	0
	0	0	0	0	0	0	0	0	0	0
	1.68463	0	0	0	0.434768	0	0.453121	0	0	0.1633
	0	0	0	0	0	0	0	0	0	0
	.0000219798	0	0	0	.0000056725	0	0.000005912	0	0	2.1613E-24
	0	0	0	0	0	0	0	0	0	0
	1.10982	0	0	0	0.28642	0	0.298511	0	0	-4.6741E-18
	0	0	0	-8.5953E-16	0	0	0	0	0	0
	1.22162	0	0	0	0.282089	0	0.342257	0	0	0.00950423
	0	0	0	0.819259	0	0	0	0	0	0
	0.111806	0	0	0	-0.00433177	0	0.0437458	0	0	0.00950423
	0	0	0	0.819259	0	0	0	0	0	0
	0.426002	0	0	0	-1.34447	0	0.7138	0	0	-3.6260E-17
	0	0	0	0	0	0	0	0	0	0
	0.469539	0	0	0	-1.48187	0	0.78675	0	0	-3.9966E-17
	0	0	0	0	0	0	0	0	0	0
	0.0138044	0	0	0	-0.0435671	0	0.0231305	0	0	-1.1750E-18
	0	0	0	0	0	0	0.6607	0	0	0

A	COL1	COL2	COL3	COL4	COL5	COL6	COL7	COL8	COL9	COL10
	COL11	COL12	COL13	COL14	COL15	COL16	COL17	COL18	COL19	COL20
	0.0339949	0	0	0	-0.107289	0	0.0569612	0	0	-2.8935E-18
	0	0	0	0	0	0	0	0	0	0
	0.00860523	0	0	0	-0.0271583	0	0.0144188	0	0	-7.3245E-19
	0	0	0	0	0	0	0	0	0.4673	0
	-0.0287957	0	0	0	0.0908799	0	-0.0482495	0	0	2.4510E-18
	0	0	0	0	0	0	0.6607	0	-0.4673	0

B	COL1	COL2	COL3	COL4	COL5	COL6	COL7	COL8
	COL9	COL10	COL11	COL12	COL13	COL14	COL15	COL16
ROW1	0.3307	0	0	0	0	0	0	0
	0	0	0	0	0	0	0	95.1237
ROW2	0	0	0	0	0	0	0	0
	1.465	0	0.8031	0	0	0	1.921	-2.3764
ROW3	0	0	0	0	0	0	0	0
	0	11.5406	0.3454	0	0	0	1.224	-0.013
ROW4	0	-1.3856E-19	-1.2962E-14	0	0	-1.8783E-15	-1.0527E-15	0
	128.098	472.253	84.3564	-5.0677E-15	0	0	218.058	-401.852
ROW5	0.252721	0	0	0	0	0	0	0
	0	213.72	6.39646	-22.823	0	0	22.6673	277.34
ROW6	-.000021734	0.000039	3.6484	0	0	0.5287	0.2963	0
	0	-0.01838	-.000550096	0.00196278	0	0	-0.00194938	-0.0355512
ROW7	0.310276	-.000854763	-79.962	0.1987	0	-11.5875	-6.49401	0
	11.6441	43.3306	7.68005	-0.0430182	0	0	19.8642	324.926
ROW8	0.893697	-.000854763	-79.962	0.1987	0	-11.5875	-6.49401	0
	139.742	729.304	98.4329	-22.866	0	0	260.589	295.537
ROW9	0.858664	-.000821256	-76.8275	0.190911	0	-11.1333	-6.23944	0
	134.264	700.715	94.5743	-21.9697	0	0	250.374	387.807
ROW10	0.567319	-.000542604	-56.7599	0.126135	0	-7.35576	-4.1224	0
	88.7085	462.963	62.4853	-14.5154	0	0	165.422	-34.3058
ROW11	0.0000067402	-7.0795E-09	-.000662277	.0000016457	0	-.000095972	-.000053786	0
	-.000013403	0.00172404	.0000442517	-.000189386	0	0	0.000165278	0.0142206
ROW12	0.373744	-.000357462	-33.4401	0.0830964	0	-4.8459	-2.71579	5.5511E-17
	18.0252	155.999	14.5502	-9.56257	0	6.3998E-13	40.1812	529.982

B	COL1 COL9	COL2 COL10	COL3 COL11	COL4 CCL12	COL5 COL13	COL6 COL14	COL7 COL15	COL8 COL16
ROW13	0.411396 24.2373	0.0303085 177.546	2835.32 18.6005	0.0952739 -9.41795	0 0	410.875 -609.989	230.267 50.612	-1.0582 708.486
ROW14	0.0376521 6.21201	0.030666 21.5472	2868.76 4.05026	0.0121775 0.144622	0 0	415.721 -609.989	232.983 10.4309	-0.0582011 178.504
ROW15	0.143461 139.742	1.34539 94.8463	125859 79.4441	0.1987 44.8871	0 0	18238.6 0	10221.5 193.298	0 -931.655
ROW16	0.158123 154.024	1.48288 104.54	173241 87.5633	0.219007 49.4746	0 0	20102.6 0	11266.1 213.053	0 -1026.87
ROW17	0.00464881 4.5283	0.0435968 3.07346	5093.28 2.57436	0.00643881 1.45455	0 0	591.016 0	331.224 6.26377	0 292.765
ROW18	0.0114482 11.1514	0.107362 7.56874	40132.6 6.33964	0.0158563 3.58199	0 0	1455.44 0	815.675 15.4252	0 420.516
ROW19	0.00289791 2.82279	0.0271768 1.9159	2542.35 1.60477	0.00401374 0.90672	0 0	368.42 0	206.474 3.90462	0 76.6647
ROW20	-0.0096973 -9.44592	-0.0909418 -6.41117	-37581.6 -5.37005	0.986569 -3.03416	1 -1	-1232.84 0	-690.925 -13.0661	0 -204.416

5.3 State Space Representation Of The Econometric Model

5.3.1 Notion of state space representation

The concept of state is one of the key notions in dynamics and arises essentially in problems of optimization over time. All dynamic models can be put into their corresponding state space representations. This implies that given the future time paths of the instruments and exogenous disturbance terms, it is possible to uniquely determine all the future time paths of the state variables of the system provided the present values of the state variables of the system are known. A state space equation thus admits a unique solution. This uniqueness is the deterministic counterpart of the Markovian property of Markov processes. A state space representation is an extremely desirable form in order to analyze dynamics since most of the techniques and concepts associated with the theory of dynamic models are available only for models in their state space representation.

We shall now outline the concept of a "dynamic model" in state space terminology. Let $\underline{x}$ be a state vector, i.e., a (minimal) collection of endogenous variables which uniquely determines the future time paths of all the endogenous variables in the model, once the future time paths of all the policy variables and exogenous disturbance terms are specified. For example, in discrete time analysis, if $\underline{x}$ is governed by a (vector) difference equation

$$x_{t+1} = f(x_t, u_t, e_t^1, t) \qquad \text{---(5.16)}$$

where $\underline{u}$ is the vector of policy variables and e^1 is a vector of disturbances, then $\underline{x}$ is a state vector under suitable regularity conditions which are given as follows:

(i) The function $f(x_t, u_t, e_t^1, t)$ is continually differentiable with respect to $\underline{x}$ and $\underline{u}$, and

(ii) The function $f(x_t, u_t, e_t^1, t)$ and all its partial derivatives with respect to $\underline{x}$ and $\underline{u}$ are bounded.

These conditions ensure the existence and uniqueness of solutions to eq.(5.16) because all future values of $\underline{x}$ are uniquely determined by its initial conditions and the future time paths of the vectors of policy variables and disturbances. The time variable $\underline{t}$ is included in the arguments of f(.) to indicate the possibility that the dynamics of the system may be time-varying. When e^1 is random, the system is known as a stochastic one. All endogenous variables of a model not represented in $\underline{x}$, denoted by $\underline{z}$, must be expressible as functions of $\underline{x}$, $\underline{u}$ and possibly some other set of disturbances, e^2,

$$z_t = g(x_t, u_t, e_t^2, t) \qquad \text{---(5.17)}$$

for $\underline{x}$ to serve as a state vector of the model. It needs to be noted that the components of the state vector may not have any direct economic significance. The dynamic equation (5.16) is derived from the structural form specification of a model. Some of these structural equations are dynamic equations representing stock-flow relations in the economic model, while the others are purely algebraic equations relating some of $\underline{x}$ and $\underline{u}$ in the form of accounting identities.

The stock-flow relations are of the form

$$x_{t+1} = F(x_t, v_t, u_t, t) \qquad \text{---(5.18)}$$

while the algebraic relations are expressed as

$$G(u_t, x_t, v_t, t) = 0 \qquad \text{---(5.19)}$$

where $\underline{x}$ and $\underline{v}$ are some endogenous variables and $\underline{u}$ is a vector of policy variables. If G(.) can be solved for $\underline{v}$ in terms of $\underline{x}$, then a vector difference equation for $\underline{x}$ of the form of eq.(5.16) is obtained by replacing $\underline{v}$, in eq.(5.18), by its solution in terms of $\underline{x}$.

We now assume, given eq.(5.16), that some of the components of the state vector $\underline{x}$ may not be explicitly available for direct measurement. In such situations, an "observation" equation specifies the amount of information available for measurement or the extent to which the state variables can be observed either directly or indirectly,

$$y_t = h(x_t, u_t, e_t^3, t) \qquad \text{---(5.20)}$$

where e^3 is the exogenous disturbance vector representing measurement inaccuracy. Eq.(5.20) indicates that the vector $\underline{y}$ is the additional information content amalgamated with the information set at time $\underline{t}$. The answers to the questions as to who acquires this information and how are likely to vary depending upon the context or framework. In our case we assume that eq.(5.20) represents the information acquisition program of a macroeconomic policy maker (see Aoki 1981).

What such a policy maker observes is not necessarily the same as the set of endogenous variables that he wishes to control or influence. In general, it is very likely that he will observe less variables than he seeks to influence but more variables than he is interested in controlling. The vector of endogenous variables that a policy maker is interested in controlling, denoted by $\underline{w}$, must be a subset of $\underline{x}$ or $\underline{z}$ and must be related to $\underline{x}$ and $\underline{u}$ by the following

$$w_t = H(x_t, u_t, e_t^4, t) \qquad \text{---(5.21)}$$

where the vector e^4 represents random slippages between the targets and the instruments. We shall adopt the general framework, set out by eqs.(5.16) and (5.20), for representing a dynamic economic model. The application of such a methodology, for the remainder of this study, will demonstrate the viability of state space representations for understanding the complex nature of dynamic stochastic economic phenomena. Unless otherwise specified, we shall regard $\underline{u}$ to include instruments, which are not explicitly endogenized by reaction equations, as well as exogenous variables. The disturbances will not be specified separately but will be assumed to be incorporated in the functional form of the equation itself.

We shall now discuss the procedure for converting the reduced form of an econometric model into its equivalent state space form using the notions just outlined.

5.3.2 Converting into a state space representation

Structural equations for economic models are often represented in a form analogous to the following equation in which $\underline{x}$ and $\underline{u}$ are vectors:

$$
\begin{aligned}
x(t) = {} & A_0x(t) + A_1x(t-1) + \ldots + A_px(t-p) \\
& + B_0u(t) + B_1u(t-1) + \ldots + B_qu(t-q) + f(t) \qquad \text{---(5.22)}
\end{aligned}
$$

where $x(t)$ is a vector of endogenous variables, $u(t)$ is an instrument vector and $f(t)$ is some exogenous disturbance vector. Kendrick (1972) and Pindyck (1973), for example, have used this form as the basis of their analysis with $B_0 = 0$ and $f(t) = 0$ for all $\underline{t}$.

If $(I - A_0)$ is nonsingular, we obtain the reduced form equation from it which is given by

$$
\begin{aligned}
x(t) = {} & A_1x(t-1) + \ldots + A_px(t-p) \\
& + B_0u(t) + B_1u(t-1) + \ldots + B_qu(t-q) + d(t) \qquad \text{---(5.23)}
\end{aligned}
$$

where some of the B's may be zero and where we redefine

$$(I - A_0)^{-1}A_i = A_i \text{ and}$$

$$(I - A_0)^{-1}B_i = B_i$$

for all $\underline{i}$ under the present context. We also specify

$$(I - A_0)^{-1}f(t) = d(t)$$

for all $\underline{t}$. This transformation could however complicate covariance computations when the A's are random (see Aoki 1976). In general, if $\dim x(t) = n$, then $\dim u(t) = m \; n$. When $(I - A_0)$ is singular, no general formula exists for the derivation of the reduced form.

Our reduced form system, in vector notation, was given by

$$x(t) = Ax(t-1) + Bu(t) \qquad \text{---(5.24)}$$

where $\underline{x}$ was a 20x1 vector of endogenous variables, $\underline{u}$ was a 16x1 vector of exogenous (including control) variables, $\bar{A}$ and B were appropriately dimensioned matrices and where the prefix $\underline{d}$ has been deleted for reasons of convenience.

We now define a 20-dimensional vector $z(t)$ by

$$z(t) = x(t) - Bu(t) \qquad \text{---(5.25)}$$

Substituting eq.(5.24) into eq.(5.25), we get

$$z(t) = Ax(t-1) \qquad ---(5.26)$$

and therefore the reason for the definition of a new state variable, given by eq.(5.25), is now apparent. From it we have the following specification for $\underline{x}$ given by

$$x(t) = z(t) + Bu(t) \qquad ---(5.27)$$

Lagging eq.(5.27) by one period and substituting the result into eq.(5.26), we obtain

$$z(t) = Az(t-1) + ABu(t-1) \qquad ---(5.28)$$

Therefore the pair of equations (5.28) and (5.27) provide us with the state space representation of the econometric model whose reduced form was given in eq.(5.24). Combining them we obtain the state space model

$$z(t) = Az(t-1) + B^*u(t-1) \qquad ---(5.29)$$

$$x(t) = Cz(t) + Bu(t) \qquad ---(5.30)$$

where $B^* = AB$ and C is an identity matrix. Eq.(5.29) is called the state transition equation while eq.(5.30) is known as the observation equation. This representation is referred to as an observable canonical form. We can also obtain an alternative representation known as a controllable canonical form wherein the dynamic matrices are transposes of each other and the roles of B^* and C are interchanged. We call a dynamic system

$$\left.\begin{aligned} z(t) &= Az(t-1) + Bu(t-1) \\ x(t) &= C'z(t) + Du(t) \end{aligned}\right) \qquad ---(5.31)$$

a dual of another dynamic system

$$\left.\begin{aligned} z(t) &= A'z(t-1) + Cu(t-1) \\ x(t) &= B'z(t) + Du(t) \end{aligned}\right) \qquad ---(5.32)$$

This correspondence $A' — A$ and $B — C$ is the precise sense in which these two systems are the duals of each other. Intertemporal minimization of a quadratic cost functional subject to linear dynamics and the optimal one-step ahead predictor (Kalman filter) are the duals in this sense.(see Aoki 1983)

5.4 Conclusions

We initiated this chapter by outlining a possible solution to the problem of dimension-reduction of an estimated macroeconometric model from the viewpoint of determing robust optimal economic policies. This was carried out through a resolution of the system which isolated a small set of loop variables - the removal of which immediately rendered the model recursive. The model was then further compacted through the elimination of the prologue and epilogue variables, as well as a few core variables, on the basis of their dynamic simulation performance. The reordered subsystem was then converted into its state space representation of the observable canonical form through the specification of artificially created state variables which provided a basis for defining an observation equation for the dynamic control system.

We shall now proceed with the estimation of the Kalman filter where this basis will be utilized in order to identify an observation mechanism for the Indian economy well in keeping with the concept of errors in information signalling which is quite prevalent within the current Indian data base context.

CHAPTER 6

OPTIMAL FILTERING AND STATE RECONSTRUCTION

6.1 Introduction

As the knowledge about the functioning of the Indian economy has been far from perfect, most policy decisions have been made in the presence of uncertainty. The principal objective of stochastic control theory is to improve system performance essentially when uncertainty is involved. The regulator mechanism of optimal control, where a quadratic penalty is placed on the deviations of the system state from the desired state, is well suited for short term economic policy problems. The economic regulation problem can therefore be formulated as a linear stochastic problem with a quadratic criterion of optimum performance.

The closed-loop control policy leads to a feedback rule for determining optimal control actions, given appropriate statistics based on available information. The determination of these statistics, namely, the conditional mean and the error covariance matrix of the system state, can take place separately. However, the relationship between the system state and the information set has to be kept in view explicitly. The observations of economic activity are assumed to comprise errors in observation as well as changing or fragmentary economic indicators.

The incorporation of such error-prone information is achieved through the Kalman filter which provides minimum-variance, unbiased estimates of the system state, conditional upon the available information. Since the filter recursively updates estimates, it has great practical advantages in as much as past information need not be repeatedly used since its effect is instantly summarized in the earlier estimates which are constantly and optimally revised with the arrival of fresh information.

This chapter will deal with the estimation of the Kalman filter which will be used for state reconstruction and the derivation of optimal economic policies given the resolved model of the last chapter.

6.2 Some Economic Applications Of The Kalman Filter

There have been numerous applications of the Kalman filter to dynamic economics as well as econometrics. This section provides a few illustrative examples of these studies - all of them reflecting breakthroughs in diverse areas using widely differing concepts and methodologies - with an idea of providing a backdrop to our ensuing analysis.

6.2.1 Interindustry demand

One example of an area where the Kalman filter has been applied is the prediction of interindustry demand. Let x_t be an nxl vector of outputs from n industries in period t; A_t an nxn matrix of input-output coefficients where the i-jth element represents the amount of output from industry i required to produce one unit of output in industry j; and d_t an nxl vector of final demands for the products of the n industries. The standard input-output relationship in period t is given by the linear equation

$$x_t = A_t x_t + d_t \qquad ---(6.1)$$

the solution of which is

$$x_t = (I - A_t)^{-1} d_t = B_t d_t \qquad ---(6.2)$$

where $B_t = (I - A_t)^{-1}$. The problem is to estimate the coefficients B_t in the linear relation (6.2) between industry outputs and the final demands under the assumption that the linear relation contains a vector e_t of random errors, i.e.,

$$x_t = B_t d_t + e_t \qquad ---(6.3)$$

In eq.(6.3) the production x_t and the final demand d_t are assumed to be directly observable. The vector e_t is assumed to have mean 0 and covariance matrix Q and to be serially uncorrelated. Vishwakarma, de Boer and Palm (1970) cast this model into its state space representation and applied the Kalman filter to the system in order to estimate the elements of B_t. The essential difficulty with this approach, especially in the case of developing countries wherein the input-output technique is extensively used,is the fact that the paucity of data makes it very unrealistic to consider B_t as a time-varying matrix and as such the possibility of applying the Kalman filter algorithm to such a system is very limited.

6.2.2 Macroeconometric forecasting

A second area where the Kalman filter has been applied is that of macroeconometric forecasting. Let the reduced form of a system of equations be given by

$$x_t = Ax_{t-1} + Bu_{t-1} \qquad ---(6.4)$$

In addition, assume that outside information on a subset of x_t is available one period ahead; for example, if x_t contains an equation for price determination then the outside information could consist of data on price expectations. If, on the other hand x_t comprises NNP or its components then this outside information could well be some preliminary estimate of NNP and/or its components. It is this set of outside information which constitutes the elements of the observation equation which is given by

$$y_t = Cx_t + Du_t \qquad \text{---(6.5)}$$

The matrix C can take the special form of

$$C = \begin{bmatrix} I & 0 \end{bmatrix} \qquad \text{---(6.6)}$$

if x_t is rearranged so as to place the subset of variables with outside measurements before the other variables in the state vector. Under the assumptions of the model x_t cannot be known accurately. Mariano and Schleicher (1972) used such a setup and applied the Kalman filter to estimate the state in the presence of such fragmentary information. For the purpose of our study, we adopted a similar approach because it preempted the need to create artificial state variables of the kind depicted in the last chapter. However, the concept of preliminary estimates notwithstanding, we felt that eq.(6.6) represented too naive a hypothesis and, as such, suitable modifications were carried out on it before we incorporated this type of observation equation into our control system.

6.2.3 Formation of expectations

A third area of application of the Kalman filter to dynamic economics is the explanation of the formation of expectations of economic agents. Several economists have written on this subject extensively. It was Muth (1960,1961) who laid much of the theoretical foundations for subsequent studies in this area and his works contained results on optimal forecasting independent of Kalman's work which was published around that time. However, the basic idea that economic behaviour is contingent on expectations or the expected value of variables was first mooted by Cagan (1956) within the framework of economic dynamics. For example, in Cagan's work on hyperinflation, when formulated in terms of discrete time, expected price in time $\underline{t}$, p_t^*, is assumed to satisfy the difference equation

$$p_t^* - p_{t-1}^* = d(p_t - p_{t-1}^*) \qquad \text{---(6.7)}$$

where p_t is the observed price. Eq.(6.7) implies

$$\begin{aligned} p_t^* &= dp_t + (1-d)p_{t-1}^* \qquad \text{---(6.8)} \\ &= dp_t + (1-d)dp_{t-1} + (1-d)^2 dp_{t-2} + \dots \end{aligned}$$

The problem is to provide a justification for the behavioural assumption of the formation of expectations in eq.(6.7). One such was provided by Taylor (1970) who assumed that the "permanent component" of an economic variable denoted by the expression y_t^1 satisfied a qth order linear stochastic difference equation. Using state variable notation it is possible to convert this equation into a first-order system of q equations

$$y_t = Ay_{t-1} + u_t \qquad ---(6.9)$$

Taylor also assumed that the observed (scalar) series p_t is generated by

$$p_t = Cy_t + e_t \qquad ---(6.10)$$

where $C = [1\ 0\ \ldots\ 0]$. Applying the Kalman filter algorithm (see Appendix A) to this problem, we have

$$y_{t/t} = y_{t/t-1} + K_t(p_t - Cy_{t/t-1}) \qquad ---(6.11)$$

where $y_{t/t}$ is the conditional expectation of y_t given all information uptil t and K_t is the Kalman gain. The quantity within brackets in eq.(6.11) is the prediction error based on information available upto one period before. In the special case when q=1, C equals one and the prediction $y_{t/t}$ of the scalar time series $y_t^1 = y_t$ is given by the sum of the prediction $y_{t/t-1}$ in the last period and the currently observed prediction error weighted by the Kalman gain for that period (under the assumption that the gain is a time-varying matrix). If $y_{t/t}$ is identified with the expected variable in eq.(6.7), then eq.(6.11) is identical with eq.(6.7) provided that C=1 and the Kalman gain converges to a steady state solution. In this context, Taylor stated conditions under which K_t would approach a steady state. He, thus, provided a rational explanation for the formation of expectations described in eq.(6.8). The explanation assumes a linear structure for the true variable, an observation equation and the rational behaviour on the part of economic agents in forming minimum-variance unbiased estimates of the true variable by means of the Kalman filter. As Chow (1975) has pointed out, if the economic variable y_t^1 satisfies a higher order linear stochastic difference equation, the solution provided by eq.(6.11) to optimally forecast still applies, but with a difference: $y_{t/t}^1$ will still be a linear function of past p_{t-k}, k=0,1,2,..., but the coefficients will no longer decline geometrically with k as in eq.(6.8). However, the result will still provide a theory of rational expectations. Nerlove (1972) also recognized the Kalman filter explicitly while providing other economic examples of the formation of expectations by economic agents.

6.2.4 Time-varying parameters

One very important area where the Kalman filter has been

extensively applied is the estimation of time-varying coefficients in linear models in econometrics. Consider the random coefficient linear regression model

$$y_t = x_t B_t + e_t^1, \qquad t = 1,\ldots,T \qquad \text{---(6.12a)}$$

where B_t is assumed to have evolved according to the relation

$$B_t = MB_{t-1} + e_t^2, \qquad t = 1,\ldots,T \qquad \text{---(6.12b)}$$

In the above formulation, y_t is a vector of observations on the dependent variable and is assumed to be directly observable. x_t is a row vector of k fixed explanatory variables. When $e_t^2 = 0$ and $M = I$, this model reduces to the standard normal regression model. This model is a special case of the state space representation of a given system, in which B_t takes the place of x_t, $A = M$ and $u_t = 0$. The Kalman filter can be applied to estimate the random coefficient B_t.

Assuming that M as well as the variances of the errors are known, we can consider the problem of estimating B_t using information I^s upto time s. Denote by $B_{t/s}$ the conditional expectation of B_t given information I^s. The evaluation of $B_{t/t}$ is known as filtering; the evaluation of $B_{t/s}$ where $s>t$ is called smoothing; while the evaluation of $B_{t/s}$ where $s<t$ is called prediction. While it is possible to obtain the filtered and smoothed estimates of B_t using the basic results due to Kalman, Sant (1977) derived the same results alternatively using the method of Aitkin's generalized least squares.

It is also possible to compute the time-varying coefficients using the method of maximum-likelihood in conjunction with the Kalman filter. The technique is to compute the maximum-likelihood estimates of the parameters in time-varying, reduced-form models. The estimated parameters can then be interpreted as time-varying regression coefficients or as observed components in an economic time series. The manner in which the estimation problem is formulated governs the interpretation of the parameters.

The estimation problem is solved in two stages. The first is concerned with the estimation of the unknown stochastic specification parameters, i.e., the coefficients and covariances which define the evolutionary behaviour of the time-varying elements. These estimates are derived using maximum likelihood. The second stage uses the results of the first stage to produce efficient estimates of the time paths of the varying parameters. It is in this context that the Kalman filter is extensively invoked.

The general limitations on the use of the Kalman filter to estimate the time-varying parameters of a model require that the regression equations be linear in the time-varying coefficients and that the observation error be zero-mean, sequentially independent and normally distributed. In addition, any random disturbances acting on the time-varying parameters must also be zero-mean, normally distributed and sequentially independent.

Extensive research in this area has been carried out by Rosenberg (1973), Athans (1974), Swamy (1974), Cooley and Prescott (1976), Harvey and Phillips (1978), Pagan (1980), amongst others.

6.2.5 Seasonal components

Yet another important area in (applied) econometrics where the Kalman filter has been applied is in the estimation of seasonal components in economic time series. Seasonal adjustment of economic time series has been a topic which has been intensively researched over the years. Of late, economists have begun to combine seasonal analysis with the estimation of an econometric model for cyclical fluctuations. While Pagan (1975) had pointed out the possibility of applying filtering and estimation methods for state space models to help in separately estimating the seasonal and cyclical components in economic time series, it is Chow (1981), amongst others, who has been actively engaged in this area.

Assume, initially, that the vector x_t of endogenous variables is the sum of cyclical (x^c), seasonal (x^s) and irregular (v) components, and is given by

$$x_t = x_t^c + x_t^s + v_t \qquad \text{---(6.13)}$$

and, also, that the cyclical component is governed by the following model:

$$x_t^c = Ax_{t-1}^c + Bu_t + e_t^1 \qquad \text{---(6.14)}$$

where u_t is a vector of exogenous variables and e_t^1 is a vector of random disturbances. The exogenous variables might or might not be seasonally adjusted but that fact does not, in any way, affect the analysis because u_t is assumed to be predetermined. Finally, an autoregressive seasonal model is assumed for the seasonal component and is given by

$$x_t^s = Cx_{t-12}^s + e_t^2 \qquad \text{---(6.15)}$$

where e_t^2 consists of random residuals. Combining eqs.(6.14) and (6.15), we can construct the vector z_t of unobserved components by specifying that

$$z_t = Mz_{t-1} + Nu_t + e_t^3 \qquad \text{---(6.16)}$$

where z_t includes both x_t^c and x_t^s as its first two subvectors as well as the necessary lagged x_{t-k}^c and x_{t-k}^s to transform the original models (6.14) and (6.15) of orders 1 (maybe higher) and 12, respectively, into a state space representation of the first order. Here, the matrix M will be determined by the matrices A and C, the matrix N will depend upon B, the vector u_t which includes all exogenous and policy variables will comprise a dummy variable to absorb the intercept term inherent in such models and e_t^3 will depend upon e_t^1 and e_t^2. Eq. (6.13) can be rewritten as

$$x_t = \begin{bmatrix} I & I & 0 \end{bmatrix} z_t + v_t \qquad ---(6.17)$$

Thus, eqs.(6.16) and (6.17) are in the standard state space form with the former explaining the unobserved state variables z_t and the latter relating the observed x_t to z_t. Given observations on x_t and u_t, the conditional expectations of the unobserved components of z_t can be estimated by the Kalman filter, provided the parameters A, B and C are known. In applications, however, the parameters A and B of eq.(6.14) are unknown. But, it is possible to employ seasonally adjusted data for x_t^c and conventional estimation techniques to obtain estimates of A and B. Incorporating these, we can then estimate the seasonal and cyclical components in z_t by invoking the Kalman filter. The revised estimates of x_t^c will then serve as fresh data during the re-estimation of the model (6.14). As such, new estimates of the seasonal components will emerge from this process. A variant of this method was used by Vishwakarma (1974) to project the values of the exogenous variables in his model as a prelude to the optimal control experiments he carried out.

6.3 The Observation Mechanism

While it would have been possible to work with the state space representation of the econometric model as provided by the observable canonical form of eqs.(5.29) and (5.30), we felt that by doing so we would be losing valuable information in as far as the effects of scanty and inaccurate observations on the state vector are concerned. Therefore, from the viewpoint of making the model reflect the existing realities of the data assimilation position inherent in the Indian economy, we adopted the approach outlined by Mariano and Schleicher (1972), albeit with a suitably modified observation mechanism well in keeping with the current Indian data base context.

A brief prelude to the problem of determining a suitable observation equation for the estimated econometric model is in order. While the data base of the Indian economy is quite sound and accurate, the one endemic feature pervasive in all the three chief information gathering agencies of the country, namely, the Central Statistical Organisation, the Reserve Bank of India and the Planning Commission, has been their inability to provide uptodate information on any of the major macroeconomic aggregates in a timely manner. Consider, for example, that even now with the financial year 1986-87 almost halfway through (it needs to be mentioned that the Indian financial year runs from April to March), firm estimates of GNP and its principal components are available only uptil 1983-84. This implies that there is an information lag of almost 3 years in the case of GNP, which can be even longer in the case of certain other less important variables. The fact that the Indian economy is a planned one makes the situation all the more piquant because nearly all the models and exercises of the Planning Commission, which is entrusted with the onerous task of building up a consistent plan framework, are constructed using the latest available data which in many cases are nothing more than tentative estimates and as such heavily contaminated by observation noise. Under the circumstances, it comes as no surprise that most of the Indian Five Year plans have gone awry because they have been based on projections using extremely 'noisy' bases. The 'settling' of the base-year estimates with the passage of time has implied the 'unsettling' of all the plan estimates. This has rendered imperative the construction of a filter mechanism which can help in the reconstruction of the true state in the presence of contaminated observations on it. It is in this context that the Kalman filter will prove to be of invaluable assistance because it is a very convenient technique to revise optimally the estimates based on past information in the light of fresh information alone. The Kalman filter updates these estimates as and when fresh information is available and thereby obviates the need to use past data again. Great economy in computation is thus achieved. The construction of such a filter will therefore go a long way in improving the accuracy of Indian plan models as the base year estimates can be sufficiently 'cleansed' before any projections are carried out.

In this regard, it needs to be mentioned that all estimates of macroeconomic aggregates in the Indian context can be chronologically classified into the following successive categories: (i) tentative, (ii) quick, (iii) provisional, (iv) revised and (v) firm. This implies that, in the Indian context, the very first estimate of, say, GNP for 1984-85 (which is the base year of the Indian Seventh Five Year Plan 1985-90) would have appeared only in early 1986. This is termed as a 'tentative estimate'. By the time it transits through all the subsequent categories and ends up as a 'firm estimate' it will be around about early 1988. By this time the Seventh Plan will be more than halfway through and if there are substantial revisions in the base (as doubtless there will be), it will be absolutely too late to make any midterm plan modifications or optimal course corrections inspite of the fact that the Annual Plan exercises of the Ministry of Finance, Government of India, do permit a modicum of flexibility *vis-à-vis* plan modifications. To preempt such a need for plan revisions, we shall adopt the following methodology to estimate an observation equation for our model which will incorporate the possibility of such incorrect observations on the state of the economy. Such an equation will be essential, because we shall be using the control system to predict optimal economic policies under the assumption that the state is incapable of being accurately observed.

6.3.1 The observation equation

From the 20x1 resolved state vector provided in the last chapter, an optimal subset of 10 variables was identified using as a basis their relative importance as well as the ease with which they could be observed (in the form of tentative estimates). Considering the fact that the restructured version of the model had only 8 sectors (the monetary and miscellaneous sectors were exogenized in the process of model modification), we took care to ensure that the observation vector had at least one representative variable from each sector.

The vector $\underline{z}$ of observed variables was given by:

1	WP
2	YNFR
3	CPR
4	s
5	SNR
6	INR
7	MR
8	YNFN
9	YNMN
10	BD

In all the cases, it was noticed that the variables were sufficiently influential enough to warrant being included in the observation vector. More important was the fact that each of them came under the class of macroeconomic variables which, from the viewpoint of the Indian information gathering agencies, merited a tentative estimate. This was the most important consideration of all as it was the estimated relationship between the initial 'tentative estimate' and the final 'firm estimate' which provided the basis for estimating the matrices C and D in the observation equation given by eq.(5.31).

6.3.2 Observing the system state

In each of the equations (the coefficients of which ultimately formed the elements of the observation matrix) we obtained estimates of the observed components of the system state in the following manner. For any given year, we regressed the tentative estimate of each variable (as and when it was first observed) on the firm estimate of the concerned variable (as and when it finally settled down to its true value) assuming, in the process, that for the year in question the variable had in fact settled down to its firm (and, consequently, final) estimate. We confined our sample subspace to only the last five years of the entire sample space, i.e., the period 1977-78 to 1981-82, because as of the time this study was carried out, firm estimates on national income and its components were available only uptil 1981-82. While we could have predated our sample subspace and taken a larger one, the considerable improvements evinced over the last few years in the methodology and techniques *vis-à-vis* data collection in India would have implied that *by doing so* the observation equation would have ended up with an unnecessarily larger variance which could have possibly contaminated the filter.

The procedure of computing the specific elements of the C matrix will be demonstrated by means of an example. Take the case of, say, the budget deficit. The data used in the estimation process is provided below:

(Rs. crores at current prices)

Year	Observed Budget Deficit	Actual Budget Deficit
1977-78	493	933
1978-79	1246	1506
1979-80	2433	2700
1980-81	1975	2577
1981-82	1392	1539

The table indicates that the observed (tentative) budget deficit, which is the preliminary estimate provided by the finance minister in his budget speech, invariably falls short of the actual (firm) budget deficit, which is the final estimate obtained after a lag of nearly two years, by a considerably big margin. If we regress the observed estimates on their corresponding actual ones, we will obtain a parameter which will indicate the extent of underestimation and which will be an element *inter alia* in the C matrix.

6.3.3 Estimating the observation equation

The observation mechanism contains, as indicated earlier, ten equations. In order of appearance, these are: (i) wholesale price level from the price sector, (ii) NDP at factor cost at constant prices from the production sector, (iii) private final consumption expenditure from the consumption sector, (iv) and (v) savings rate and net domestic savings from the savings sector, (vi) net domestic capital formation from the investment sector, (vii) imports of goods from the trade sector, (viii) and (ix) NDP at factor cost at current prices and NDP at market prices at current prices from the income sector and (x) budget deficit from the government sector. The estimation of each equation is presented below.

In all the equations that follow, the figures in parentheses denote the elasticities of the coefficients beneath which they appear. Also reported are the standard error of the estimate (S.E.E.), the mean of the dependent variable (Mean) and the coefficient of variation (C.V.). The bar over the dependent variable in each case indicates that it is the initial observed value of the concerned variable that is being regressed on its true value.

(i) Wholesale price level (WP): The estimated equation was

$$\overline{WP} = \underset{(0.96)}{0.9784}\ WP \qquad \text{---(6.18)}$$

S.E.E. = 0.0449 Mean = 1.9865 C.V. = 2.2623

It is thus noticed that observations on the implicit national income deflator (which was proxied by the wholesale price index) are always an underestimate of the actual price level. This is well in keeping with the nature of the structure of errors inherent in an observation mechanism. The equation translating the true system state into its corresponding observation indicates that there is an approximate 2.2 percent downward bias in perceiving the nature of inflation in India. This is an important result and indicates a possible reason as to the high 'noise' content in policy responses to such types of error-prone early-warning information signalling.

(ii) NDP at factor cost (YNFR): The estimated equation was

$$\overline{YNFR} = \underset{(1.16)}{1.0118}\ YNFR \qquad ---(6.19)$$

S.E.E. = 897.195 Mean = 46890.2 C.V. = 1.9134

As before, the results of the estimation bear out theory. It is noticed that the preliminary estimates of national income are always overestimated by as much as 1.2 percent of their actual true values. Therefore, in this case, the observation mechanism induces an upward biased error in the information content. Considering the fact that the true estimate of NDP at factor cost for 1981-82 was Rs. 49802 crores ($ 40 billion), an error of 1.2 percent would imply a discrepency of the magnitude of nearly Rs. 600 crores ($ 0.5 billion), a figure large enough to render inoperative some of the consistency planning exercises of the Planning Commission. Thus, it is in this context that the estimated filter should be capable of reconstructing the true state of the system from the given observations.

(iii) Private consumption expenditure (CPR): The estimated equation was

$$\overline{CPR} = \underset{(1.04)}{1.0089}\ CPR \qquad ---(6.20)$$

S.E.E. = 263.527 Mean = 38615.2 C.V. = 0.6824

The equation indicates that the preliminary estimate of consumption expenditure is an overestimate and that the actual estimate, as and when it appears, would be lower by approximately 0.9 percent. Once again this is in keeping with the true nature of the observation error mechanism and serves to bear out the plausibility of the hypothesis governing our procedure.

(iv) Savings rate (s): The estimated equation was

$$\bar{s} = \underset{(1.19)}{1.0172}\ s \qquad ---(6.21)$$

S.E.E. = 0.0033 Mean = 0.2012 C.V. = 1.6455

This is an extremely important parameter from the viewpoint of Indian planning because in most applications almost the entire plan framework has hinged on a successful forecast of the propensity to save in the economy. The equation indicates that there is an upward bias to the extent of 1.7 percent in perceiving the true nature of the savings rate. This *per se* may look marginal but when translated into preempted resources, by multiplying it with a similarly upward biased estimate of national income, the resulting downward revision in the actual quantum of savings has been quite substantial to warrant large plan modifications.

(v) Net domestic savings (SNR): The estimated equation was

$$\overline{SNR} = \underset{(1.22)}{1.0266}\ SNR \qquad \text{---(6.22)}$$

S.E.E. = 298.640 Mean = 9421.20 C.V. = 3.1699

The result is basically a 'spin-off' from eqs.(6.19) and (6.21). The equation translating the true system state into its corresponding observation indicates that there is a 2.7 percent upward biased error in estimating the actual quantum of savings in India.

(vi) Net domestic capital formation (INR): The estimated equation was

$$\overline{INR} = \underset{(1.03)}{1.0828}\ INR \qquad \text{---(6.23)}$$

S.E.E. = 428.706 Mean = 8951.40 C.V. = 4.7893

This equation is clearly a critical one and serves to highlight, in no uncertain manner, the nature and magnitude of observation errors that can vitiate macroeconomic planning in India. The equation indicates that there is an upward bias of almost 8.3 percent in perceiving the quantum of investible resources in India. This is an error of quite a serious magnitude and typifies the manner in which observation errors can cumulate.

(vii) Imports of goods (MR): The estimated equation was

$$\overline{MR} = \underset{(0.99)}{0.9937}\ MR \qquad \text{---(6.24)}$$

S.E.E. = 16.1759 Mean = 3046.40 C.V. = 0.5310

It is noted that there is an error of less than one percent in observing the true state. This is not surprising in view of the fact that most imports are contracted well in advance and while the value of imports may show a marked difference between their preliminary and final estimates due to unanticipated price increases there is not likely to be much of a revision as far as the estimates of the quantum of imports are concerned.

(viii) NDP at factor cost at current prices (YNFN): The estimated equation was obtained by using actual preliminary estimates of YNFN as reported by the three agencies and not by using derived preliminary estimates of YNFN (i.e., those constructed by multiplying the preliminary estimates of WP with the preliminary estimates of YNFR). The estimated equation was

$$\bar{YNFN} = \underset{(0.99)}{1.0035}\ YNFN \qquad \text{---(6.25)}$$

S.E.E. = 148.401 Mean = 94882.4 C.V. = 0.1564

The equation serves to indicate that there is a discrepency of less than 0.5 percent between the preliminary and the final estimates of NDP at factor cost at current prices. This is not as reassuring as it might look at first glance if we pause to consider the fact that the downward biased errors in the observed price level have just about served to offset the upward biased errors in observed real national income thereby leading to this spurious accuracy.

(ix) NDP at market prices at current prices (YNMN): The estimated equation was

$$\bar{YNMN} = \underset{(1.00)}{1.0055}\ YNMN \qquad \text{---(6.26)}$$

S.E.E. = 459.101 Mean = 107699.0 C.V. = 0.4264

As in the preceding case, it is noticed that the observation mechanism records an upward biased error of about 0.5 percent only while estimating the true system state. Once again this is due to the presence of counteracting errors in the price and real income observation structures of the economy.

(x) Budget deficit (BD): The estimated equation (in passing, we note that this equation was estimated by making use of the sample subspace of observations provided in section 6.3.2) was

$$\bar{BD} = \underset{(0.85)}{0.8302}\ BD \qquad \text{---(6.27)}$$

S.E.E. = 197.526 Mean = 1507.80 C.V. = 13.1003

The equation, which translates the true system state into its observations, indicates that the actual budget deficit is underestimated by as much as 17 percent within the context of the Indian economy. This is an error of a very high magnitude and serves to underscore the unreliability of the preliminary estimates of the budget deficit especially from the viewpoint of using them as a basis for countercyclical policy.

6.3.4 Deviations model of the estimated observation structure

In eqs.(6.18)-(6.27) above, all the variables in the estimated equations were specified in actual level terms. Thus, they are incompatible with the state variables of the system which, from the consideration of the reduced form deviations model of eq.(5.15), are all specified as deviations from an operating point. In order that the observation structure be conformable with the overall model structure, we have to specify these variables as deviations from the same (in the case of similar variables) set of reference values used earlier.

Let us consider, say, eq.(6.27) and convert the observation equation for the budget deficit into its deviations form structure. As before, we replace each variable in the equation by the sum of its reference value and the deviation from this reference value. Therefore, eq.(6.27) can be replaced by

$$\hat{\bar{BD}} + d\bar{BD} = 0.8302(\hat{BD} + dBD)$$

where, as before, ^ designates the reference value of the concerned variable and $\underline{d}$ designates the difference between the value of the variable and its reference value. The original equation indicates that the observed budget deficit was approximately 0.8302 times the actual one. Carrying this analogy one step further we can assume that even in the reference year (1970-71) the preliminary estimate of the budget deficit for that year was 0.8302 times the actual budget deficit in that year, which was Rs. 285 crores. Therefore, the reference value of $\bar{BD}$ would be Rs. 237 crores (0.8302x285). Thus, in terms of deviations from their respective reference values, we have the following equation

$$237 + d\bar{BD} = 0.8302(285 + dBD)$$

which reduces to the original equation but for the fact that now both the variables are specified in terms of deviations from their respective operating points which are assumed to be different.

We provide below the 10 observation equations in their deviations form structure:

$$d\bar{WP} = 0.9784\ dWP$$

$$dY\bar{N}FR = 1.0118\ dYNFR$$

$$dC\bar{P}R = 1.0089\ dCPR$$

$$d\bar{s} = 1.0172\ ds$$

$$dS\bar{N}R = 1.0266\ dSNR$$

$$dI\bar{N}R = 1.0828\ dINR$$

$$d\bar{M}R = 0.9937\ dMR$$

$$dY\bar{N}FN = 1.0035\ dYNFN$$

$$dY\bar{N}MN = 1.0055\ dYNMN$$

$$d\bar{BD} = 0.8302\ dBD$$

The state space representation of our model is therefore given by the following pair of equations

State equation : $dx(t) = Adx(t-1) + Bdu(t)$ ---(6.28)

Observation equation: $dz(t) = Cdx(t)$ ---(6.29)

Having provided the matrices A and B in Table 5.2 earlier, we present the matrix C in Table 6.1.

Table 6.1
The Observation Matrix Of The Model

C	COL1 COL11	COL2 COL12	COL3 COL13	COL4 COL14	COL5 COL15	COL6 COL16	COL7 COL17	COL8 COL18	COL9 COL19	COL10 COL20
	0 0	0 0	0 0	0 0	0 0	0.9784 0	0 0	0 0	0 0	0 0
	0 0	0 0	0 0	0 0	0 0	0 0	0 0	1.0118 0	0 0	0 0
	0 0	0 0	0 0	0 0	0 0	0 0	0 0	0 0	0 0	1.0089 0
	0 1.0172	0 0	0 0	0 0	0 0	0 0	0 0	0 0	0 0	0 0
	0 0	0 1.0266	0 0	0 0	0 0	0 0	0 0	0 0	0 0	0 0
	0 0	0 0	0 1.0828	0 0	0 0	0 0	0 0	0 0	0 0	0 0
	0 0	0 0	0 0	0 0.9937	0 0	0 0	0 0	0 0	0 0	0 0
	0 0	0 0	0 0	0 0	0 1.0035	0 0	0 0	0 0	0 0	0 0
	0 0	0 0	0 0	0 0	0 0	0 1.0055	0 0	0 0	0 0	0 0
	0 0	0 0	0 0	0 0	0 0	0 0	0 0	0 0	0 0	0 0.8302

6.4 The Error Covariance Matrices

6.4.1 The transition errors

Of the 20 equations in the model, 14 are determined as behavioural equations and the remaining 6 are identities. In all these reaction equations, random disturbances appear. These disturbances are introduced to account for errors caused by overlooking causal factors and to make the model reflect the uncertainty in the transition of the economy from year to year as indicated by it. These random disturbances by themselves are all assumed to be uncorrelated with each other and over time. In addition, they are assumed to have zero mean and fixed finite variances. The error covariance matrix of the state transition equation, denoted by Q, is presented in Table 6.2.

6.4.2 The observation errors

Our 10x1 observation vector dz was given by

1	$d\overline{WP}$
2	$d\overline{YNFR}$
3	$d\overline{CPR}$
4	$d\overline{S}$
5	$d\overline{SNR}$
6	$d\overline{INR}$
7	$d\overline{MR}$
8	$d\overline{YNFN}$
9	$d\overline{YNMN}$
10	$d\overline{BD}$

As all its components were determined by behavioural equations, it was assumed that random disturbances appeared in each of them. These disturbances were introduced to account for errors caused by observing wrong signals of the concerned state variables and to make the observation mechanism of the model reflect the uncertainty in the detection of noise from signals as well as the risk involved in the choice of an unsuitable time lag between the first observed contaminated signal and the final, presumably cleansed, signal. As before, these random disturbances by themselves are all assumed to be uncorrelated with each other and over time. In addition, they are all assumed to have zero mean and fixed finite variances. The error covariance matrix of the observation equation, denoted by R, is also presented in Table 6.2.

In passing, we note that our observation equation has no D matrix of the form indicated by the observable canonical state space representation in eq.(5.31). Therefore the instrument vector *per se* cannot exert any direct influence over the target (observation) vector at any given instant in time. Rather it will have to exploit the dynamics of the model via the state transition equation to affect the instantaneous equilibrium values of the observation vector.

Table 6.2

The Error Covariance Matrices

Q COL1 COL11	COL2 COL12	COL3 COL13	COL4 COL14	COL5 COL15	COL6 COL16	COL7 COL17	CCL8 COL18	COL9 CCL19	COL10 COL20
2307.4 -0.0757389	73.4824 0	16.6089 0	2395.7 -1065.72	-6.77392 0	0.166213 0	-197.675 -8202.33	0 2071.06	-4653.9 1643.4	-2805.41 0
73.4824 -0.0248046	25.1332 0	-3.18089 0	163.779 -171.44	131.343 0	0.00999302 0	-9.10065 1787.85	0 910.962	107.297 321.558	546.909 0
16.6089 0.00492576	-3.18089 0	4.33498 0	60.9185 21.7827	64.8656 0	0.00252321 0	-11.2871 -686.166	0 4.26072	46.7813 -126.731	-249.18 0
2395.7 -0.4836	163.779 0	60.9185 0	12717.2 1892.11	-2715.8 0	1.7282 0	-5222.75 -16167.8	0 4301.21	497.466 12470.7	1522.98 0
-6.77392 0.201814	131.343 0	64.8656 0	-2715.8 -4513.97	13974.7 0	-2.10787 0	3467.42 10414	0 13143	3553.8 -10491.6	-9930.5 0
0.166213 -.000120681	0.00999302 0	0.00252321 0	1.7282 0.8901	-2.10787 0	0.000646076 0	-1.34399 -3.67805	0 -2.11558	-0.383273 3.08349	2.50145 0
-197.675 0.647618	-9.10065 0	-11.2871 0	-5222.75 -5680.85	3467.42 0	-1.34399 0	7309.83 28627.9	0 13456.6	1131.57 -9491.49	-15766.9 0
0 0	0 0	0 0	0 0	0 0	0 0	0 0	0 0	0 0	0 0
-4653.9 -0.061565	107.297 0	46.7813 0	497.466 937.478	3553.8 0	-0.383273 0	1131.57 34399.8	0 12235.4	22840.7 -3082.74	-2816.13 0
-2805.41 -1.90788	546.909 0	-249.18 0	1522.98 10114	-9930.5 0	2.50145 0	-15766.9 11234.9	0 -18742.4	-2816.13 24631.4	85575.2 0
-0.0757389 0.0000976	-0.0248046 0	0.00492576 0	-0.4836 -0.34972	0.201814 0	-.000120681 0	0.647618 1.10224	0 0.993041	-0.061565 -1.06199	-1.90788 0
0 0	0 0	0 0	0 0	0 0	0 0	0 0	0 0	0 0	0 0
0 0	0 0	0 0	0 0	0 0	0 0	0 0	0 0	0 0	0 0
-1065.72 -0.34972	-171.44 0	21.7827 0	1892.11 6551.7	-4513.97 0	0.8901 0	-5680.85 -30198.2	0 -16554.1	937.478 5040.63	10114 0
0 0	0 0	0 0	0 0	0 0	0 0	0 0	0 0	0 0	0 0
0 0	0 0	0 0	0 0	0 0	0 0	0 0	0 0	0 0	0 0
-8202.33 1.10224	1787.85 0	-686.166 0	-16167.8 -30198.2	10414 0	-3.67805 0	28627.9 373107	0 138011	34399.8 2967.47	11234.9 0

COL1 COL11	COL2 COL12	COL3 COL13	COL4 COL14	COL5 COL15	COL6 COL16	COL7 COL17	COL8 COL18	COL9 COL19	COL10 COL20
2071.06	910.962	4.26072	4301.21	13143	-2.11558	13456.6	0	12235.4	-18742.4
0.993041	0	0	-16554.1	0	0	138011	92522	-2011.76	0
1643.4	321.558	-126.731	12470.7	-10491.6	3.08349	-9491.49	0	-3082.74	24631.4
-1.06199	0	0	5040.63	0	0	2967.47	-2011.76	26377.3	0
0	0	0	0	0	0	0	0	0	0
0	0	0	0	0	0	0	0	0	0

R

COL1	COL2	COL3	COL4	COL5	COL6	COL7	COL8	COL9	COL10
0.00151681	-17.0535	-3.45945	-.000057462	-5.56544	10.3943	0.235646	-0.540954	-14.031	0.12599
-17.0535	455045	20408.7	0.481405	117594	-128730	-7521.51	70.5759	248569	69644.8
-3.45945	20408.7	48804.8	0.154549	13243.1	-72089.7	1250.1	10736.2	17856.3	-18433.5
-.000057462	0.481405	0.154549	0.000003572	0.267624	-0.333543	-0.00353186	0.0362495	0.520806	-0.024949
-5.56544	117594	13243.1	0.267624	38675.9	-41383.5	-1588.29	2385.97	72968.9	14889.2
10.3943	-128730	-72089.7	-0.333543	-41383.5	136455	139.076	-14230.6	-88052.2	15487.2
0.235646	-7521.51	1250.1	-0.00353186	-1588.29	139.076	182.342	372.257	-4071.84	-1733.46
-0.540954	70.5759	10736.2	0.0362495	2385.97	-14230.6	372.257	2478.01	1497.82	-4480.13
-14.031	248569	17856.3	0.520806	72968.9	-88052.2	-4071.84	1497.82	164455	26984
0.12599	69644.8	-18433.5	-0.024949	14889.2	15487.2	-1733.46	-4480.13	26984	24175.7

6.5 The Initial Conditions

If conditions (1)-(4) set out by the Kalman-Bucy theorem (see Appendix A) are all shown to be satisfied, then every solution of the recurrence relationships of the Kalman filtering algorithm

$$V\bar{x}_{t+1/t} = AV\bar{x}_t A' + Q \qquad \text{---(6.30)}$$

$$K_{t+1} = V\bar{x}_{t+1/t}C'(CV\bar{x}_{t+1/t}C' + R)^{-1} \qquad \text{---(6.31)}$$

$$V\bar{x}_{t+1} = (I - K_{t+1}C)V\bar{x}_{t+1/t} \qquad \text{---(6.32)}$$

which results from initializing with <u>any</u> symmetric non-negative matrix V_0 tends uniformly to a unique finite positive-definite matrix V* (Here, A and C are the system matrices, Q and R are the error covariance matrices, $V\bar{x}_{t+1/t}$ is the <u>à priori</u> variance, K_{t+1} is the Kalman gain and $V\bar{x}_{t+1}$ is the <u>à posteriori</u> variance. Thus, the initial conditions refer to the value assigned to $V\bar{x}_t$, t=0, in eq.(6.30) to start up the filtering algorithm). However, in the event that even one of these conditions is not satisfied, then we will have to initialize the algorithm with the covariance matrix of the state variables, i.e., $\text{cov}(x_t|I_t)$, where I_t denotes the information set available at time <u>t</u>. It needs to be noted that the calculation of these covariance matrices and hence of the filter gain matrices are independent of the actual measurement data and therefore can be carried out separately.

With respect to condition (1) of the theorem, as we employed the operating point concept to linearize the model, all the system matrices were finite and time-invariant. Therefore, the model was stationary over the entire period. With regards to condition (2) of the theorem, it was established that the error covariance matrices were finite, positive-definite and time-invariant. Thus, the preliminary conditions for the stability and uniqueness of the filter were established.

As far as conditions (3) and (4) of the theorem were concerned, i.e., the linear system should be completely controllable and perfectly observable, respectively, we realized that to actually ascertain whether they are satisfied or not, via an application of the controllability and observability criteria (see Appendix C), would have involved an inordinate amount of computation which did not seem worthwhile.

Thus, as there was no way of proving that all the conditions of the theorem were actually satisfied, we could not directly apply the results of the Kalman-Bucy theorem in our case. With this complication in mind, it was decided to execute two alternative filtering runs. In one, we initialized the procedure with an identity matrix whilst in the other we set $V\bar{x}_t$, t=0, equal to the state covariance matrix.

This variance-covariance matrix of the state, denoted by $V\bar{x}_0$, is provided in Table 6.3.

Table 6.3
The State Variance-Covariance Matrix

$V\bar{x}_0$ COL1 COL11	COL2 COL12	COL3 COL13	COL4 COL14	COL5 COL15	COL6 COL16	COL7 COL17	COL8 COL18	COL9 CCL19	COL10 COL20
218155 11.4709	4869.27 870574	4722.24 625720	556702 280211	630262 11904194	242.583 13738765	970134 1112175	2375948 1183017	2399378 443953	1684914 259124
4869.27 0.294735	178.982 22863.2	136.929 16681.7	19833.9 7253.16	15095.2 271860	5.19068 313124	22663.2 27007.5	62476.3 27606.1	63582.3 10638.2	44565.8 4155.44
4722.24 0.279041	136.929 21100.2	134.486 16144.3	15874.4 6913.3	14279.7 273055	5.43294 313955	21726.8 24892.8	56622.2 26734.6	58103.5 9832.97	40376.1 4457.89
556702 34.2082	19833.9 2610185	15874.4 1882297	2231167 795455	1714758 31045989	600.644 35708891	2557051 3068807	7061444 3157113	7191255 1204256	5019701 457292
630262 35.4243	15095.2 2660603	14279.7 1925511	1714758 819581	1894420 34130497	675.938 39389544	2841839 3338224	7083235 3448793	7172260 1283006	5035351 759203
242.583 0.0116619	5.19068 914.202	5.43294 701.089	600.644 343.32	675.938 13940.4	0.292338 16056.7	1084.78 1181.48	2604.83 1324.35	2640.26 496.804	1850.32 277.024
970134 50.3337	22663.2 3903790	21726.8 2928319	2557051 1353992	2841839 53969717	1084.78 62310002	4408131 4971790	10780376 5290459	10927345 1993130	7704873 1194567
2375948 131.471	62476.3 10047840	56622.2 7364031	7061444 3250408	7083235 131093130	2604.83 151196516	10780376 12494218	27308642 13083339	27698285 4925771	19450402 2670765
2399378 132.631	63582.3 10174564	58103.5 7493819	7191255 3327176	7172260 133159931	2640.26 153594564	10927345 12613370	27698285 13250022	28176639 4978969	19752686 2670662
1684914 90.3638	44565.8 7042395	40376.1 5237766	5019701 2392100	5035351 93583014	1850.32 107965037	7704873 8817686	19450402 9258484	19752686 3517850	13951043 1986051
11.4709 0.000834763	0.294735 55.9514	0.279041 38.0994	34.2082 11.7898	35.4243 591.806	0.0116619 682.045	50.3337 65.282	131.471 64.307	132.631 22.4628	90.3638 10.6408
870574 55.9514	22863.2 3990666	21100.2 2827922	2610185 1054252	2660603 46487935	914.202 53616205	3903790 4795807	10047840 4849888	10174564 1754266	7042395 905760
625720 38.0994	16681.7 2827922	16144.3 2256640	1882297 935086	1925511 35722742	701.089 41150911	2928319 3425554	7364031 3537861	7493819 1277431	5237766 739060
280211 11.7898	7253.16 1054252	6913.3 935086	795455 526396	819581 17192536	343.32 19821530	1353992 1372370	3250408 1548156	3327176 605487	2392100 371080
11904194 591.806	271860 46487935	273055 35722742	31045989 17192536	34130497 680068208	13940.4 783934362	53969717 59352419	131093130 65138184	133159931 24518803	93583014 13937339
13738765 682.045	313124 53616205	313955 41150911	35708891 19821530	39389544 783934362	16056.7 903895168	62310002 68488952	151196516 75133885	153594564 28289810	107965037 16151932
1112175 65.282	27007.5 4795807	24892.8 3425554	3068807 1372370	3338224 59352419	1181.48 68488952	4971790 6041805	12494218 6131665	12613370 2244658	8817686 1351516

COL1 COL11	COL2 COL12	COL3 COL13	COL4 COL14	COL5 COL15	COL6 COL16	COL7 COL17	COL8 COL18	COL9 COL19	COL10 COL20
1183017 64.307	27606.1 4849888	26734.6 3537861	3157113 1548156	3448793 65138184	1324.35 75133885	5290459 6131665	13083339 6508682	13250022 2403202	9258484 1351089
443953 22.4628	10638.2 1754266	9832.97 1277431	1204256 605487	1283006 24518803	496.804 28289810	1993130 2244658	4925771 2403202	4978969 921848	3517850 522866
259124 10.6408	4155.44 905760	4457.89 739060	457292 371080	759203 13937339	277.024 16151932	1194567 1351516	2670765 1351089	2670662 522866	1986051 568350

6.6 The Covariance Matrix

It was noticed that numerical convergence via both the runs, using alternative initializations, yielded exactly identical steady-state *à posteriori* covariance matrices. This steady-state covariance matrix, denoted by V*, has been provided in Table 6.4. This matrix preserves the minimum-variance property of the Kalman filter.

It should be noted that the error covariance matrix $V\bar{x}_t$ does not vanish as $t \rightarrow \infty$; rather, it approaches a finite positive-definite matrix. This is in direct contrast with what happens in some other estimation methods, like the least squares method, in which the error covariance of the estimates of the regression coefficients tends to zero asymptotically. The finite error covariance resulting from the filter implies that the true values of the state vector can never be learned from the observations z_t. This is well in keeping with the true nature of the stochastic process we are dealing with.

6.7 The Predictor-Corrector Matrix

As the matrix V* is obtained by numerical convergence in the manner described above, the steady-state Kalman gain which is the predictor-corrector matrix of the control system is obtained simultaneously as it forms one link of the recursion used. This matrix converts the differences between observations and predictions into corrections of the predictions of the state vector via

$$x_t^* = \hat{x}_t + K(z_t - C\hat{x}_t) \qquad \text{---(6.33)}$$

where x_t^* is the optimal state at time $\underline{t}$ and K is the steady-state Kalman gain, which is provided in Table 6.5.

The term 'filter' for the kind of algorithm provided by eqs.(6.30)-(6.32) arose in communications engineering where signals are to be separated from noise. The algorithm furnishes Bayesian estimates x_t^* as linear combinations of the set of observation data. The quantity x_t^* is the orthogonal projection of x_t on the Hilbert space of observations $\underline{z}$ (see Kalman 1960, Kalman and Bucy 1961, Deutsch 1965). It is a linear, minimum-variance, unbiased estimator of a system state over a sample of observation data $\underline{z}$. This implies that it is 'optimal for any positive-semidefinite quadratic optimality criterion' (see Luenberger 1969). The optimality of the Kalman prediction-correction scheme has also been demonstrated, albeit from an alternative standpoint, by Krasnakevich and Haddad (1969).

Table 6.4

The Steady-State A Posteriori

Covariance Matrix

V* COL1 COL11	COL2 COL12	COL3 COL13	COL4 COL14	COL5 COL15	COL6 COL16	COL7 COL17	COL8 COL18	COL9 COL19	COL10 COL20
2615.16	9.00116	23.1804	2059.2	-343.102	0.200449	109.002	2226.45	-2622.58	-296.543
-.000143175	931.104	1008.11	-20.5726	5.33772	5.88324	-9212.14	101.72	2655.41	1546.25
9.00116	4.97328	0.876492	86.4939	23.0069	0.000983458	0.150683	10.9289	161.243	-1.45032
-6.9578E-07	4.57047	4.94048	-0.100838	0.0272729	0.0300602	546.011	481.363	96.8433	7.58628
23.1804	0.876492	3.13102	50.6951	85.8196	0.00166447	-14.3781	18.502	49.7335	-2.4502
-1.1712E-06	7.73755	8.35627	-0.170573	0.0472032	0.0520274	-351.906	81.4413	-73.0337	12.8395
2059.2	86.4939	50.6951	7415.19	1598.06	0.21627	-711.968	2403.99	2860.28	-318.384
-.000152205	1005.35	1085.78	-22.1634	6.12813	6.75442	-650.588	10482.8	5593.27	1668.28
-343.102	23.0069	85.8196	1598.06	5374.2	0.0286948	-1120.21	318.932	2327.44	-42.2706
-.000020234	133.377	144.095	-2.94123	0.806648	0.889088	-6311.31	3357.22	-1270.78	221.348
0.200449	0.000983458	0.00166447	0.21627	0.0286948	.0000424098	0.150685	0.471422	0.428611	-0.0624278
-2.9838E-08	0.197149	0.21291	-0.00434604	0.00120315	0.00132612	0.195884	-0.00328526	0.368845	0.327144
109.002	0.150683	-14.3781	-711.968	-1120.21	0.150685	1331.79	1676.66	3120.21	-220.372
-.000103928	701.179	754.744	-15.4115	4.61429	5.08587	6304.53	-1051.23	457.892	1162.35
2226.45	10.9289	18.502	2403.99	318.932	0.471422	1676.66	5240.33	4764.45	-693.893
-.000331606	2191.51	2366.63	-48.3091	13.3853	14.7532	2176.54	-36.6903	4099.4	3636.49
-2622.58	161.243	49.7335	2860.28	2327.44	0.428611	3120.21	4764.45	25979.8	-630.874
-.000301485	1992.49	2151.71	-43.9219	12.1708	13.4146	45473.8	15733	2065.91	3306.25
-296.543	-1.45032	-2.4502	-318.384	-42.2706	-0.0624278	-220.372	-693.893	-630.874	91.9353
.0000439808	-290.186	-313.456	6.39829	-1.76147	-1.94149	-289.105	4.68786	-543.495	-481.56
-.000143175	-6.9578E-07	-1.1712E-06	-.000152205	-.000020234	-2.9838E-08	-.000103928	-.000331606	-.000301485	.0000439808
2.1078E-11	-.000138678	-.000149867	.0000030589	-8.3256E-07	-9.1765E-07	-.000138921	.0000020963	-.000260303	-.000230166
931.104	4.57047	7.73755	1005.35	133.377	0.197149	701.179	2191.51	1992.49	-290.186
-.000138678	916.488	989.725	-20.2029	5.59772	6.1698	910.229	-15.3439	1714.37	1520.78
1008.11	4.94048	8.35627	1085.78	144.095	0.21291	754.744	2366.63	2151.71	-313.456
-.000149867	989.725	1068.94	-21.8195	6.02861	6.64473	984.319	-16.3139	1852.38	1642.36
-20.5726	-0.100838	-0.170573	-22.1634	-2.94123	-0.00434604	-15.4115	-48.3091	-43.9219	6.39829
.0000030589	-20.2029	-21.8195	0.445389	-0.123094	-0.135675	-20.0897	0.333551	-37.8098	-33.5248
5.33772	0.0272729	0.0472032	6.12813	0.806648	0.00120315	4.61429	13.3853	12.1708	-1.76147
-8.3256E-07	5.59772	6.02861	-0.123094	0.0363981	0.040118	5.37754	-0.128156	10.3344	9.28093
5.88324	0.0300602	0.0520274	6.75442	0.889088	0.00132612	5.08587	14.7532	13.4146	-1.94149
-9.1765E-07	6.1698	6.64473	-0.135675	0.040118	0.044218	5.92712	-0.141253	11.3905	10.2294
-9212.14	546.011	-351.906	-650.588	-6311.31	0.195884	6304.53	2176.54	45473.8	-289.105
-.000138921	910.229	984.319	-20.0897	5.37754	5.92712	181030	53116.4	20589.8	1511.03

COL1 COL11	COL2 COL12	COL3 COL13	COL4 COL14	COL5 COL15	COL6 COL16	COL7 COL17	COL8 COL18	COL9 COL19	COL10 COL20
101.72 .0000020963	481.363 -15.3439	81.4413 -16.3139	10482.8 0.333551	3357.22 -0.128156	-0.00328526 -0.141253	-1051.23 53116.4	-36.6903 48754.9	15733 10254.7	4.68786 -25.3413
2655.41 -.000260303	96.8433 1714.37	-73.0337 1852.38	5593.27 -37.8098	-1270.78 10.3344	0.368845 11.3905	457.892 20589.8	4099.4 10254.7	2065.91 12873.2	-543.495 2845.23
1546.25 -.000230166	7.58628 1520.78	12.8395 1642.36	1668.28 -33.5248	221.348 9.28093	0.327144 10.2294	1162.35 1511.03	3636.49 -25.3413	3306.25 2845.23	-481.56 2523.54

Table 6.5
The Steady-State
Predictor-Corrector Matrix
(Kalman Gain)

K COL1	COL2	COL3	COL4	COL5	COL6	COL7	COL8	COL9	COL10
-162.247	0.0567243	-0.0328127	-2064.22	0.159383	0.0722074	-0.188366	0.26828	-0.108265	-0.12398
2.82006	-0.00211638	0.00499323	-333.117	0.00384916	0.00223381	-0.0526661	-0.00129508	0.00223065	-.000501551
5.80127	0.000086942	-0.00343593	24.3879	0.00239987	-0.00179284	0.0094089	0.00102079	-0.0010156	-.000766558
1839.78	-0.0801556	-0.121655	-4898.02	0.24855	-0.1187	-0.016672	-0.185249	0.150727	-0.0999988
-3529.43	0.140591	-0.0824592	-5009.57	0.278878	0.11098	-0.363615	0.395492	-0.563459	-0.0137778
0.903808	-.000034578	.0000084701	0.102519	-1.6465E-07	-.000030022	.0000112352	-.000119142	0.000116699	-.000019494
-992.723	0.0983252	0.00626185	3731.84	0.0198024	0.119783	-0.535093	0.502898	-0.201996	-0.0381807
-1314.65	0.204278	0.0941491	1139.5	0.237478	0.282186	0.124875	0.485449	-0.354864	-0.215815
-2391.84	0.230341	-0.0135749	-1879.21	0.320045	0.266108	0.41374	0.67283	-0.50794	-0.196134
174.105	0.0661441	0.978708	-150.991	-0.641284	0.575027	-0.0165505	-0.469383	0.400529	0.029454
.0000832253	.0000050436	-5.9634E-09	0.983019	-.000012708	7.9196E-07	-7.9223E-09	.0000051827	-4.7157E-06	1.4816E-08
-549.789	0.0854292	0.0393732	476.537	0.0993135	0.11801	0.0522227	0.203015	-0.148404	-0.090254
-593.762	0.0935827	0.042526	514.777	0.102101	0.130665	0.0564188	0.239011	-0.158226	-0.0987843
12.1201	0.017853	-.000868052	-10.5076	-0.046439	0.0142369	1.00519	-0.0313571	0.0272474	0.00201366
-3.35272	0.0400333	0.000239617	2.88911	-0.239609	0.124445	0.000315873	0.878734	0.104181	-.000374059
-3.69536	0.0441247	0.000264106	3.18438	-0.264097	0.137164	0.000348155	0.96854	0.114828	-.000412288
-319.848	-0.321573	1.09032	12398.1	-0.713539	0.223361	-5.61264	-0.113006	0.3929	1.25851
593.749	-0.0335945	0.187707	5114.07	-0.108905	0.0588944	-2.63709	0.0314695	0.0793363	-0.00124832
2713.99	-0.106641	0.151342	-4311.4	0.115829	0.00454058	-0.0345775	-0.321788	0.346682	-0.0215925
-912.314	-0.264877	0.0653372	790.819	-0.00243612	-0.234397	0.0866667	0.0561779	0.0172288	1.05415

6.8 Observations On The Filter

Certain more interesting elements of the filter will now be discussed.

The first column indicates that the observed inflation rate z(1) strongly affects the estimated inflation rate WP: every 1% inflation observed below the predicted rate leads to an upward correction of the estimated inflation rate by 0.90%. It is interesting to note that the filter automatically reduces YNFR to the extent of Rs. 1315 crores in the presence of such a phenomenon.

The second column shows that the observed national income z(2) does not really affect the estimated YNFR. This is well in keeping with reality as the Indian data base has been quite prone to overestimating NDP in the past, much to the detriment of the Indian plans which have had to rely upon such overestimates. The filter indicates that every Rs. 1 crore observed above the predicted rate leads to an upward correction in the estimated national product by only Rs. 0.20 crores. The filter reduces the share of agriculture (YAR) by Rs. 0.08 crores, while allocating the overall increase between manufacturing (Rs. 0.14 crores), transport (Rs. 0.05 crores) and services (Rs. 0.09 crores).

The third column notes that the observed consumption z(3) very strongly affects the estimated C: every Rs. 1 crore observed above the predicted rate leads to an upward correction of the estimated consumption by Rs. 0.98 crores. This is due to the very small observation noise assumed in the consumption statistics. It is noted that this leads to an increase in non-developmental expenditure (NDE) by Rs. 1.09 crores and in the budget deficit (BD) by Rs. 0.07 crores.

The fourth column alludes to the fact that the observed savings rate z(4) also very strongly affects the estimated s, due to an equally small observation error in the savings statistics. It is seen that every 1% savings rate observed above the predicted rate leads to an upward correction in the estimated savings rate by 0.98%. The filter reduces consumption by Rs. 151 crores, while increasing net domestic savings (SNR) by Rs. 477 crores and net investment (INR) by Rs. 515 crores, in the presence of such a phenomenon. National income (YNFR) is also increased by Rs. 1140 crores. All these clearly highlight the importance of accurately measuring the savings rate from the viewpoint of its repercussions upon the estimates of some key macroeconomic aggregates in the model.

The fifth column points out that the observed quantum of savings z(5) does not affect the estimated SNR in any relevant manner at all: every Rs. 1 crore observed above the predicted rate leads to an upward correction in the estimated level of savings by only Rs. 0.10 crores. This damped correction factor is due to the large level of observation noise in the national income statistics which serves to reduce the overall impact of the observed increase in the level of savings. The filter, at the same time, increases investment by approximately the same amount (Rs. 0.10 crores) while decreasing consumption by as much as Rs. 0.64 crores which serves to decrease imports by Rs. 0.05 crores.

The sixth column highlights the fact that the observed level of investment z(6) bears very little relationship to the estimated INR: every Rs. 1 crore observed above the predicted level leads to an upward correction in the estimated level of investible resources by barely Rs. 0.13 crores. This is due to the large volume of observation noise contaminating the investment statistics. The filter indicates that this excess of observed investment over predicted investment increases NDP by Rs. 0.28 crores and imports by Rs. 0.01 crores.

The seventh column linking the observed level of the quantum of imports z(7) to the predicted level is very interesting. Although the observation noise in imports statistics was minimal, it is noted that for every Rs. 1 crore observed below the predicted level there is an upward correction in the quantum of imports by almost Rs. 1.005 crores. This over-compensatory correction factor can be attributed to the fact that one of the major components of imports, consumption, had a very low level of observation noise in its measurement statistics. The filter increases investment by Rs. 0.06 crores in the presence of such a phenomenon.

The eighth column shows that the observed level of national income at factor cost at current prices z(8) is closely related to the estimated YNFN. It is seen that every Rs. 1 crore observed above the predicted level leads to an upward correction in current national income by Rs. 0.88 crores. This leads to an increase in tax revenue (TR) by Rs. 0.03 crores which is more than offset by an increase in the budget deficit to the extent of Rs. 0.06 crores.

The nineth column alludes to the fact that there is very little correlation between the observed level of current national income at market prices z(9) and the estimated level of YNMN: every Rs. 1 crore observed above the predicted level leads to an upward correction in national income at current market prices by only Rs. 0.11 crores. This is also an interesting phenomenon because there is practically no observation noise in the measurement of this variable. This damped correction factor can however be attributed to the fact that any perceived increase in the observed over the predicted level of YNMN brings about a reduction in real national income to the extent of Rs. 0.35 crores (for every Rs. 1 crore excess in the observed over the predicted value of YNMN) and this causes a dilution in national income at market prices. The filter indicates that, under the circumstances, the budget deficit rises by about Rs. 0.02 crores.

The last column shows that the observed budget deficit z(10) is far removed from the estimated BD and, indeed, this is what we were led to believe in view of the large amount of measurement noise contaminating the budget statistics. It is noted that for every Rs. 1 crore observed below the predicted level there is an upward correction in the budget deficit estimates by more than Rs. 1.05 crores - incidently, the largest correction factor invoked by the filter. This is reflected in an increase in non-developmental expenditure (NDE) by Rs. 1.26 crores and a fall in real national income by Rs. 0.22 crores. This fall in YNFR is approximately shared by YAR (Rs. 0.10 crores) and YTR (Rs. 0.12 crores).

6.9 Conclusions

The use of the Kalman filter enables one to deal appropriately with scarce and conflicting data. As it updates optimally the previous estimates each time fresh information is available, the filter is very convenient to use. Extreme economy in computation is thereby achieved in view of its recursive nature as the contribution of all earlier information is summarized succintly in the current estimates. Moreover, the Kalman filter estimates have the extremely desirable properties of being minimum-variance linear unbiased estimators of the system state.

Taking into consideration the extremely illuminating and innovative results obtained in the last section, the Kalman filter estimates will be utilized in the determination of optimal control actions for macroeconomic regulation in India. It will be shown that the filter weights the latest and the previous statistical data in a proportional-plus-integral scheme so that not only the latest but also the past information is appropriately incorporated. The resulting control actions, as a consequence of such a proportional-plus-integral control policy, will therefore be much smoother than in the case when only the latest information is taken into consideration and will thereby add to the general stability and continuity of system performance. It is to such a type of macroeconomic regulation that we will now address ourselves in the concluding part of this study.

CHAPTER 7

OPTIMAL CONTROL AND MACROECONOMIC PLANNING: THE INDIAN CONTEXT

7.1 Introduction

Most applications of optimal control techniques that have appeared in the literature have been generally concerned with western market economies rather than with eastern planned economies. However, considering the fact that the application of optimal control presupposes that it is possible to construct and operate a plausible econometric model of the economy, there is every reason to use it in the solution of the planning problem in developing economies (planned or otherwise) which are beset by the very essence of the optimal, as opposed to the automatic, control problem which is not knowing in advance what exactly one would like to happen. Under the circumstances, in an economy where the government is actively engaged in development planning, such as in India, the techniques of optimal control should be fully availed of, if the paramount problem faced by all developing economies, i.e., the optimal allocation of limited resources over time to achieve a set of competing (and at times, conflicting) objectives, is to be satisfactorily solved. It is with this objective in mind that we now illustrate the possibility of applying control and filtering techniques within the context of Indian macroeconomic planning.

7.2 Proportional-Plus-Integral Control

Any model that is to be used for operational control of the economy must be a 'control model', i.e., a model which attempts to specify those mechanisms which operate between the control variables and the state (or target) variables.

Consider the familiar discrete-time linear state space representation of an econometric model given by

$$x_t = Ax_{t-1} + Bu_t \qquad \text{---(7.1)}$$

$$z_t = Cx_t \qquad \text{---(7.2)}$$

where $\underline{x}$ is an $\underline{n}$ dimensional state vector, $\underline{u}$ is an $\underline{r}$ dimensional control vector, $\underline{z}$ is an $\underline{s}$ dimensional observation vector, and A, B and C are appropriately dimensioned time-invariant system matrices.

The optimal Linear-Quadratic-Deterministic (LQD) formulation of the control problem leads mathematically (see Appendix B) to a feedback rule in which the optimal values of the control variables (u_t^*) are determined as linear functions of the optimal values of the state variables (x_t^*) by means of the control law given by

$$u_t^* = G_t x_{t-1}^* + J_t \qquad \text{---(7.3)}$$

where G_t is an rxn state variable feedback (SVF) matrix and the rx1 vector J_t is a function of the desired trajectories specified in the cost functional.

The conditional mean of the system state is determined by the relation (see Appendix A)

$$x_t^* = \bar{x}_t + K(z_t - C\bar{x}_t) \qquad \text{---(7.4)}$$

where $\bar{x}_t$ are the predictions of the state based upon eq.(7.1) and K is the steady-state solution of the Kalman gain. The quantity in brackets in eq.(7.4) is the error in prediction based on information available upto one interval before. Let $\bar{e}_t$ denote the one-interval ahead prediction error at time t, i.e.,

$$\bar{e}_t = z_t - C\bar{x}_t \qquad \text{---(7.5)}$$

Because the one-interval ahead conditional prediction of the state vector is given by

$$\bar{x}_t = Ax_{t-1}^* + Bu_t^* \qquad \text{---(7.6)}$$

eq.(7.4) can be written as

$$x_t^* = Ax_{t-1}^* + Bu_t^* + K\bar{e}_t \qquad \text{---(7.7)}$$

Substituting the feedback control rule given by eq.(7.3) into the above equation yields

$$x_t^* = K\bar{e}_t + (A + BG_t)x_{t-1}^* + BJ_t \qquad \text{---(7.8)}$$

Recursion and substitution leads to

$$x_t^* = K\bar{e}_t + BJ_t + (A + BG_t)(K\bar{e}_{t-1} + (A + BG_{t-1})x_{t-2}^* + BJ_{t-1}) \qquad \text{---(7.9)}$$

Successive recursion and substitution yields the following wherein we have set $\bar{x}_0 = 0$ only for convenience

$$x_t^* = K\bar{e}_t + \sum_{j=0}^{t-1} \left[\prod_{k=j+1}^{t} (A + BG_k) \right] K\bar{e}_j$$

$$+ (BJ_t + \sum_{j=0}^{t-1} \left[\prod_{k=j+1}^{t} (A + BG_k) \right] BJ_j) \qquad \text{---(7.10)}$$

The optimal control action at time t+1 is then

$$u^*_{t+1} = G_{t+1}K\bar{e}_t + G_{t+1}\sum_{j=0}^{t-1}\left[\prod_{k=j+1}^{t}(A + BG_k)\right]K\bar{e}_j$$

$$+ G_{t+1}(BJ_t + \sum_{j=0}^{t-1}\left[\prod_{k=j+1}^{t}(A + BG_k)\right]BJ_j) + BJ_{t+1} \quad \text{---(7.11)}$$

The first term on the right-hand side of eq.(7.11) represents an action which is proportional to the current one-interval ahead prediction error, while the second term denotes an action which is proportional to the weighted sum of the past errors in prediction. These two govern the transient responses of the system controlling, as it does, the closed-loop eigen value location. The third and fourth terms combine together to form the intercept which, in effect, acts as the input command signal to help guide the system along the specified nominal trajectories, given its transient responses.

In control engineering, such a scheme is referred to as proportional-plus-integral control. This nature of the feedback control policy results from the minimization of a quadratic function, subject to model constraints. Because the control actions are linear functions of some weighted average of current and past prediction errors, they will not react so sharply to recent prediction errors as compared to a policy based on recent errors alone. Thus, when prediction errors are caused largely by uncorrelated random exogenous disturbances (as is the case in the Indian economy), the appropriate control action can be smoother than it would have been otherwise. Such a scheme is also conducive to general continuity and stability.

The proportional-plus-integral feature resembles the concept of exponentially weighted moving averages developed by Holt *et al* (1960) which is well known in the literature on time series analysis and forecasting. It has also been referred to as 'exponential smoothing' by Brown (1963). Vishwakarma (1970) reported that an optimal scheme of exponential smoothing for the special case of a single time series with only a trend, as developed by Theil and Wage (1964) and Nerlove and Wage (1964), can be shown to be a particular case of the Kalman filter.

The present framework of a linear stationary stochastic control problem is, of course, much more general and takes care of not merely prediction but generalized policy making in a multivariable context, contaminated by inaccurate measurements of the observed variables which are, in general, fewer in number than the system state variables.

7.3 System State Estimation And Control Action Determination

The quantities G_t and J_t, appearing in eq.(7.3), that determine the optimal control actions are independent of the random disturbances, the observations, as well as the past control actions. They are non-random quantities and can be determined before the planning period begins. The random processes affect only the conditional means x_t^*. Since, the optimal control actions are completely specified when the means are known, the actual determination of the optimal control policy comprises two separate phases:

1. Estimation of the conditional mean x_t^* of the state x_t based on a set of observation data z_t and the steady-state solution of the Kalman gain.
2. Determination of the quantities G_t and J_t that provide the control action as a linear transformation of the estimated state. The linear transformation G_t, called the 'feedback coefficient' in engineering, is identical to the transformation that would have been applied if there had been no disturbances affecting the model and if the system state had coincided with its conditional mean. Thus, the control actions can be determined from the corresponding non-random system wherein all the random variables are equated with their expected values. This is, in essence, the operational relevance of the Certainty Equivalence principle.

The separation of the (statistical) estimation phase and the (deterministic) optimization phase in the solution of the stochastic control problem thus forms the crux of the matter. Moreover, the execution of the optimization recursively implies immense computational advantages as compared to solving the problem simultaneously at all stages. It needs to be noted that apart from the computational simplicity inherent in the nature of the Kalman filter algorithm, one of its greatest merits lies in the optimality property discovered by Luenberger (1969) who showed that the filter is a linear, minimum-variance unbiased estimator of the system state over a sample of observation data $\underline{z}$, and that it is optimal for any positive-semidefinite quadratic optimality criterion. Thus, regardless of the objective of the control policy, the estimation of the existing state, under the present framework, can, and should, be done only by means of the Kalman filter.

7.4 Econometric Models And Macroeconomic Policy Formulation

Assuming, as we have done so far, that econometric models can be used for policy analysis, this section presents a systematic approach of how to actually go about applying the techniques of optimal control in order to improve upon the formulation of macroeconomic policies and their subsequent predictions. The recommended procedure for the determination of optimal economic policies consists of seven steps, based on Chow (1981), and provides a useful guidepost of the generalized format adopted in this study.

Step 1. Insert the actual values of the policy variables and the exogenous variables not subject to control into the linearized econometric model to obtain simulated *ex-post* projections of the key economic variables over the entire planning period.

Step 2. Modify the econometric model through the suitable

incorporation of add or mul factors if the simulated values from step 1 differ radically from their actual historical counterparts. Many econometricians prefer to adjust the constants in their model utilizing information on the equation residuals in the recent years. Others insist on forecasting without any adjustment of the model in which case the final economic projections will need to be suitably scaled. Whatever adjustments are carried out, the essence of step 2 is to arrive at a set of simulated values of the important target variables which are as close as possible to their actual historical levels without in any way distorting the dynamics of the model.

Step 3. Set target values for only the very important economic variables of the system which represent the actual values desired by the central planner. The motivation here is to determine a set of mutually consistent maximal targets that can be all contemporaneously attained via the use of optimal control calculations. This presupposes the existence of a quadratic loss function into which the nominal levels of the target variables can be incorporated.

Step 4. Change the elements of the weighting matrices in the loss function systematically and re-optimize in order to trace out the policy and target constellations attainable under alternative strategies. By varying these penalties, one can compute solutions to figure out, at least heuristically, the so-called 'Pareto Optimum' situations wherein it would be impossible to improve upon the optimal values attained by a certain target variable without suffering a deterioration in the optimal values of some other target variable, no matter what policy mix is selected.

Step 5. If the solution obtained in step 4 represents a totally dominant solution *vis-à-vis* that obtained in step 1, then the choice made and the corresponding optimal policy constitute a set of intermediate guideposts for the target and control variables which could possibly require further analysis and improvement.

Step 6. If the solution paths for the target variables in step 4 drift very far away from their planned trajectories in step 3 or if the control paths of the instruments show severe fluctuations quite out of tune with reality, impose higher penalties on the errant variables in the loss function and optimize once again. Trial and error optimization runs are required in this step to obtain reasonable solution paths for the control variables, at the same time ensuring that the targets are not, in any way, compromised. If no such strategy emerges, it would imply that the constraints placed on the model are too great to allow a simultaneous attainment of the targets (at least, around a reasonable neighbourhood). This would require a scaling down of, at least, some of the target variables.

Step 7. Extract the optimal policies from the model in this manner and examine the consequences, upon the target variables, of the simultaneous occurrence of alternative combinations of exogenous factors in the presence of such optimal policy mixes. This can be done by the construction of cross-impact matrices which can help in the detection of robust economic policies, if they exist. If no such robust policy is found to dominate, then one can adopt a sort of Bayesian strategy and assign probabilities to the occurrence of certain patterns in the exogenous variables and thereby choose that policy mix which minimizes the expected loss with respect to these probabilities.

Why should such a procedure be adopted? Because one position frequently articulated is that existing econometric models are unreliable and it would be dangerous to base short-term policy decisions on them. However, it needs to be countered that decision makers in the government do, in fact, consider the reactions of the economy while making decisions, however imperfect their knowledge regarding these reactions may be. The use of an econometric model, by aiding in the quantification of these dynamic reactions, forces a decision maker to explicitly take a stand while considering an issue. It is more dangerous for decision makers to mask their assumptions, regarding the economy, on which they base their decisions, than for econometricians to quantify the economy by means of a mathematical model and advise policy makers on the correct strategy. Most policy decisions, in practice, are undertaken in the face of uncertainty, but this uncertainty does not warrant discarding a rational explicit quantitative approach. The use of the suggested methodology in determining the optimal strategy, given an econometric model, will render yet more explicit the underlying rationale in the formulation of macroeconomic policies.

7.5 Setting Up The Experiment

A quick glance at the algorithm contained in Appendix B will confirm that computing the numerical optimal control policies from the model involves a series of intricate matrix manipulations. For this reason, the computer program written to calculate the optimal control and state trajectories incorporating the Kalman filter consisted, apart from the main program, of a sequence of nested subroutines for matrix operations. The entire solution, which was effected on an IBM 3033, of the stochastic control problem was obtained in less than ten seconds which was quite rapid considering the fact that we were dealing with a 20-state-variable, 10-observation-variable and 5-time-period dynamic optimization problem.

All the optimal control experiments described in this chapter were designed to be run for five years, beginning from 1977-78 and ending in 1981-82. A dynamic simulation of the resolved model was carried out over this time period using the historical values of the policy and exogenous variables. It was noted that the simulated values of nearly all the major macroeconomic variables in the model captured most of their turning points and, as such, no attempt was made to adjust the system via any fine tuning exercises.

Reverting back to Appendix B, we realize that the solution of the optimal control algorithm requires the following inputs: (1) the system matrices, (2) the initial conditions, (3) the nominal state trajectories for the target variables, (4) the historical values of the control and exogenous variables and (5) the cost functionals.

7.5.1 The system matrices

As no adjustments were made in any of the elements of the system matrices, these matrices *per se* remained unchanged. However, from the point of simplicity of the control computations, we rewrote our B matrix so that it included only the three major controllers, i.e., government developmental expenditure (GDE), money supply (M1) and the tax rate (T).

Our control vector was therefore formally represented by:

$$\begin{bmatrix} GDE \\ M1 \\ T \end{bmatrix}$$

All other remaining control and quasi-control variables were clubbed under the category of exogenous variables and classified as uncontrollable variables.

7.5.2 The initial conditions

The initial conditions of the state variables, x(0), are all shown in Table 7.1. The '0' refers to 1976-77. All of these pertain to the actual historical levels attained by the variables in that year and are in terms of deviations from their respective reference values.

Table 7.1
Initial Conditions Of The State Variables

YTR(0) = 755	YNFR(0) = 6046	MR(0) = 191
QF(0) = 2.8	YPDR(0) = 5827	YNFN(0) = 32701
QNF(0) = 9.1	CPR(0) = 3584	YNMN(0) = 37703
YAR(0) = 552	s(0) = 0.0642	NDE(0) = 3730
YMR(0) = 2044	SNR(0) = 3403	TR(0) = 4130
WP(0) = 0.766	INR(0) = 1530	NTR(0) = 1267
YSR(0) = 2695		BD(0) = -154

7.5.3 The nominal state trajectories

Although we had twenty endogenous, and therefore potential target, variables, it was decided, in the interests of expositional clarity, to confine ourselves to the two main problems confronting Indian planners today, i.e., maximizing the national income growth rate and minimizing the inflation rate in the face of adverse exogenous conditions. As such, it was decided to include only these two variables in our target set and optimize with respect to them and the three major controllers. While it could be argued that the model possessed other equally important variables which should have been included for consideration, our basic concern in this study was to prescribe a 'broad-spectrum' optimal compensatory policy in the face of uncertainty and to this end the present framework, with its emphasis on just a few selected variables, seemed quite apt.

Before setting the target rates for these two variables, we computed their simulated growth rates in an effort to measure how closely they approximated reality. It was noticed that the model predicted an average annual inflation rate of 9.1% over the period 1977-82 as against its historical record of

9.7%. As far as the growth rate of national income was concerned, the model predicted an average annual growth rate of 5.8% over the same period as against the historical growth rate of 4.2%. As mentioned earlier, no effort was made to adjust the model because we felt that this would effect its dynamic structure. The planned levels of these two target variables were therefore constructed from the viewpoint of these simulated values. As the Planning Commission of the Government of India had set the desired growth rate of national income for the period 1977-82 at 5.2%, which was one percentage point over the historical growth rate, we set the nominal growth rate for YNFR to be 7%, thereby specifying an increase, over the simulated value, which was of about the same order of magnitude. The nominal annual inflation rate was specified to be 5% which was slightly more than half of the average annual price rise recorded by the model. We have provided in Table 7.2 the nominal trajectories for these two target variables using their planned growth rates and translating the subsequently predicted values into deviations from their respective reference values.

Table 7.2
Nominal Trajectories For The Target Variables

Year	WP	YNFR
1977-78	0.854	8886
1978-79	0.947	11924
1979-80	1.044	15175
1980-81	1.147	18653
1981-82	1.254	22376

7.5.4 The control trajectories

The control trajectories of the three controllers are shown in Table 7.3. They are all set at their historical levels and are specified in terms of deviations from their respective reference values.

Table 7.3
The Control Trajectories

Year	GDE	M1	T
1977-78	4138	10652	0.0150
1978-79	4923	13935	0.0277
1979-80	5989	16284	0.0360
1980-81	7774	20116	0.0287
1981-82	9117	21974	0.0373

7.5.5 The exogenous variables

The values taken by the exogenous variables over the planning period are shown in Table 7.4. As before, they all represent historical levels and are specified in terms of deviations from their respective reference values.

Table 7.4
Trajectories For The Exogenous Variables

Variable	1977-78	1978-79	1979-80	1980-81	1981-82
1. ITR	136	236	182	350	419
2. DEF	1434	1668	2073	2600	3000
3. WPF	0.704	0.726	0.854	1.167	1.374
4. WPNF	0.780	0.704	0.946	1.177	1.395
5. XR	2555	2621	2325	2722	3404
6. AF	3.2	4.3	0.9	1.5	4.3
7. ANF	1.1	1.5	0.6	1.0	1.8
8. R	2.0	1.0	-11.5	0.5	2.5
9. MDL	4.81	1.95	24.29	2.37	0.96
10. CR	3065	4414	3629	5415	6481
11. PM	1.487	1.597	2.593	2.384	2.375
12. t	7	8	9	10	11

7.6 Optimal Policies: The Results

In all the following six experiments, numerical solutions are obtained for the optimal policy program using specific cost functionals. In each successive run, the cost functionals were suitably modified in accordance with the results of the previous experiments, so as to iteratively extract an optimal set of weights from the dynamic responses of the model.

One very important point that needs to be noted at this juncture is that although the budget deficit was not included in the target set, we were fully aware that this variable being the last link in our recursive model structure incorporated, in itself, practically the entire ramifications of the system. It was also true that it reflected the eccentricities exhibited by the controllers because it served as an important measure of their efforts. Thus, while BD was directly affected by GDE, it was indirectly affected by M1 (via YNMN and consequently by NDE and NTR) as well as by T (via YNFN and thereafter by TR).Hence, any marked variations in the instruments would have shown up in the form of radical changes in the budget deficit. With this countervailing presence in mind, we constructed a 90% confidence interval for the budget deficit around its simulated values and ensured that the cost coefficients on the three controllers were so assigned that the budget deficit was restricted to lie approximately within this zone.

When actually carrying out the experiments and interpreting the results it is important to realize the fact that the cost functional accumulates penalties only over the finite planning horizon. As a result, many researchers have noticed a type of anomaly that has generally manifested itself in the form of a peculiar behaviour pattern in the optimal paths of certain control variables towards the end of the planning horizon, regardless of the consequences stemming from such actions thereafter. For example, if changes in the money supply affect the interest rate with a lag of two quarters (in a quarterly model) and both the interest rate as well as money supply are included in the cost functional, then the optimal change in the money supply will always be equal to its nominal change during the last two quarters of the planning period. This will occur because whatever happens to the interest rate beyond the planning period, i.e., when the time-to-go is less than two quarters, is overlooked by the monetary instrument and, therefore, does not enter into the cost functional.

The practical solution to this problem has been to extend the planning horizon beyond the period of actual interest by as many time periods as the maximum control delay. However, as in our model all the control variables were assumed to affect the target variables instantly, such an anomaly would not have arisen and therefore there was no need to extend the time horizon of the experiment.

However, as we wanted the feedback coefficient, G_t, to attain a steady-state solution, we solved the matrix Riccati equation (29) in Appendix B iteratively until it converged. It needs to be noted that the solution of this nonlinear difference equation depends only upon the system matrices and the weighting coefficients in the cost functional and, as such, does not involve any of the nominal trajectories that are to be tracked. Consequently, the number of iterations required for convergence was not limited by any data constraint and far exceeded the number of planning periods in the horizon.

7.6.1 Run 1

In this first run, we attempted to obtain a general idea of the optimal time paths that the control and target variables followed in an effort to track their respective trajectories. Both the target variables, i.e., national income and the price level, were weighted equally as we wanted to identify which of the two variables responded easily to the control effort. Although the remaining 18 state variables were assigned zero weights in all the experiments, their responses, especially those of the budget deficit, were carefully monitored in order to examine the overall policy implications of the control strategy adopted.

All the three instruments, i.e., government developmental expenditure, money supply and the tax rate, were equally weighted as we wanted to spot that controller which would stray the furthest from its historical trajectory so that we would have some idea of the control costs to be re-assigned in the next run.

Numerical values for the cost functionals, i.e., the diagonal elements of the Q and R matrix, are also provided with each experiment. We should remember that we are penalizing for percent deviations from the nominal in each case and because the variables are scaled differently, their assigned weights will have different scales. Thus, for example, the average value of the historical trajectory for the tax rate was 0.0360. Therefore, a 1% mean deviation from its path would have a magnitude of 3.6×10^{-4} and when this is squared, as it is in our quadratic cost functional, we end up with a number of magnitude 1.3×10^{-7}. Thus, the weighting coefficient in the R matrix that corresponds to the tax rate will be much higher than the coefficients of either government spending or money supply although technically speaking, the effective weights are the same. For all the experiments that follow, the figures in parentheses, if any, immediately after the optimal values of the state or control variables refer to their contemporaneously corresponding nominal or historical values, respectively.

The cost functionals for this run are given below. We have provided the actual as well as the relative weights, in each case.

Q matrix:	Actual weights	Relative weights
WP	2.0×10^8	1:1
YNFR	1.0	
R matrix;		
GDE	7.0	1:1:1
M1	1.0	
T	2.0×10^{11}	

Observations: The results, in the case of a few important targer and/or control variables, are presented graphically so as to facilitate a quick comprehension of the general form and characteristics of the optimal policies. The selected results of this experiment are shown in Figures 7.1.1-7.1.3. The optimal policy, as far as government spending and the tax rate were concerned, invoked was one of very close tracking of the nominal trajectories. The former ended the run in 1981-82 at an optimal level of Rs. 10631 crores (Rs. 10594 crores). The latter ended up at an optimal level of 13.81% (13.95%). However, a considerable divergence was discernible between the optimal and historical trajectories of money supply. The optimal annual growth rate of narrow money over the planning period worked out to be around 11.6% (14.1%). This considerable reduction in the growth rate of money supply and the relative inflexibility of the fiscal controllers clearly demonstrate that the target variables of the model respond more strongly to monetary policy rather than to fiscal policy.

It is interesting to note the effect of this optimal policy on the target variables. The optimal annual inflation rate slips down to 8.2% (9.1%). On the other hand, the average annual growth rate of national income rises to 6.5% (5.8%). As both, government spending and the tax rate were quiescent, this improvement in the basic characteristics of the economy is directly attributable to the reduced rate of growth of money supply. However, it needs to be stressed that one very important outcome of this policy is the consequent rise in the rate of savings. The optimal savings rate in the last year of the experiment worked out to be nearly 1.5 percentage points higher than its simulated counterpart. Thus, this optimal policy does seem to offer some scope for increasing a key planning parameter in the Indian economy. It also provides an answer to a question posed earlier regarding the effect of the inflation rate on the rate of savings and indicates that a reduction in the rate of inflation helps to increase the savings rate. As both the fiscal controllers were tightly leashed to their nominal trajectories, no untoward increase in the budget deficit was evinced.

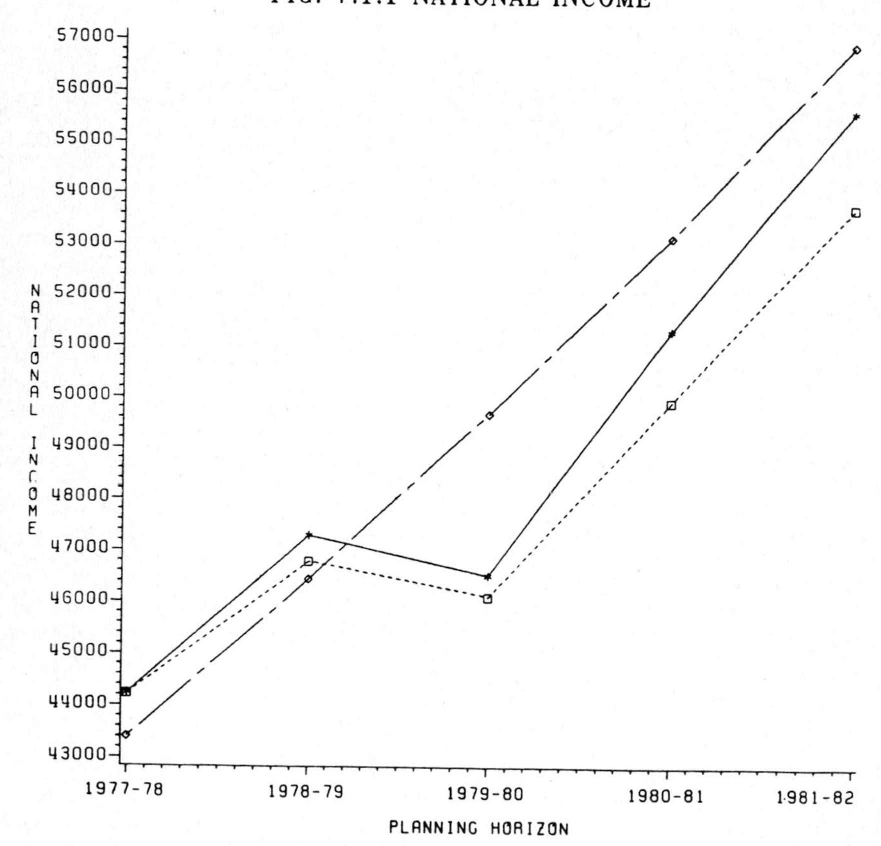

HISTORICALLY SIMULATED PATH (DOTTED LINE)
NOMINAL PATH (BROKEN LINE)
OPTIMAL PATH (SMOOTH LINE)

FIG. 7.1.2 PRICE LEVEL

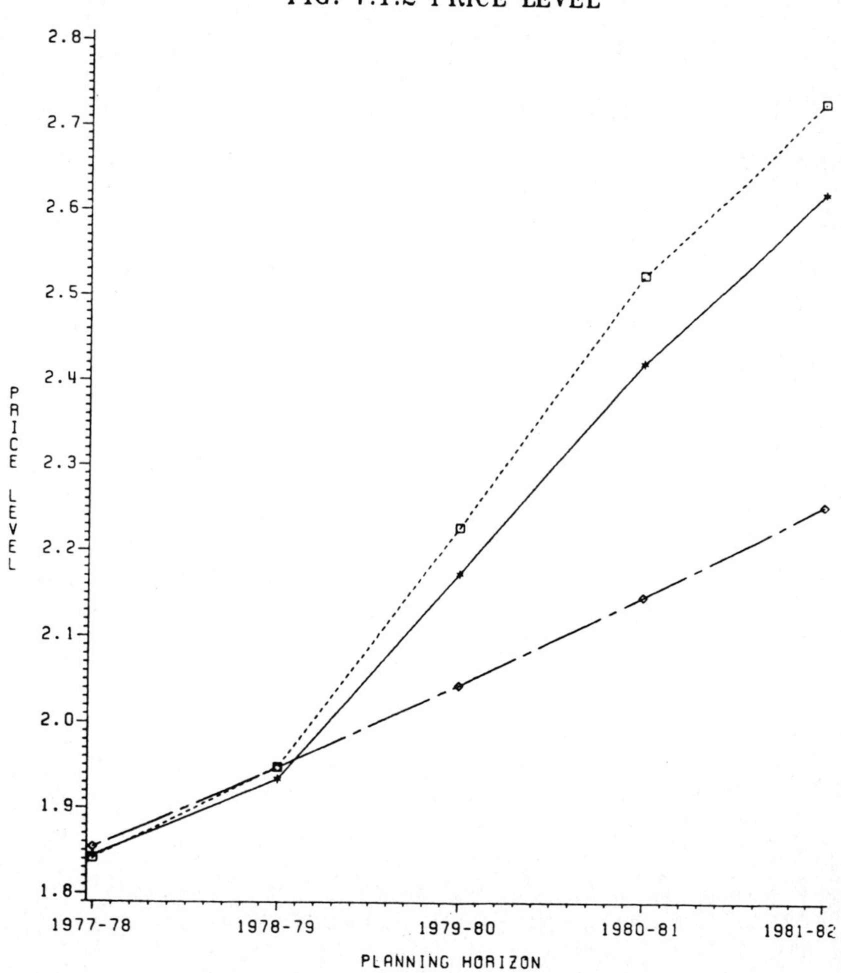

FIG. 7.1.3 MONEY SUPPLY

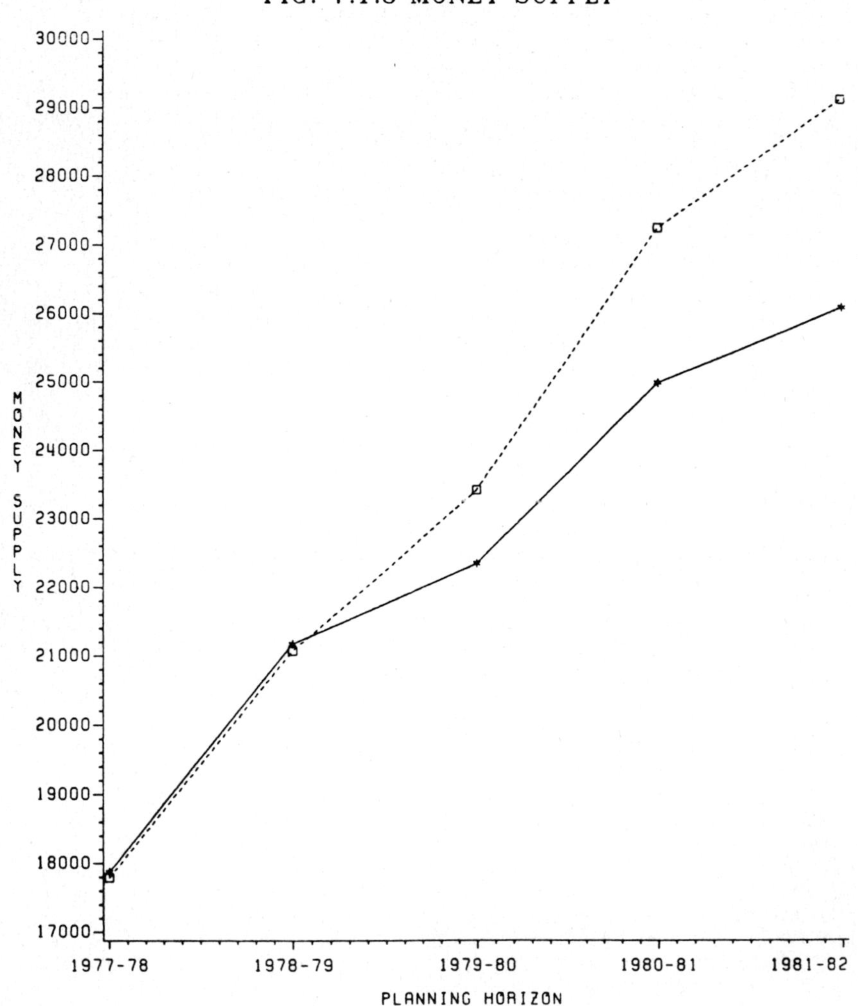

HISTORICAL PATH (DOTTED LINE)
OPTIMAL PATH (SMOOTH LINE)

7.6.2 Run 2

The results of the earlier run demonstrated that monetary policy has clearly a greater role to play than fiscal policy in controlling the economy. Although all the three instruments were equally weighted, it was money supply that was most extensively used in order to control the target variables. This formally seems to justify the earlier (Mundellian) hypothesis established in chapter 4 regarding the tracking ability of monetary policy.

In this run, we reduced the cost coefficients of government spending and the tax rate by half in an effort to encourage their use. As a reverse measure, we increased the effective cost of monetary policy by a factor of ten. We also increased the penalties imposed on the target variables by an equivalent factor in order to improve their tracking abilities. The cost functionals are:

Q matrix:	Actual weights	Relative weights
WP	2.0×10^9	1:1
YNFR	10.0	
R matrix:		
GDE	3.5	1:20:1
M1	10.0	
T	1.0×10^{11}	

Observations: The selected results of this experiment are shown graphically in Figures 7.2.1-7.2.3. The first thing to be noticed is that although deviations from the nominal were heavily penalized in the case of monetary policy, it was just as extensively invoked as in the last run. The optimal annual growth rate of money supply over the planning period was 11.8% (14.1%) which, interestingly enough, was slightly more than before. This time, however, both the fiscal instruments were utilized in an effort to control the economy. There was an impulse response in the control trajectory of government spending in 1979-80 which was a disastrous year for the Indian economy. Government spending that year reached a level of Rs. 10256 crores (Rs. 7466 crores). The spike did not damp down and the optimal values of this controller remained consistently above their historical levels throughout. However, it was the tax rate which was the most extensively invoked. It practically levelled out in 1979-80 and thereafter plunged sharply to reach a terminal rate of 11.39% (13.95%).

The overall annual inflation rate over the planning period worked out to be 8.0% (9.1%) while the average annual rate of growth of national income was 6.5% (5.8%). The assignment of much higher penalties on the target variables did not seem to provoke an overly enthusiastic response on their part, leading one to believe that there could be a limit to the optimal performance of the target variables that can be coaxed out from the system. The cumulative budget deficit over the entire planning period was Rs. 10225 crores (Rs. 9255 crores) and the predicted values of this variable just about lay in the 90% confidence interval constructed for it.

FIG. 7.2.1 GOVERNMENT EXPENDITURE

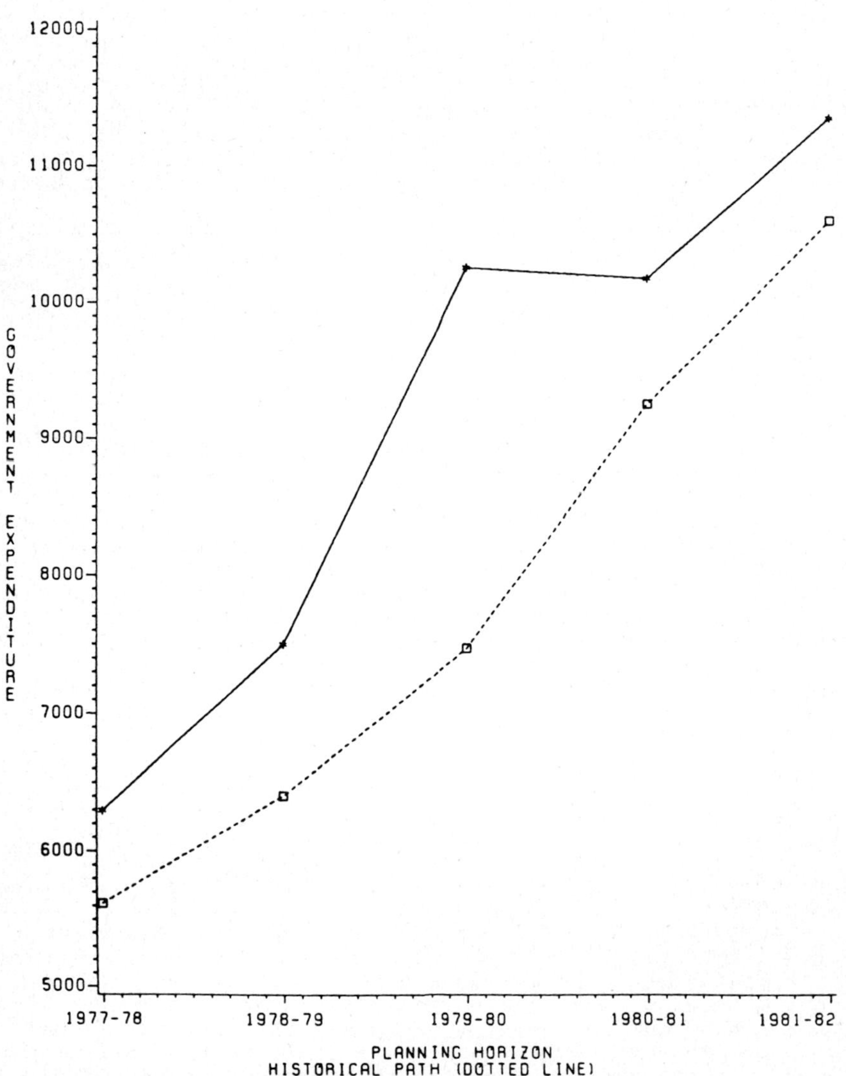

FIG. 7.2.2 MONEY SUPPLY

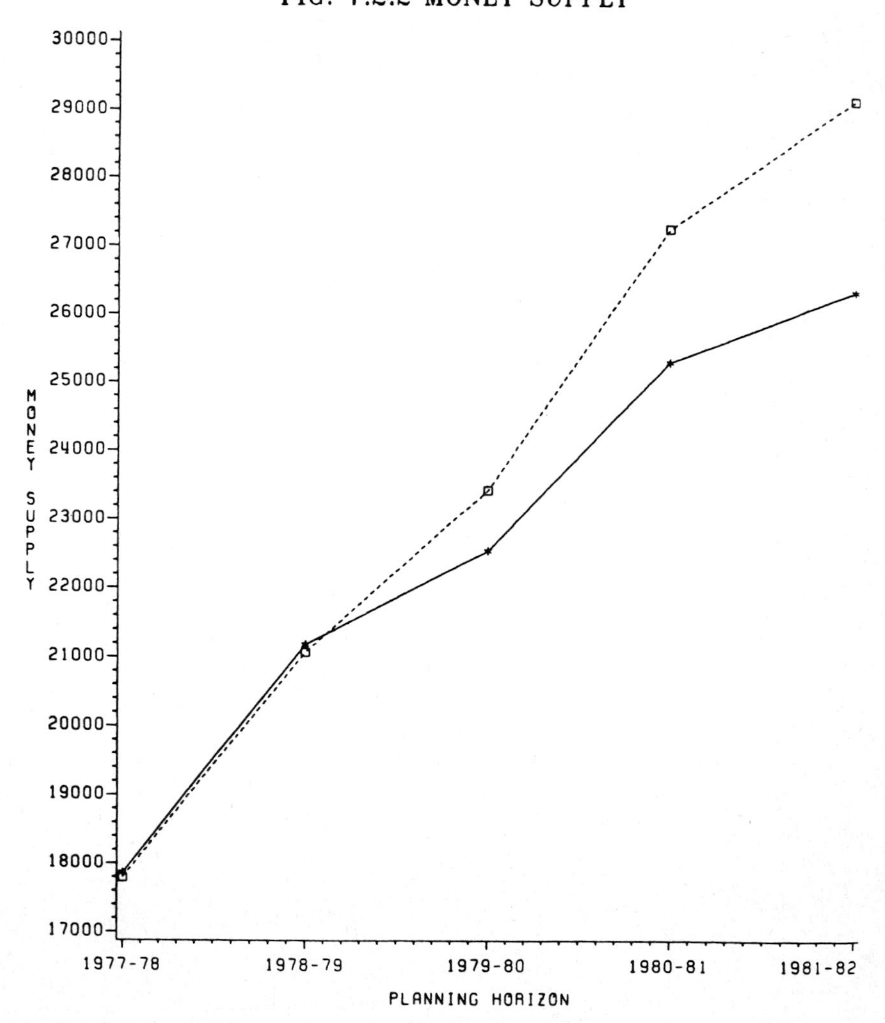

HISTORICAL PATH (DOTTED LINE)
OPTIMAL PATH (SMOOTH LINE)

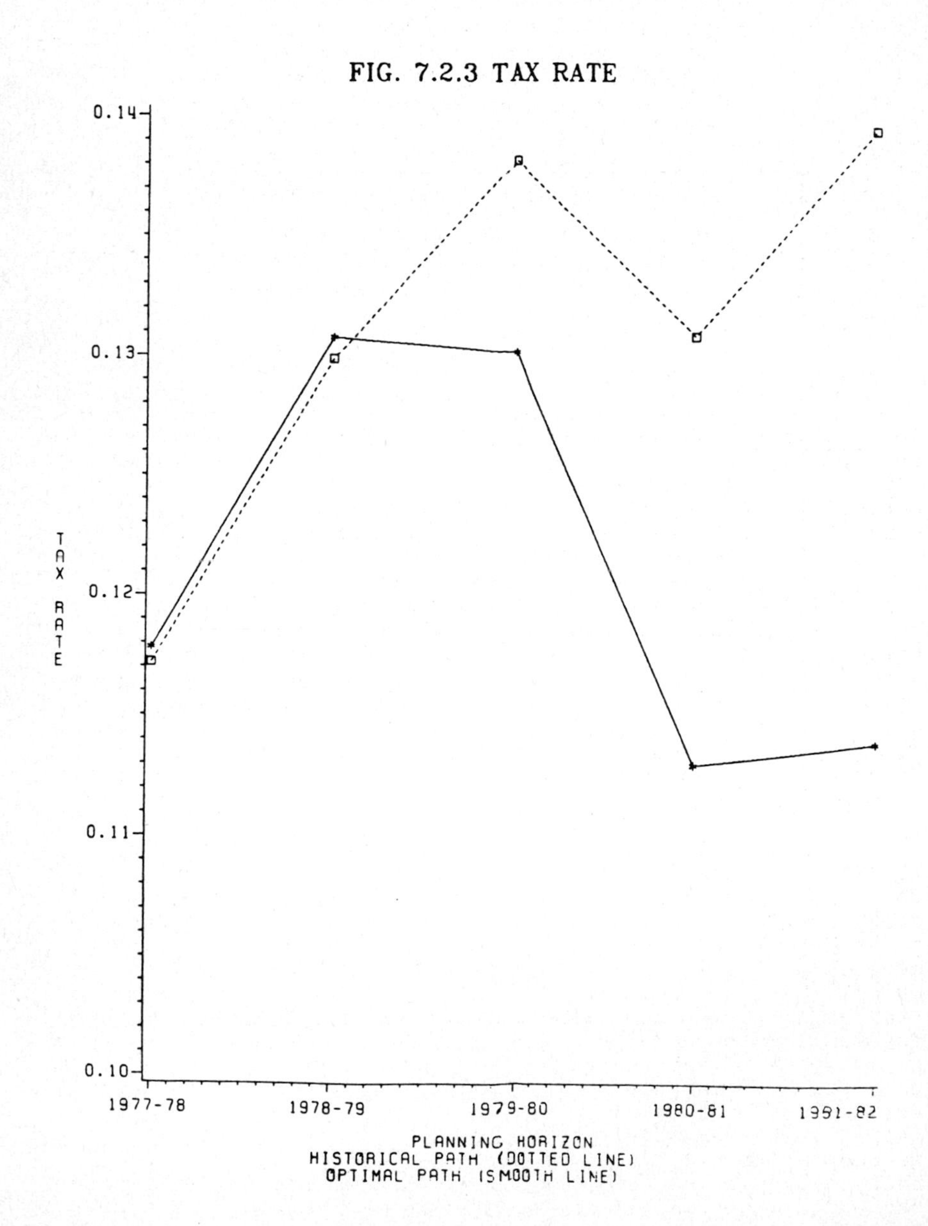
FIG. 7.2.3 TAX RATE
0.14
0.13
0.12
0.11
0.10
TAX RATE
1977-78
1978-79
1979-80
1980-81
1981-82
PLANNING HORIZON
HISTORICAL PATH (DOTTED LINE)
OPTIMAL PATH (SMOOTH LINE)

7.6.3 Run 3

The downward plunge in the optimal values of the tax rate in the previous experiment led us to increase its penalty in this run by a factor of half, so that we could arrest its decline. In an effort to coax a yet greater response from government spending, we further reduced its cost coefficient by a factor of half. A certain amount of flexibility was accorded to monetary policy by reducing its effective cost by an equivalent factor. There was no change in the costs assigned to the target variables. The cost functionals are:

Q matrix:	Actual weights	Relative weights
WP	2.0×10^9	1:1
YNFR	10.0	
R matrix:		
GDE	1.7	1:20:3
M1	5.0	
T	1.5×10^{11}	

Observations: The selected results of this experiment are shown graphically in Figures 7.3.1-7.3.3. The increase in the penalty imposed on the tax rate led to an upward correction in its optimal trajectory in an effort to track its historical values more closely. While the basic profile was similar,- a levelling off during 1979-80, a sharp reduction in the very next year and an increase thereafter - all the optimal values remained consistently higher than their Run 2 counterparts. The terminal value of the tax rate was 12.32% in 1981-82 as against 11.39% obtained in the previous experiment. Government spending once again exhibited an impulse response - of a much sharper type - in 1979-80. It reached a level of Rs. 11563 crores (Rs. 7466 crores) that year. Although the spike died down, the terminal optimal level of government spending remained much higher than its historical counterpart. The optimal annual rate of growth of money supply which worked out to be 10.4% was lower than last time.

The optimal annual rate of inflation is controlled at 7.8% (9.1%) what with the low money supply growth rate compensating for the increase in the tax rates. The overall annual rate of growth of national income over the planning period worked out to be 6.6% (5.8%) because of a higher level of real government developmental expenditure. The cumulative budget deficit over the planning horizon worked out to be Rs. 10717 crores (Rs. 9255 crores) and the predicted values of this variable lay fractionally beyond the 90% confidence interval constructed for it. Inspite of the relatively higher tax rates predicted in this run, the increasingly higher levels of government spending served to vitiate this constraint. The terminal level of the savings rate was nearly 1.7 percentage points higher than its corresponding simulated value. The results seem to indicate that, although India has one of the highest savings rates in the world, this rate is below that which is needed to achieve the desired targets under the circumstances. This result is well in keeping with those obtained by Rao (1984).

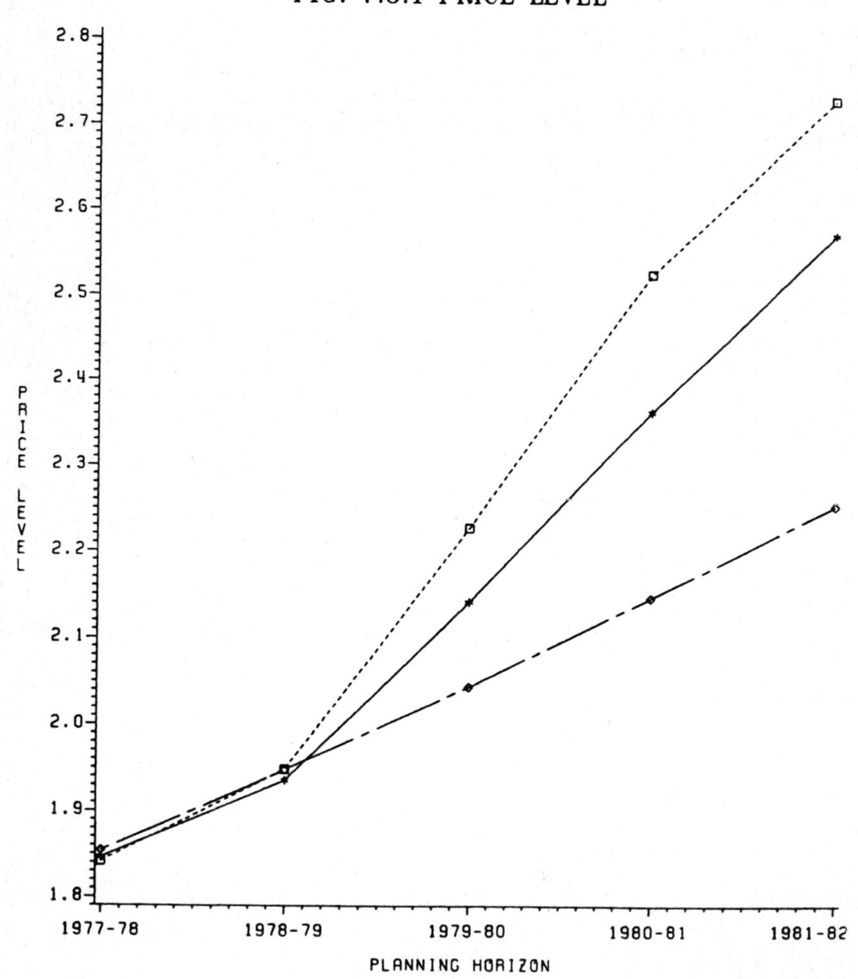

HISTORICALLY SIMULATED PATH (DOTTED LINE)
NOMINAL PATH (BROKEN LINE)
OPTIMAL PATH (SMOOTH LINE)

FIG. 7.3.2 GOVERNMENT EXPENDITURE

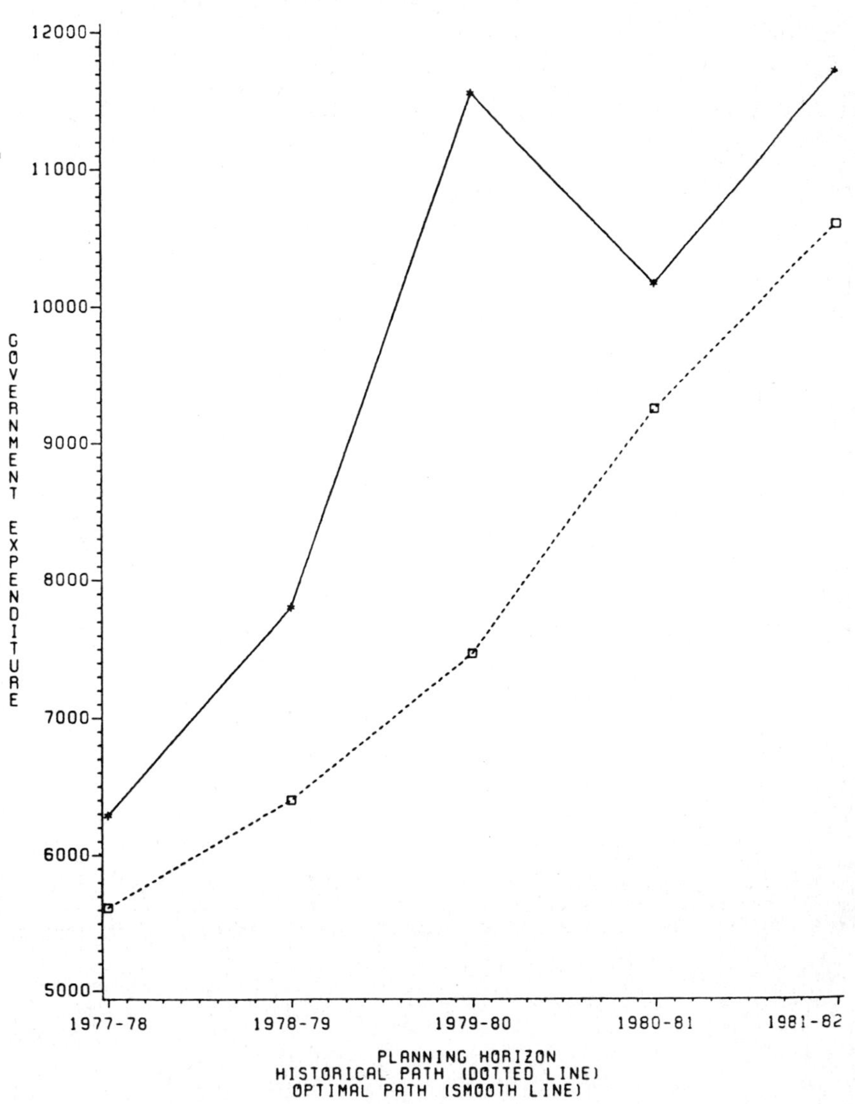

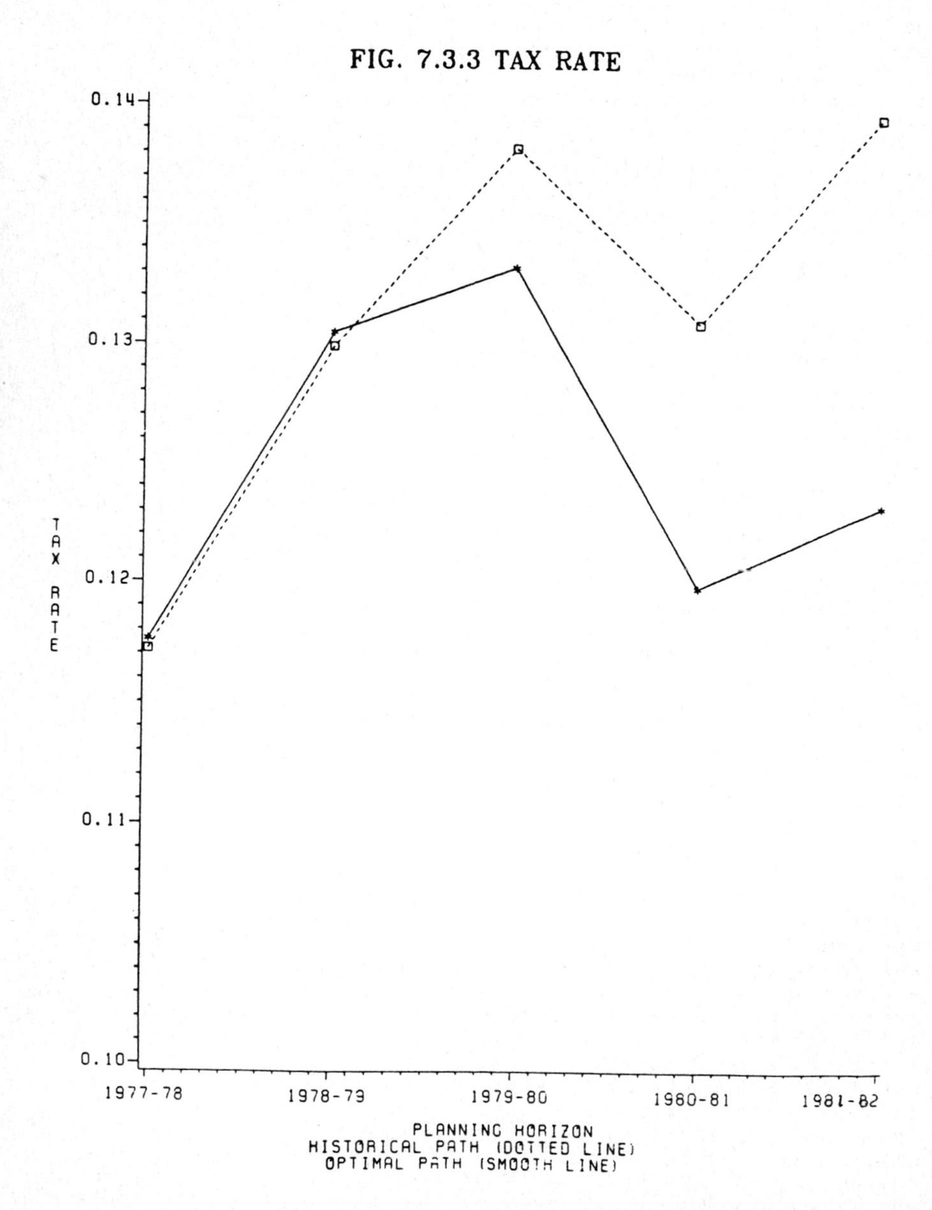
FIG. 7.3.3 TAX RATE
0.14
0.13
0.12
0.11
0.10
TAX RATE
1977-78
1978-79
1979-80
1980-81
1981-82
PLANNING HORIZON
HISTORICAL PATH (DOTTED LINE)
OPTIMAL PATH (SMOOTH LINE)

7.6.4 Run 4

The less-than-satisfactory performance of the inflation rate in tracking its nominal trajectory prompted us to increase its cost coefficient by a factor of ten in this run. There was no change in the penalty imposed on real national income. As far as the controllers were concerned, we increased the cost coefficient imposed on money supply by a factor of six, thereby emphasizing a restrictive role for monetary policy. Similarly, the penalty on deviations from the nominal for the tax rate was increased by a factor of five. There was no change in the effective cost of government spending. The cost functionals are:

Q matrix:	Actual weights	Relative weights
WP	2.0×10^{10}	10:1
YNFR	10.0	
R matrix:		
GDE	1.7	1:120:15
M1	30.0	
T	7.5×10^{11}	

Observations: The selected results of this experiment are shown graphically in Figures 7.4.1-7.4.3. As before, there is a surge in the level of government spending in 1979-80. It rises to Rs. 11618 crores (Rs. 7466 crores) that year. The intensity of this shock was quite similar to the one encountered in the previous run. In order to track the extremely low nominal rate of inflation, it became necessary for the growth rate of money supply to be reduced even further. The optimal annual growth rate of narrow money was just 9.0% (14.1%) over the planning period. The tax rate displayed a similar profile as in the earlier runs. It levelled off in 1979-80, a characteristic feature of all the experiments, and then dipped sharply, only to rise again in the final year of the plan. The tax rate in 1981-82 was, albeit, much lower at 11.37% as compared to its terminal level of the earlier run.

The annual rate of inflation is now controlled at 7.2% (9.1%). This clear reduction in the inflation rate by almost two percentage points can be attributed to the slackening in the growth rate of money supply and a reduction in the tax rates. The combined efforts of the instruments help to increase the annual growth rate of national income to 6.7% (5.8%) over the planning period. This was ensured even without assigning it a higher penalty in the cost functional. We now come to the price we have to pay for this overexertion in the controllers which was provoked as a result of such a high weight being assigned to the inflation rate in the cost functional. The budget deficit now starts straying well outside its 90% confidence interval what with the cumulative budget deficit being about Rs. 10866 crores (Rs. 9255 crores) over the planning period. While this feature does not, by itself, vitiate the optimal policy, which otherwise looks promising, we have to take into consideration the trade-offs involved between the short-term gains (of increasing income) and the long-term losses (of increasing indebtedness) before adopting such a policy.

FIG. 7.4.1 NATIONAL INCOME

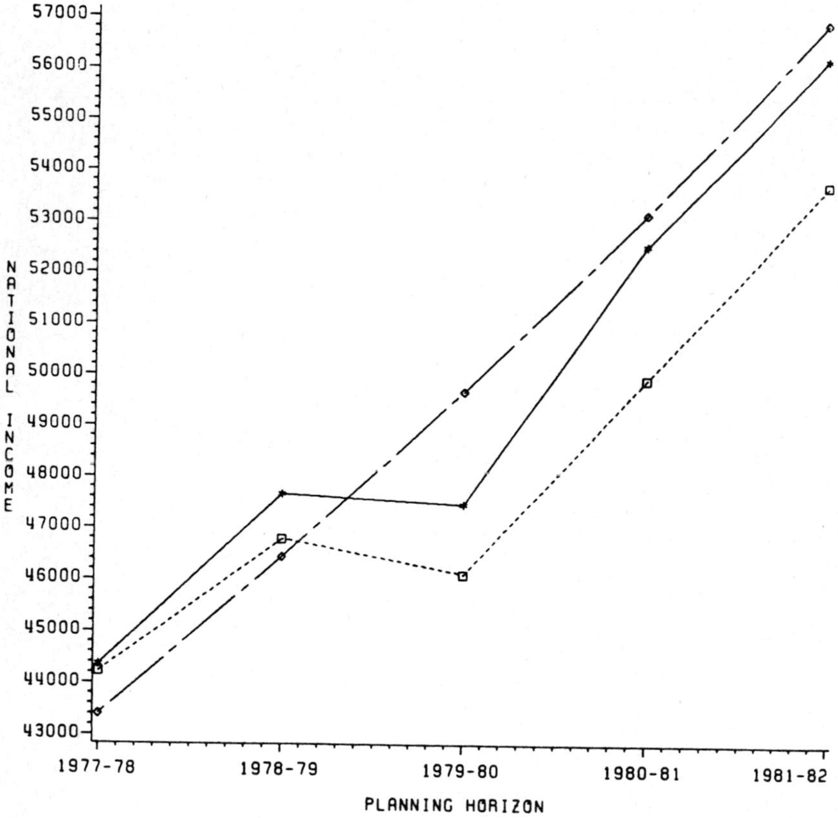

HISTORICALLY SIMULATED PATH (DOTTED LINE)
NOMINAL PATH (BROKEN LINE)
OPTIMAL PATH (SMOOTH LINE)

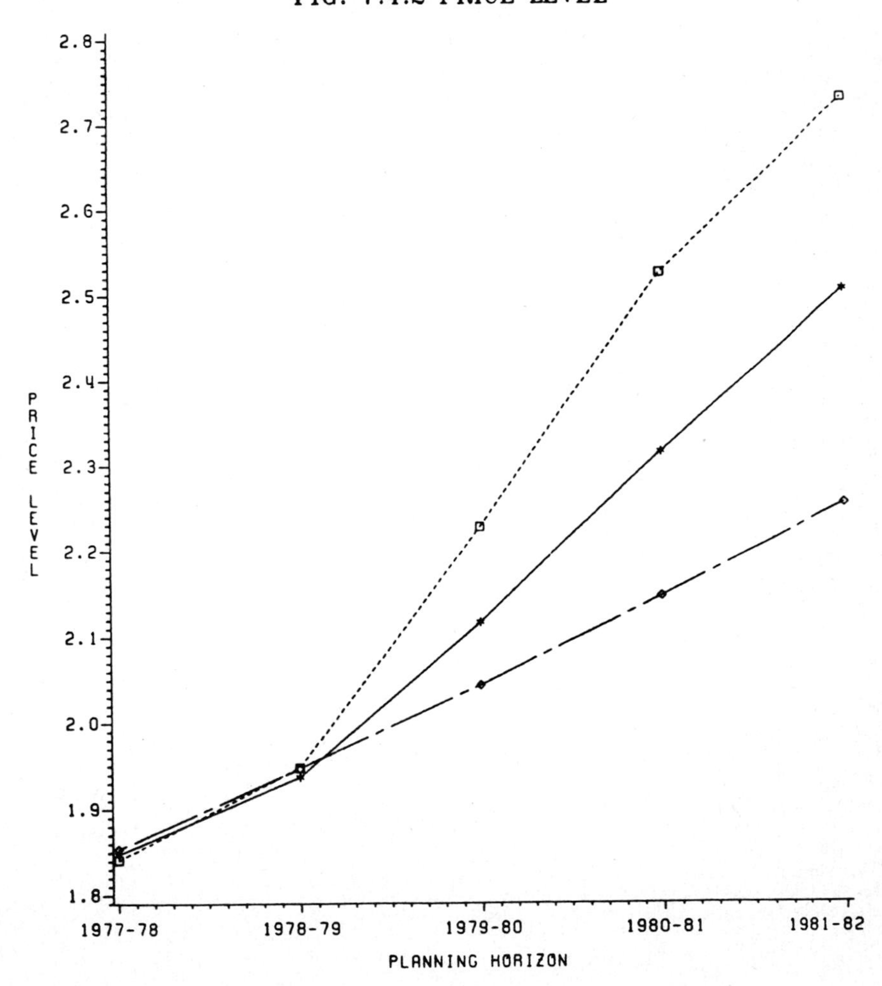

HISTORICALLY SIMULATED PATH (DOTTED LINE)
NOMINAL PATH (BROKEN LINE)
OPTIMAL PATH (SMOOTH LINE)

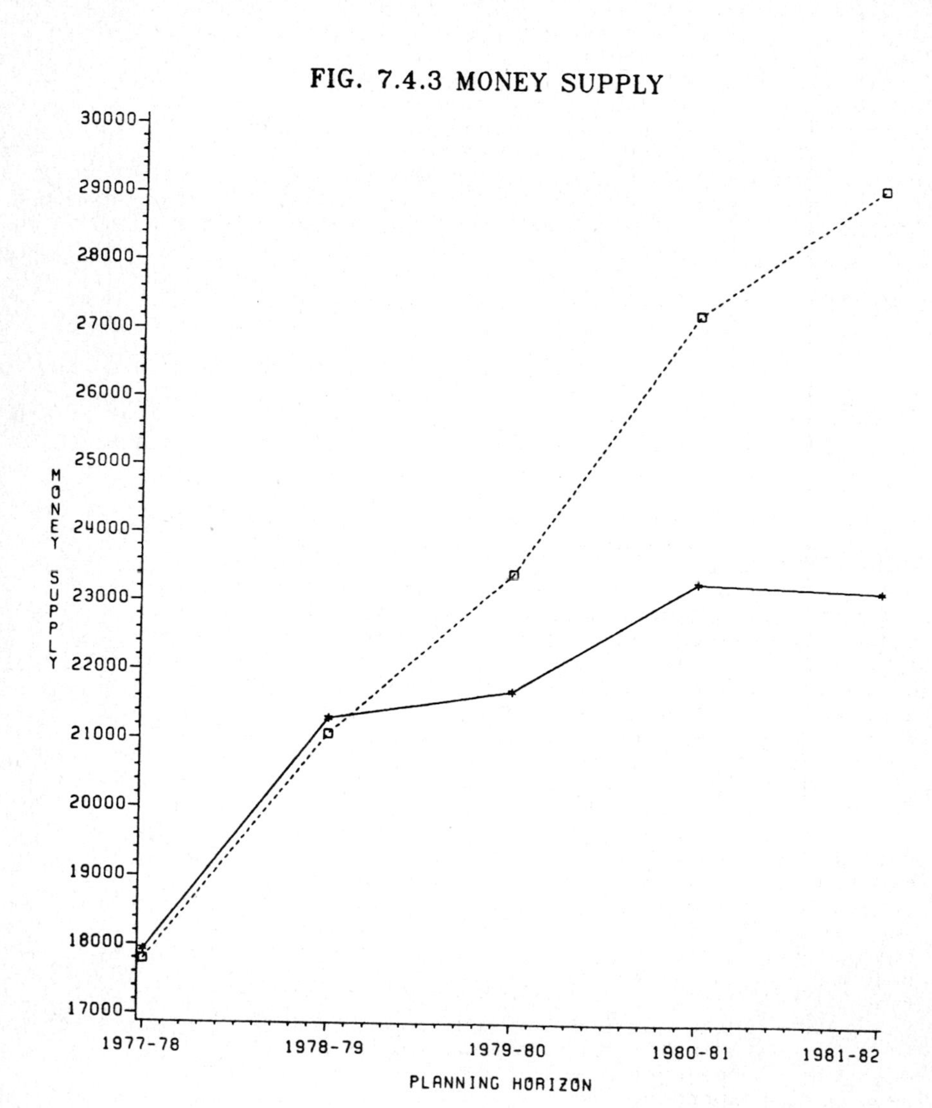
FIG. 7.4.3 MONEY SUPPLY
30000
29000
28000
27000
26000
25000
24000
23000
22000
21000
20000
19000
18000
17000
MONEY SUPPLY
1977-78
1978-79
1979-80
1980-81
1981-82
PLANNING HORIZON
HISTORICAL PATH (DOTTED LINE)
OPTIMAL PATH (SMOOTH LINE)

7.6.5 Run 5

In this run, we switched our priorities vis-a-vis the target variables. Herein, the emphasis was on tracking the growth rate of real national income which, hitherto, had not shown any marked improvement, unlike the inflation rate which seemed to respond fairly well to the controllers. The cost coefficient on national income was increased by a factor of ten while that of the price level was decreased by an equivalent factor. As far as the instruments were concerned, we decreased the penalties imposed on money supply and the tax rate by a factor of half and increased that of government spending by a factor of eight. Our objective was, therefore, to achieve a high growth rate of income through the use of fiscal instruments only. The cost functionals are:

Q matrix:	Actual weights	Relative weights
WP	2.0×10^9	1:10
YNFR	100.0	
R matrix:		
GDE	14.0	1:8:1
M1	15.0	
T	3.7×10^{11}	

Observations: The selected results of this experiment are shown graphically in Figures 7.5.1-7.5.3. As far as the optimal time profiles of the instruments are concerned, we notice that the switching of target priorities induces no shift in the control priorities. Government spending once again exhibits an impulse response in 1979-80, but this time of a much sharper variety because of the relative flexibility accorded to it. It rose to Rs. 12384 crores (Rs. 7466 crores) that year. The tax rate ended the run at 13.18% (13.95%). The annual growth rate of money supply was 12.3% (14.1%). Thus, the relatively high cost placed on this variable ruled out any considerable deceleration in its growth rate as was evident in the last few experiments.

The results, as far as target-tracking is concerned, are very interesting. The annual growth rate of national income was 6.4% while the annual rate of inflation was 8.5%. It is thus noticed that the growth rate of real income (rate of inflation) is lower (higher) than in the last run. This clearly seems to have been brought about by a wrong target-instrument pairing because we emphasized the role of fiscal policy in tracking real output when, in reality, we should have left this variable to be tracked by monetary policy. This result has very important planning implications for the Indian economy. Rather than emphasizing the excessive use of fiscal policy and thereby needlessly expending control effort, we must adopt a restrictive monetary policy to help control the rate of inflation. This will, by the very nature of the dynamics of the model, help us to achieve a high growth rate of national income as well. The cumulative budget deficit over the planning period was Rs. 11690 crores (Rs. 9255 crores) and the predicted values of this variable once again lay well outside the 90% confidence interval constructed for it.

FIG. 7.5.1 PRICE LEVEL

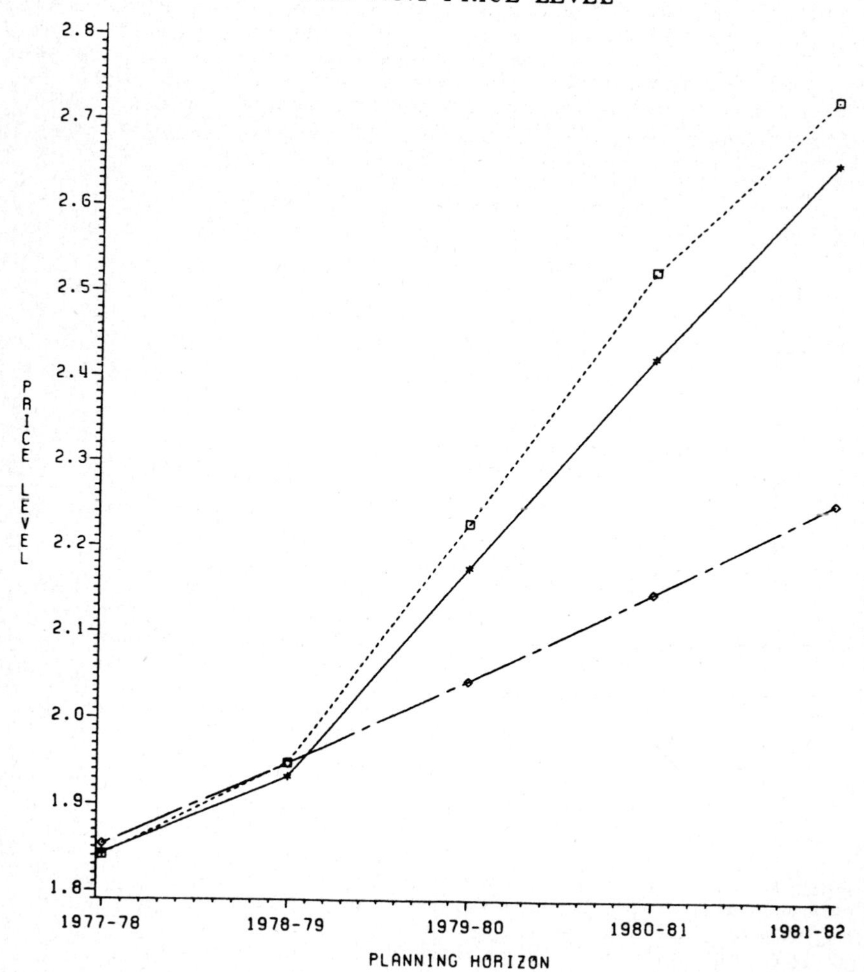

HISTORICALLY SIMULATED PATH (DOTTED LINE)
NOMINAL PATH (BROKEN LINE)
OPTIMAL PATH (SMOOTH LINE)

FIG. 7.5.2 GOVERNMENT EXPENDITURE

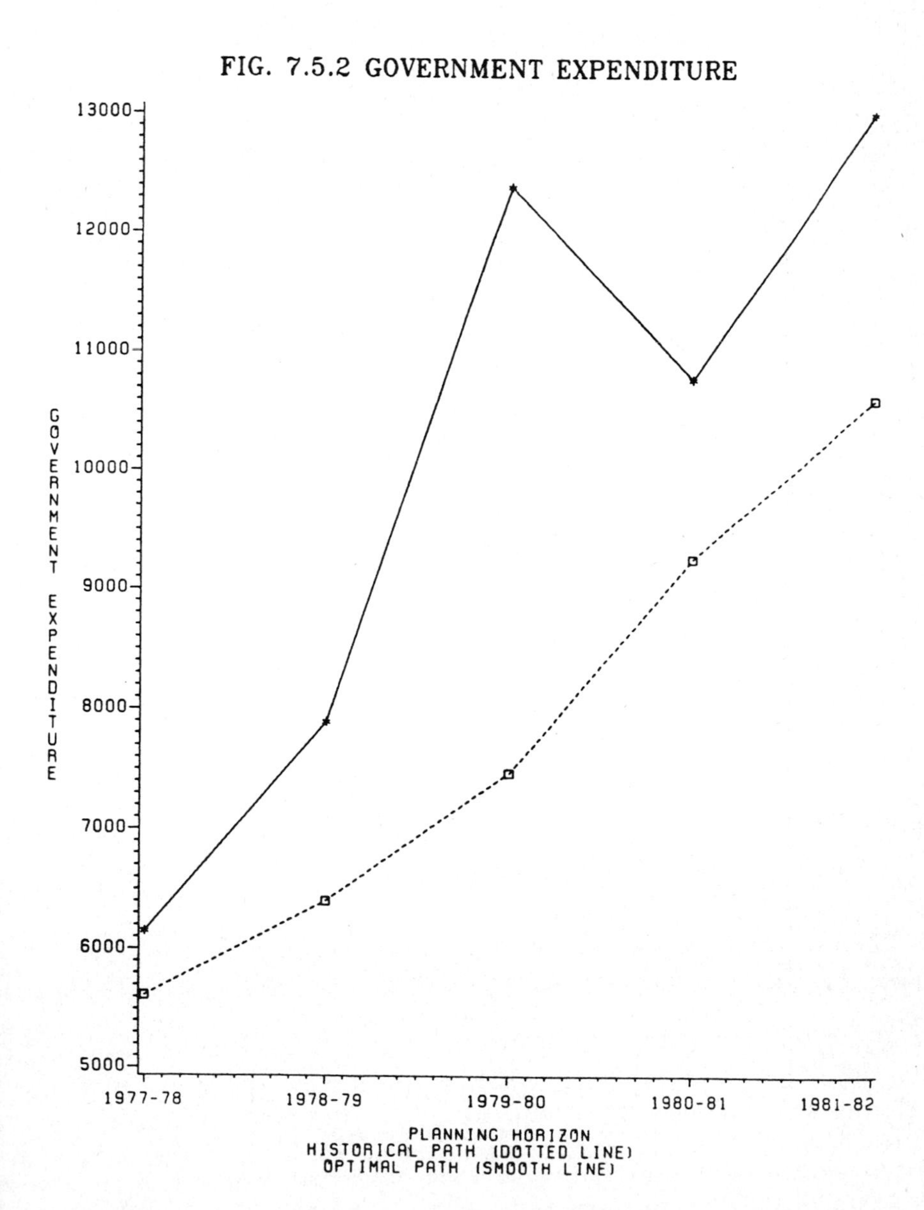

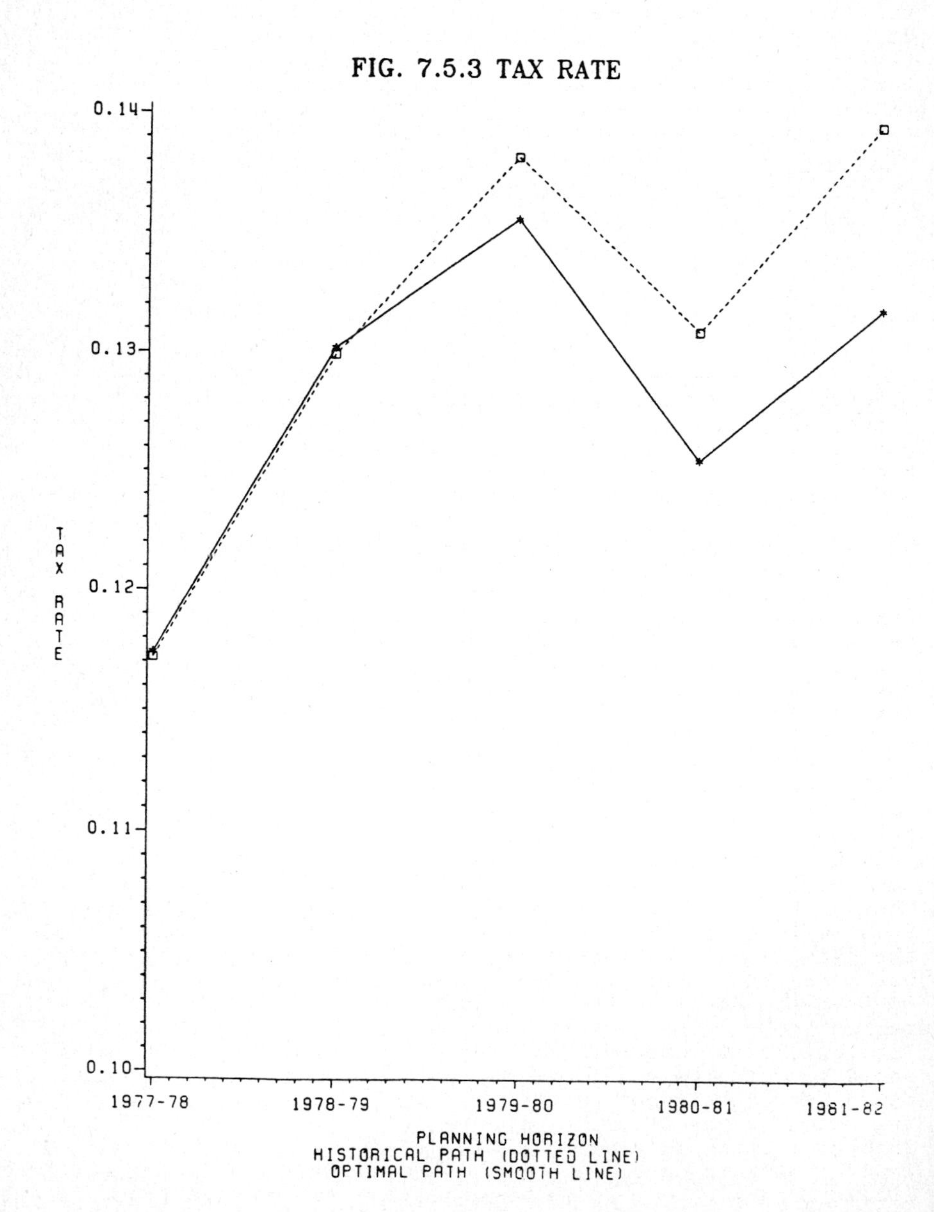
FIG. 7.5.3 TAX RATE
0.14
0.13
0.12
0.11
0.10
TAX RATE
1977-78
1978-79
1979-80
1980-81
1981-82
PLANNING HORIZON
HISTORICAL PATH (DOTTED LINE)
OPTIMAL PATH (SMOOTH LINE)

7.6.6 Run 6

In this last experiment, we attempted to modify the results of the last run through an optimal weighting pattern which, while still emphasizing fiscal policy as the main controller, did delegate a certain role to monetary policy. We reduced the cost coefficients on government spending and the tax rate by a factor of half and that on money supply by a factor of four. There was no change in the penalties imposed on either of the two target variables. The cost functionals are:

Q matrix:	Actual weights	Relative weights
WP	2.0×10^{9}	1:10
YNFR	100.0	
R matrix:		
GDE	7.0	1:4:1
M1	3.7	
T	2.0×10^{11}	

Observations: The selected results of this experiment are shown graphically in Figures 7.6.1-7.6.3. The extreme flexibility permitted to government spending allowed it to rise to Rs.14138 crores (Rs. 7466 crores) in 1979-80. This provides us with an approximate clue as to the extent of fiscal effort necessary to achieve our target of a high growth rate of national income. The terminal value of the optimal tax rate was 12.48% (13.95%). The comparitive flexibility accorded to money supply permitted a rapid deceleration in its growth rate to 9.0% per year which, interestingly enough, was the optimal annual growth rate recorded in Run 4. In that run, while we had singled out prices as the main target, we had set clamps on monetary policy (through the assignment of a relatively high cost coefficient to it) which, however, were overridden by the optimal policy by invoking an immediate application of monetary brakes.

The optimal annual rate of inflation in this run was 7.5% while the optimal annual growth rate of national income was 6.4%. On both counts it is noticed that the optimal policy invoked in Run 4 was marginally superior. This can be attributed, as mentioned before, to the isolation of real national income (instead of prices) as the primary target. To highlight this aspect, a quick comparison of the results of both these runs is in order: While the monetary effort exercised in both the experiments was identical, the fiscal effort in this run was definitely more pronounced what with the terminal level of government spending (the tax rate) being Rs. 2277 crores (1.1 percentage points) higher than its Run 4 counterpart. This 'recycling' of resources served no constructive purpose which seems to definitely suggest that, within the Indian context, there are certain optimal levels of government spending and the tax rate and any effort made to exceed these, as a result of an incorrect assignment between targets and instruments, could lead to serious repercussions in the form of a reduced real income growth rate coupled with a higher dose of inflation. Thus, our priorities should lie in selecting an appropriate primary target and then pairing it with the proper instrument.

FIG. 7.6.1 PRICE LEVEL

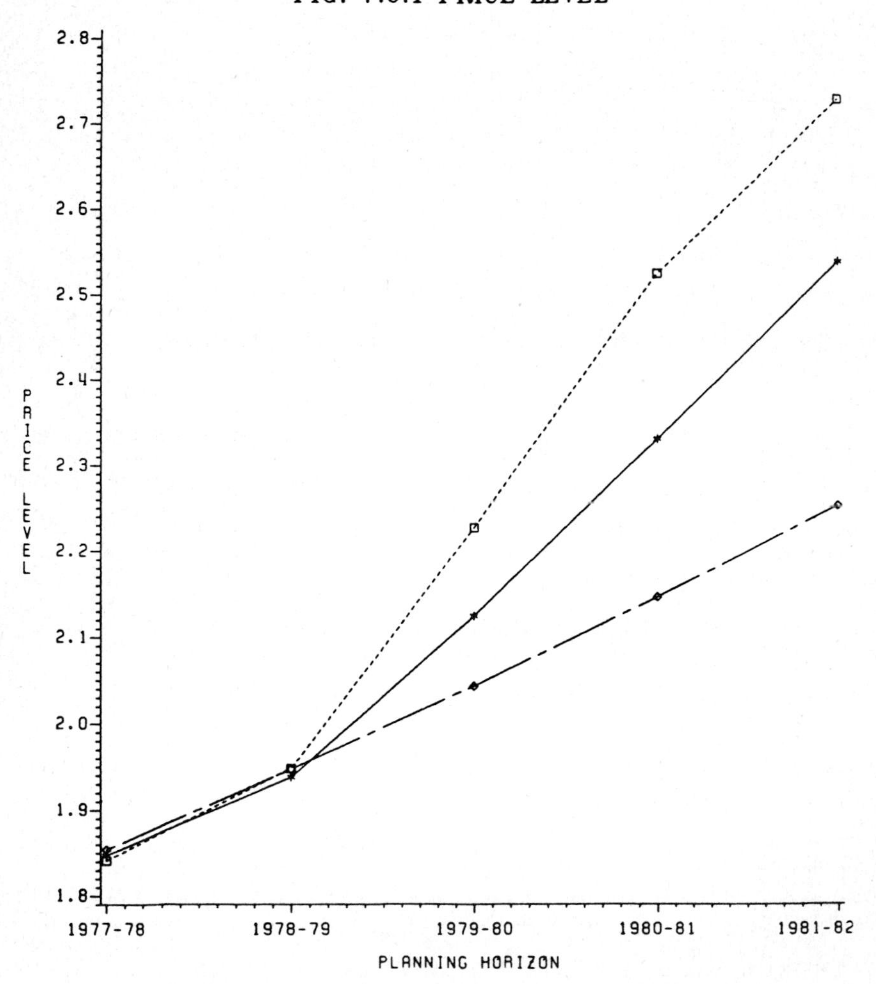

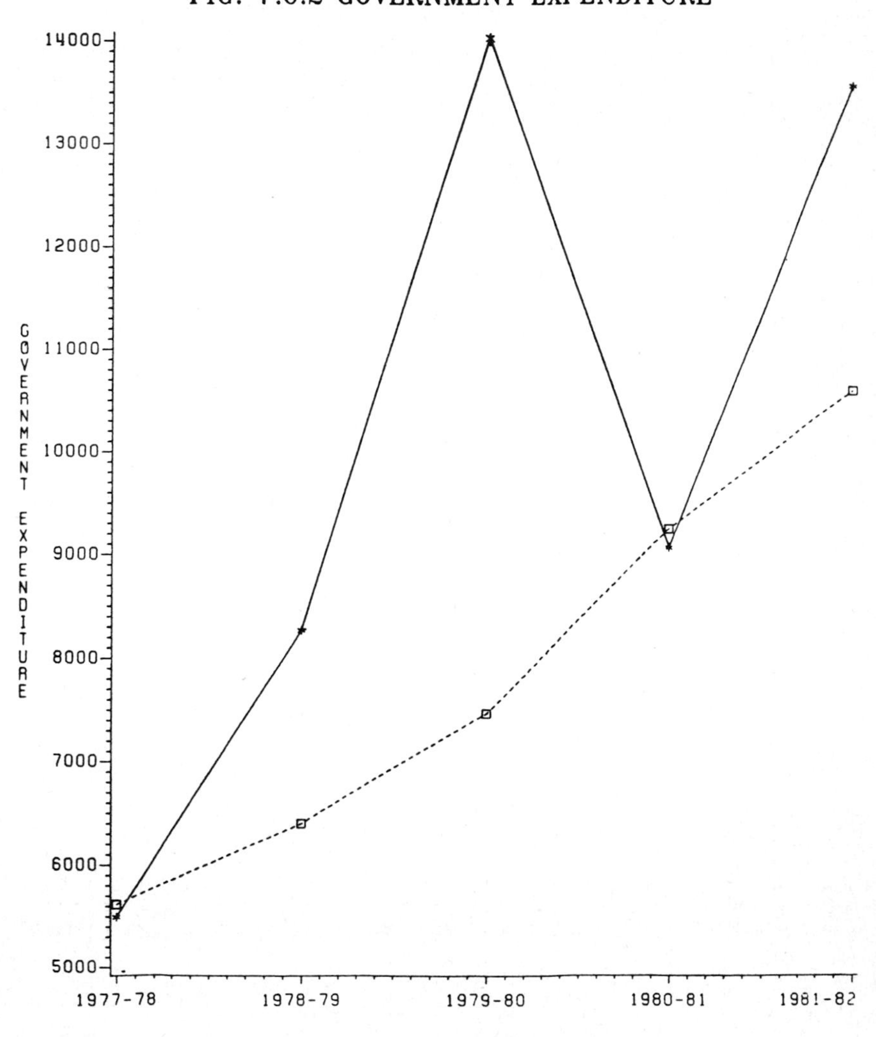
FIG. 7.6.2 GOVERNMENT EXPENDITURE
14000
13000
12000
11000
10000
9000
8000
7000
6000
5000
GOVERNMENT EXPENDITURE
1977-78
1978-79
1979-80
1980-81
1981-82
PLANNING HORIZON
HISTORICAL PATH (DOTTED LINE)
OPTIMAL PATH (SMOOTH LINE)

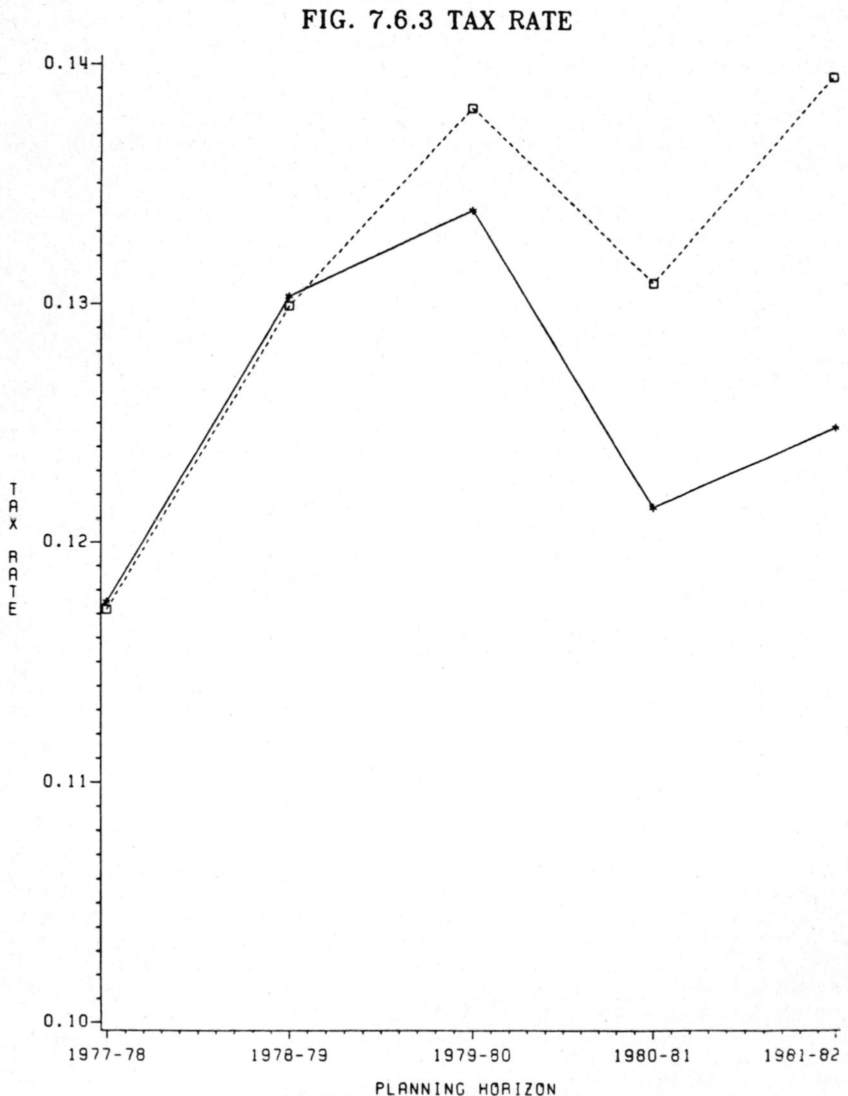
FIG. 7.6.3 TAX RATE
0.14
0.13
0.12
0.11
0.10
TAX RATE
1977-78
1978-79
1979-80
1980-81
1981-82
PLANNING HORIZON
HISTORICAL PATH (DOTTED LINE)
OPTIMAL PATH (SMOOTH LINE)

7.7 A Summing-Up

Inspite of the restricted size of the econometric model, the optimal control experiments performed, using the Kalman filter, helped us to understand several important facets of behaviour in the control variables when forced to track the growth rate of real national income as well as the rate of inflation in India. These results therefore do seem to provide valuable guidelines for Indian short-term countercyclical policy planning under uncertainty.

There were certain patently distinguishing features of, both, fiscal as well as monetary policy. The Mundellian hypothesis that there ought to be a proper target-instrument pairing seemed to be borne out what with monetary policy proving to be superior to fiscal policy both with regards to the control effort expended as well as the proximity of the targets realized. In all the six experiments it was established that there were certain well defined time-profiles for the instruments. The following observations based upon all the results need to be stressed: (1) Government spending followed an expansionary optimal path with a time-profile which was consistently higher than its historical counterpart. Moreover, the existence of a sharply defined impulse response in 1979-80, which was an extremely disastrous year for the Indian economy, seems to bear out the fact that fiscal policy should be activated during such times to stimulate the economy. The optimal policy indicated an increase in the level of government spending by at least 35 percent in 1979-80 as against the historical increase of barely 17 percent. The total quantum of developmental expenditure over the planning period was indicated at around Rs. 45000 crores ($ 36 billion) as against Rs. 39000 crores ($ 31 billion) which was actually spent. (2) The optimal profile of the tax rate was equally revealing. While the historical and the optimal paths meshed for the first two years of the plan, i.e., 1977-78 and 1978-79, the optimal path remained consistently below the historical one thereafter. The optimal policy indicated an immediate levelling off in the tax rate in 1979-80 with a marked reduction in the tax rate the very next year, i.e., 1980-81, with a view of diluting the inflationary impact of the earlier period. The tax rate subsequently drifted upward in the last year of the plan. The optimal path indicated an average tax rate of about 11.9% over the period 1979-82 as against the historical average tax rate of 13.6%. (3) While both the fiscal instruments exhibited clear profiles with well-defined optimal 'course-corrections', in the case of money supply there was no such indication. In all the runs, the optimal monetary policy indicated a definitive deceleration in the annual growth rate of money supply suggesting a range between 9-12 percent per year as against the historical record of 14 percent per year. The results, by and large, favoured the lower bound of the range thereby implying that the optimal annual growth rate of money supply ought to have been nearly 5 percentage points lower than what it actually was over the period 1977-82. This considerable reduction in the money supply growth rate advocated by the optimal policy clearly vindicates many Indian economists (see Brahmananda 1980) who have been strongly recommending a restrictive monetary policy for the Indian economy.

Under this control setting, the average annual growth rate of real national income for the Indian economy, over the period 1977-82, would have ranged between 4.8-5.1 percent, while the average annual rate of inflation over this same period would have ranged between 7.8-9.1 percent (These are appropriately scaled estimates after taking into consideration the adjustments made in view of the differences between the actual historical values of the target variables and their corresponding simulated values). The actual annual growth rate of national income for the Indian economy over this period was 4.2 percent while the actual annual inflation rate was 9.7 percent. It is thus seen that the use of stochastic control theory could have helped the central planners in selecting an optimal short-term trajectory for each of the controllers, based on all the inherent uncertainties in the system, which could have helped in improving the dynamic short-term performance of the Indian economy over the period of the (old) Indian Sixth Plan.

7.8 Conclusions

One of the major intricacies in the practical designing of optimal policies is to determine the proper phasing, i.e., the magnitude and timing, of fiscal and monetary policy under conditions of uncertainty. This is especially true in the case of developing countries which are usually plagued by problems of a cyclical as well as a structural nature. It is not merely a question of how much monetary expansion or fiscal effort is necessary, but also how much when. The proper phasing of policies is therefore of critical importance otherwise the end results of these policies could well be disastrous. Simulation, through the mechanical use of intuitive policies, can never be successful in designing optimal policies and that is why stochastic control theory has to be invoked so as to provide us with some clue as to how exactly the phasing should be executed. Beyond the shadow of a doubt, the value of optimal control as an aid in the determination of policy will depend largely (if not entirely) on the robustness, accuracy and dynamic nature of the estimated econometric model. We have shown that, given such a model, this approach is probably the best one, not only for short-term policy planning but also for a better comprehension of the dynamic trade-offs that a policy maker will eventually be confronted with in the course of macroeconomic planning.

CHAPTER 8

CONCLUSIONS

8.1 Optimal Control: Some Findings

The regulation of a macroeconomic process is an extremely complex task and highly sophisticated techniques are usually needed to answer the intricate questions posed thereof. Years of applied econometric research have led economists to believe that our insights into the functioning of the economy and the effects of control actions upon it are far from perfect. For all practical purposes, economic phenomena are best considered as stochastic in nature. This coupled with the fact that the observation of economic activity itself is haphazard and that decision making involves a considerable amount of approximation, error and bias, leads one to conclude that an economy should be viewed as a stochastic control system and that its regulation should be considered as a problem in control engineering. This is exactly what we have done in the present study.

It is interesting to note that, in 1976, the Financial Secretary to the Treasury of the United Kingdom, by command of Her Majesty, created a "Committee on Policy Optimisation" which was chaired by Professor R. J. Ball, whose terms of reference were:

> " To consider the present state of development of optimal control techniques as applied to macroeconomic policy. To make recommendations concerning the feasibility and value of applying these techniques within Her Majesty's Treasury."

The Committee, which sought and obtained oral and written evidence from some of the most eminent specialists in the area of economic forecasting, modelling, policy formulation and control theory, submitted its report in 1978. The findings indicated that (see Ball 1978):

> " The application of optimal control techniques to economic problems presupposes that it is possible to construct and operate a plausible mathematical model of the economy. There is a *prima facie* case for applying optimal control to such a model of the economy, based on the analogy between the structures of certain economic and physical problems. In the macroeconomic system, policy makers are well aware of the constraints that affect their choices. Thus, in formulating policy, account must be taken of 'trade-offs' among different policy objectives. It therefore seems

permissible to adopt a systematic way of exploring and comparing different feasible paths, obtained by different settings of the instruments of economic policy. In particular, one might hope to construct an ideal compromise between competing objectives, the 'optimum path' and its associated 'optimal policies'. This is one of the most important advantages of considering macroeconomic policy-making within an optimal control framework. "

As the essence of the optimal, as opposed to the automatic, control problem is that one does not know in advance what exactly one would like to happen; and since economic analysis is concerned for the most part with the exercise of choice, given constraints, the possibility of transferring optimal choice technology from physical to economic systems via control theory seems very appealing.

8.2 Economic Significance Of The Results: Robust Policies

One of the major challenges facing policy planners concerned with the exercise of such (target) choices, given (instrument) constraints, is the design of robust optimal economic policies and not just optimal economic policies. This implies, within our context, that the optimal policies extracted using control techniques should be robust enough to withstand alternative model specifications.

Thus, we have to ascertain whether the optimal economic policies, derived on the basis of our stochastic 20-equation model incorporating the Kalman filter, are actually robust in the sense of being able to control the target variables optimally even under alternative model formulations.

If our target variables, i.e., the price level and real output, are viewed as a part of the many endogenous variables which are jointly determined by a system of alternative simultaneous econometric equations, their relationship can be derived only by solving the entire (alternative) system using specified values for the policy variables. Because the relationship between these two variables in any econometric model is not rigid, an arbitrary choice of policy settings in a simulation approach will ordinarily produce combinations of these two variables that can be improved upon. Therefore, the policy variables must be set optimally, rather than arbitrarily, in order to derive the most favourable relationship (or choice) between the targets.

To this end, we incorporated the generalized optimal trajectories of the instruments (derived on the basis of optimal control) back into the original 40-equation deterministic model and carried out a dynamic simulation in order to test the inherent robustness of the optimal policy package suggested by the (resolved) 20-equation stochastic model.

An outline of the procedure actually adopted is in order. Of the three major controllers in the 20-equation model, i.e., government spending, the tax rate and money supply, both the fiscal instruments were exogenous even in the 40-equation model. Thus, when carrying out the dynamic simulation, we incorporated the optimal trajectories of these two instruments based on the results of Run 4 of the last chapter which was, incidently, superior to all the other runs in the exercise (A quick glance at the results of this run will confirm this statement what with the growth rate of real national income being 6.7 percent per year and the annual inflation rate being 7.2 percent).

The hitherto dominating role played by monetary policy in the control of the economy prompted us to examine the existing implications (on income and prices) of pre-selecting a desired growth rate of money supply. We were basically interested in determining via simulation the extent of relationship between the target variables, i.e., the rate of growth of real output and the inflation rate, and the rate of growth of money supply and comparing the results with those obtained via optimal control theory.

However, considering the fact that money supply was an endogenous variable in the 40-equation model, we carried out the dynamic simulation by varying the only monetary controller in this model, i.e., the bank rate. Needless to say, different settings of this instrument prompted different annual rates of growth of money supply. Thus, we were able to gauge the repercussions on the economy (now represented by an alternative fully simultaneous model) as a result of varying growth rates of money supply. As alternative money supply "rules" would give rise to alternative solutions of national income and the price level, under optimally controlled conditions of government spending and the tax rate, an analysis of the results would ascertain whether the derived relationships between the targets and the instrument were in any way compatible with our earlier control solutions.

The simultaneous solutions of both income and prices for each policy setting of money supply (via the bank rate) were recorded and we were able to derive the implied relationship between: (i) money supply growth and the rate of inflation and (ii) money supply growth and the real growth rate. Both these plots are graphed in Figure 8.1. They depict that, given a fixed fiscal effort, any increase in money supply would increase prices and decrease real income.

The results of the dynamic simulation bear out our earlier optimal control solutions. They indicate that, given the optimal fiscal effort, an annual growth rate of money supply ranging between 9-12 percent (the optimal range suggested by the control runs of the last chapter) would imply an annual inflation rate ranging approximately between 7.5-8.3 percent and an annual growth rate of real national income ranging approximately between 6.5-6.9 percent,essentially the same bounds as before.Thus, using the optimal policies derived from the stochastic 20-equation model as a policy constellation in the deterministic 40-equation model still preserved its essential optimal characteristics, thereby testifying to its robustness. The consequent robustness of the optimal policy package therefore has considerable economic significance from the viewpoint of <u>short-term</u> macroeconomic planning in India. We have deliberately stressed 'short-term' because the long-term implications of such an optimal policy have not yet been examined. This will be the focus of our next section.

Figure 8.1
Money, Prices And Income

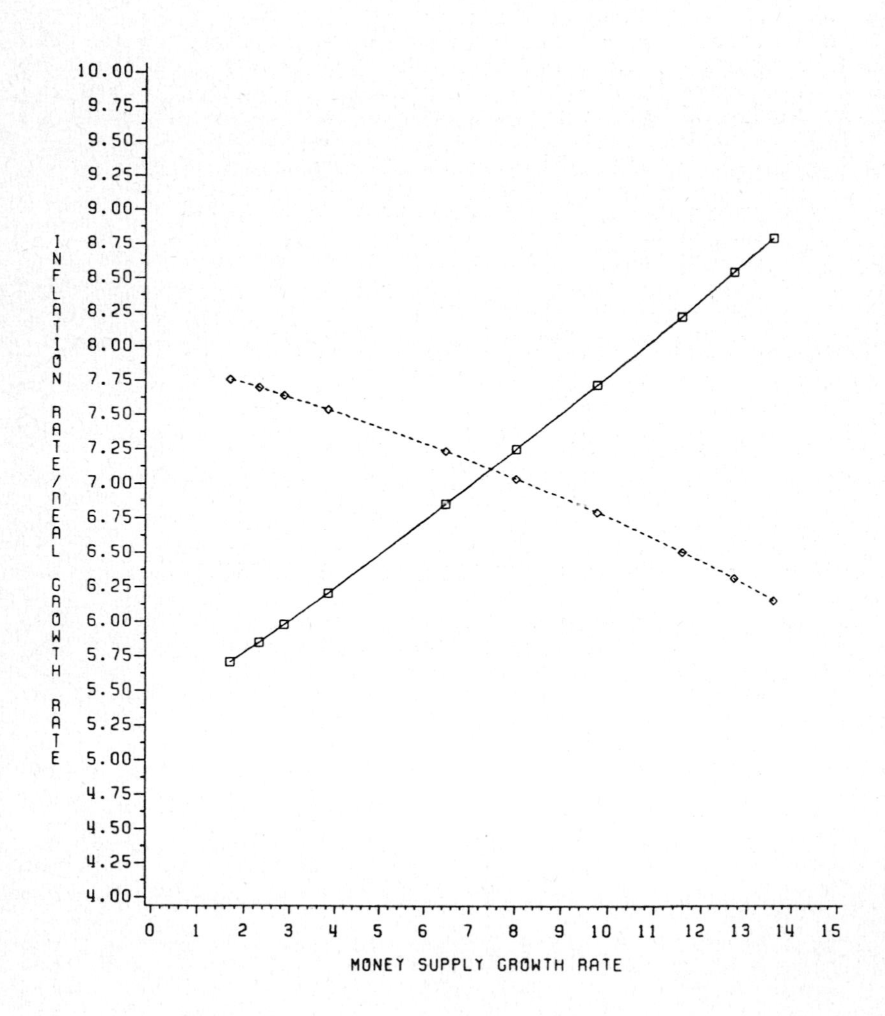

RATE OF INFLATION (Smooth line)
RATE OF REAL INCOME GROWTH (Dotted line)

8.3 Long-Term Economic Planning in India: Control Settings

Considering that both our models were essentially short-term models, any optimal policy indicated by them would be intrinsically optimal only in the short run. To demonstrate this point and to obtain some ideas regarding the long-term (perspective) implications of an optimal short-term policy, we carried out the following exercise.

We initially fixed the control trajectory for government spending at its optimal setting indicated in Run 4 of the last chapter. We then applied optimal control within the framework of the entire 40-equation model incorporating the following tracking features. We only penalized deviations from the nominal trajectory for the price level and set its weight, in the cost functional, at a level which exceeded its corresponding cost coefficient in Run 4 by a factor of 100. All other endogenous variables (including real income) in the model were assigned zero weights. In a similar manner, we assigned extremely heavy weights to all the exogenous policy variables of the model (including government spending) and thereby stabilized them around their historical trajectories (Needless to say, government spending was now stabilized around its optimal trajectory of Run 4). Instrument flexibility was accorded only to the tax rate and the bank rate and, as such, alternative weighting patterns (with varying relative weights) assigned to these two controllers yielded alternative optimal tax rates as well as money supply growth rates (the latter via bank rate changes) with each solution pair tracking a 5% annual inflation rate (on the basis of which our nominal trajectory for the price level was constructed). This short-term _iso-price_ line is depicted by means of the smooth line in Figure 8.2. This iso-price line is downward sloping indicating thereby that increasing (decreasing) tax rates have to be compensated by decreasing (increasing) money supply growth rates to maintain the rate of inflation at a constant given level.

The results reveal that in order to track a 5% inflation rate over the period 1977-82, both, the optimal tax rates as well as the money supply growth rates have to be set at very low levels, quite out of tune with their current historical rates.

We now have to consider the long-term implications of such an 'optimal' short-term policy. Given the fact that government spending is stabilized at a high (optimal) level and that the optimal predicted tax rates are very low, the underlying implications of this optimal policy are that the budget deficits are bound to increase. Such increasing budget deficits coupled to a low growth rate of money supply imply that the bank rate must be stepped-up drastically beyond the planning period as this is the only antidote to force down the growth rate of money supply in the face of a swelling budget deficit. While this _per se_ seems feasible enough, considering that many Indian economists have indeed been advocating an increase in the bank rate, we have to understand the deeper ramifications of such a policy for the Indian economy. As one of the prime determinants of currency expansion is _net_ domestic borrowings, i.e., gross borrowings less repayments, any drastic increases in interest rates (proxied by the bank rate) would lead to ever increasing interest repayments which would serve to decrease net borrowings and consequently increase currency expansion (given the fiscal effort) and money supply. As a result, the bank rate should be hiked up even further to reduce the growth rate of money supply and maintain prices at their pre-assigned levels. Ultimately, in the long run something has got to 'give'. This could be in the form of a sudden increase in the tax rates (to decrease the budget deficits) or a sudden increase in money supply (to permit a fall in the bank rate). Both these would instantly increase the rate of inflation. Thus, the short-term optimal policy could be suboptimal in the long-term.

In order to obtain a heuristic idea of an optimal policy setting (in terms of the tax rates and the rates of growth of money supply) to ensure reasonable long-term stability (in terms of a constant bank rate), we carried out the following exercise.

We referred back to our linearized structural-form deviations model of our 40-equation system (see chapter 4) and obtained the structural form equation for M1 in terms of CC and BR alone by substituting out SCBR and TBR from the concerned equation. This was given by

$$dM1 = -296.133 + 2.5864\, dCC - 352.714\, dBR$$

We then endogenized dBR and exogenized dM1 yielding

$$dBR = -0.8396 + 0.007333\, dCC - 0.002835\, dM1$$

A very *ad hoc* procedure, no doubt, and hence our earlier qualifying statement: "... heuristic idea of an optimal policy setting ..." The above formulation can be theoretically justified if we consider CC to be the demand for currency in which case it can be specified as a function of national income.

This reformulated equation was incorporated into the system which was now reduced to a 38-equation model considering that, both, SCBR and TBR had been eliminated in the process of substitution. An optimal control experiment was now performed within the framework of this model and we tracked the bank rate by assigning it an extremely heavy weight in the cost functional. Its nominal trajectory was fixed at a constant 15% over the planning period 1977-82. This is the approximate level suggested for it by many Indian economists. All other endogenous variables were assigned zero weights in this run. As far as the policy variables were concerned, the tracking features were identical to the earlier experiment, the only exception being that instrument flexibility was now accorded to the tax rate and money supply (which had been exogenized in the present framework).

Thus, alternative optimal pairs of tax rates and money supply growth rates yielded the same level of the bank rate. This *iso-interest* line is depicted by means of the dotted line in Figure 8.2. The iso-interest line is generally downward sloping indicating thereby that increasing (decreasing) tax rates would imply decreasing (increasing) budget deficits and

Figure 8.2
Control Setting

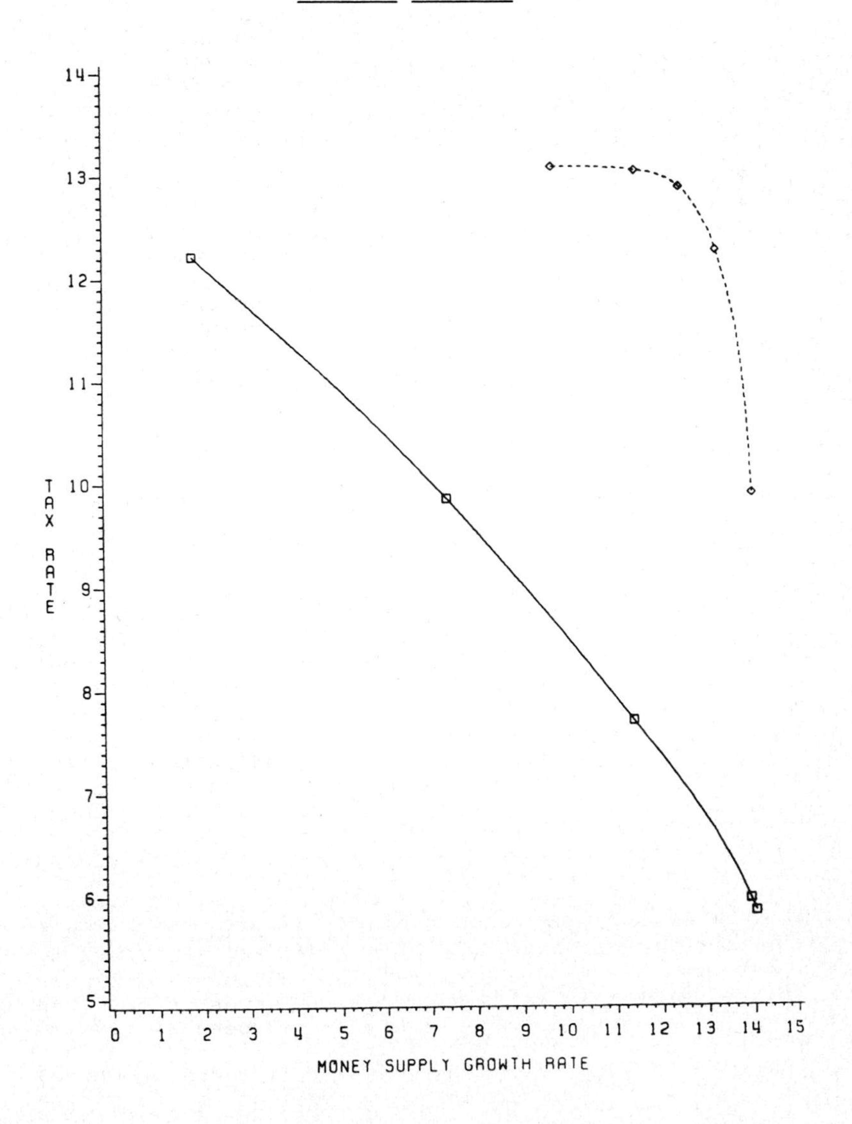

ISO-PRICE LINE (Smooth line)
ISO-INTEREST LINE (Dotted line)

thereby decreasing (increasing) levels of currency and consequently money supply should be decreased (increased) to maintain the bank rate at its given level.

The results reveal that in order to track a 15% bank rate, both, the optimal tax rates as well as the money supply growth rates should be much higher than those required to track a 5% annual inflation rate. This seems quite reasonable because, as mentioned earlier, to track a 5% annual inflation rate indefinitely would have implied that one, or both, of the instrument values would have been forced to increase when the pressure to decrease the (rising) bank rate became too great. Thus, in order to maintain the bank rate at a reasonable long-term level both the instruments have to be set at considerably higher levels than those indicated on the basis of short-term considerations alone. As there is no common meeting point between these two settings of short-term optimal policy and long-term optimal policy, it seems logical to conclude that it would be very difficult to reconcile short-term optimality (*vis-à-vis* tracking a 5% annual inflation rate) and long-term optimality (*vis-à-vis* tracking a 15% bank rate) within the Indian context.

More generally, there has to be some trade-off between these two types of optimal policies and, in effect, we should try and fix instrument values at some intermediate optimal setting. Thus, an ideal compromise between short-term optimality and long-term optimality is essential in order to improve economic planning in India.

8.4 Control Theory And Economic Policy Planning In India

Planning efforts in India have mostly concentrated on the theoretical side with building consistency models which fit into a framework of specified overall targets. Thus, the basic concern of the Indian Planning Commission has been invariably to specify short-run five-yearly targets of most of the major macroeconomic variables, all of which have then been dovetailed into a long-run perspective plan. The planning exercises have therefore consisted of working within highly disaggregated and a consistent static input-output model, given these targets. However, thirty-five years of planning have indicated that the actual performance of the Indian economy has almost always fallen short of these targets. This has led to a growing awareness amongst many Indian economists that consistency plans need to be supplemented by short-run macroeconometric models which, by incorporating the behavioural, technological and institutional relationships existing in the Indian economy, can help in identifying the bottlenecks, constraints and lags operating in the system, besides generating the actual achievable growth rate of the economy (see Desai 1973).

From this point of view, the econometric model need not be large; indeed, the idea of a possible 'Law of Large Systems' together with the associated concept of 'dominant mode' dynamic behaviour (Young 1977) would seem to suggest that the size of a development planning control model could be quite modest in comparison to most practical forecasting models. Thus, it would be possible to specify a reasonably compact development planning model for the Indian economy and obtain general guidelines regarding the feasibility of applying control theory techniques within its framework. It will be our contention in this section to show that even within a consistency planning framework, one which is invariably adopted by Indian central planners, it is possible to use optimal control techniques in order to try and improve economic policy planning in India.

Consider the following 5-equation consistency model which is basically a synthesis of the Harrod-Domar and the two-gap models and which is supposed to capture a few essential elements of the Indian economy. Such a framework usually provides an underpinning to most of the macroeconomic forecasts which underlie Indian economic planning.

The time-path of national income over the planning period is usually computed by

$$Y(t+1) = Y(t) + b\Delta K(t) \qquad \text{---(8.1)}$$

where $Y(t)$ is GNP at time t, $K(t)$ is capital stock, b is the inverse of the ICOR (incremental capital-output ratio) and Δ is the forward difference operator.

The capital-stock generating forward recursion equation is given by

$$K(t+1) = (1-d)K(t) + I(t) \qquad \text{---(8.2)}$$

where $I(t)$ is gross investment and d is the fraction of capital stock depreciated each period.

Foreign trade is brought into the model via the identity that capital inflows, i.e., the difference between imports, $M(t)$, and exports, $X(t)$, add to investible resources, $I(t)$, given gross savings, $S(t)$, so that the savings-investment balance becomes

$$I(t) = S(t) + M(t) - X(t) \qquad \text{---(8.3)}$$

If investment has a marginal import share of h and production of a unit of GNP requires imports of the amount m, then

$$M(t) = hI(t) + mY(t) \qquad \text{---(8.4)}$$

Finally, the residual level of consumption, $C(t)$, is given by

$$C(t) = Y(t) - S(t) \qquad \text{---(8.5)}$$

It should be borne in mind that all these variables are measured at constant prices which implies that price variations are conspicuously absent in all Indian plan models.

To reduce the dimension of the system, we combine eqs. (8.1) and (8.2) to form

$$Y(t+1) = (1-abd)Y(t) + bI(t) \qquad \text{---(8.6)}$$

where a is the ACOR (average capital-output ratio).

Considering the fact that we are modelling an economy through a set of equations comprising only accounting identities, i.e., eqs.(8.3) and (8.5), and technological relationships, i.e., eqs.(8.4) and (8.6), we realize that there are no reaction equations to estimate and, as such, there is no error-term structure in the state-transition equation of the system. This is one of the primary reasons as to why Indian planners have been unable to incorporate the concept of transition errors in the modelling process.

In order to apply optimal control theory techniques to the above model, we need numerical estimates for the parameters of the system. It needs to be noted that the capital elasticity of output $\underline{e}$ is given by

$$e = ab = \frac{K(t)}{Y(t)} \cdot \frac{\Delta Y(t)}{\Delta K(t)} \qquad \text{---(8.7)}$$

It is this measure that is usually used in Indian planning rather than the ACOR. Using the planned estimates provided in 'A Technical Note On The Sixth Plan Of India', Planning Commission, Government of India (1981), we set e = 0.7608, b = 0.2381, d = 0.0212, h = 0.4305 and m = 0.0936. The state-transition equation of the system was thus given by

$$x(t) = \begin{bmatrix} 0.9839 & 0.2381 & 0 & 0 \\ 0.1617 & 0.0391 & 0 & 0 \\ 0.1617 & 0.0391 & 0 & 0 \\ 0.9839 & 0.2381 & 0 & 0 \end{bmatrix} x(t-1) + \begin{bmatrix} 0 & 0 \\ -1.7559 & 1.7559 \\ -0.7559 & 0.7559 \\ 0 & -1 \end{bmatrix} u(t) \qquad \text{---(8.8)}$$

where

$\underline{x} = (Y\ I\ M\ C)^T$ and $\underline{u} = (X\ S)^T$

Calculation of the open-loop eigen values and eigen vectors of the system yields

Eigen value:	1.0230	0, 0, 0
Corresponding eigen vector:	$\begin{bmatrix} 0.6977 \\ 0.1149 \\ 0.1149 \\ 0.6977 \end{bmatrix}$	$\begin{bmatrix} 0.2352 \\ -0.9719 \\ 0 \\ 0 \end{bmatrix}$

The result demonstrates the well-known instability of the Harrod-Domar open dynamic model. In order to stabilize the system, we use the theorem due to Wonham (1974) which tells us that if the system is controllable, it is possible to choose a feedback gain matrix G which will place the closed-loop poles at any arbitrarily specified positions in the complex plane. In practice, this result is meaningful provided the system remains linear under closed-loop control.

In order to apply this fundamental result, we invoked the controllability criterion (see Appendix C) and determined that the controllability matrix

$$(B,\ AB,\ A^2B,\ A^3B)\ :\ 4x8 \text{ matrix}$$

indeed had rank equal to 4. Thus, in order to stabilize the model, we have to design an appropriate feedback gain matrix G which will relocate all the eigen values within the unit circle.

In view of our earlier comments regarding a possible trade-off between short-term and long-term optimality within the Indian context, we shall now examine the implications of trying to obtain an optimal allocation of output (over time) between consumption (being a proxy for short-term optimality) and investment (being a proxy for long-term optimality). Thus, we would be tracking only these two target variables from the viewpoint of determining optimal 'intermediate' policy. As exports are usually considered uncontrollable within the Indian context, we shall accord instrument flexibility only to the level of savings. Hence, our objective would be to use savings, one of the most important control variables in the Indian planning context, in order to steer the economy on an optimal course between the conflicting goals of simultaneously trying to maximize,both, consumption as well as investment.

Our cost functionals were therefore given by

$$Q = \begin{bmatrix} 0 & 0 & 0 & 0 \\ 0 & 1 & 0 & 0 \\ 0 & 0 & 0 & 0 \\ 0 & 0 & 0 & 1 \end{bmatrix} \quad \text{and} \quad R = \begin{bmatrix} 1 & 0 \\ 0 & 0.1 \end{bmatrix}$$

We therefore have a 2 target-1 instrument tracking case. When the number of target variables exceeds the number of instruments, not all target values can be contemporaneously satisfied by the solution of the deterministic system given by eq.(8.8). Thus, the trade-offs between short-term and long-term optimality will be jointly determined by the sequence of SVF gain matrices, G(t), and the intercept vectors, J(t), specified in eq.(21) of Appendix B. The former is more important because it also determines the closed-loop characteristic polynomial which defines the poles of the closed-loop system.

Recursive solutions, backward in time, of the matrix Riccati eq.(29) of Appendix B yielded the following steady-state SVF gain matrix:

$$G = \begin{bmatrix} 0.6725 & 0.1628 & 0 & 0 \\ 0.3114 & 0.0754 & 0 & 0 \end{bmatrix}$$

It is this solution in conjunction with the time-varying J(t), which will depend upon the planned trajectories of the target and control variables <u>inter alia</u>, that will determine the optimal level of savings <u>for the Indian</u> economy which would be a compromise optimal solution between short-term and long-term targets.

The new state transition matrix is now given by

$$(A+BG) = \begin{bmatrix} 0.9839 & 0.2381 & 0 & 0 \\ -0.4725 & -0.1143 & 0 & 0 \\ -0.1113 & -0.0269 & 0 & 0 \\ 0.6725 & 0.1628 & 0 & 0 \end{bmatrix}$$

The closed-loop eigen values are seen to be 0.8696, 0, 0 and 0. Thus, the choice of the SVF gain matrix has helped us to uniquely relocate the poles of the system within the unit circle. The closed-loop pole assignment problem has thereby yielded a design for the robust optimal control of a consistency planning model for the Indian economy. It is in this context that we advocate the extensive use of optimal control analysis for economic policy planning in India.

Needless to say, the use of such techniques presupposes the existence of a large output gap for the Indian economy. Considering the preoccupation of Indian planners with increasing the growth rate of real national income, one of the most challenging problems confronting them has been the measurement of potential output for the Indian economy. This is so because when decision makers contemplate the use of optimal monetary and/or fiscal policies in order to bring the economy closer to the full-employment level of output, they must need to know where exactly to aim at. In control terminology, this amounts to specifying a nominal trajectory for real national income.

It is in this context that Rao and Fernandes (1986) constructed a "locked" variant of the consistency model just described and applied control theory, in particular stability theory, within its framework in order to obtain an idea of the rate of growth of potential GDP for the Indian economy. The model predicted a 4.1 percent annual growth rate of potential GDP which was considerably higher than the 3.6 percent annual growth rate of actual GDP evinced over the period 1950-84.

The results are graphed in Figure 8.3. It indicates that the output gap grows during a recession as a result of more resources becoming unemployed. Conversely, during an expansion the gap declines, although for the Indian economy, unlike, say, the U.S. economy, it has never actually become negative. A negative gap implies that there is overemployment, overtime of workers and more than the usual rate of capacity utilization - phenomena practically unheard of in the Indian economy.

Much of the Indian economic track record can be read from the output gap. The most important amongst them is that this gap was fairly stable and comparitively negligible during the period 1950-64, except for a recession in 1957 which was immediately followed by a long expansion during the first half of the 1960s. The output gap at the end of this striking phase was barely 0.9 percent. However, the gap witnessed a quantum jump as a result of the two successive drought years of 1965-66 during which period the output gap surged forward to 14.3 percent. Every successive recession thereafter, namely, the oil price crunch of 1971-72, the droughts during 1974 and 1976, and the oil price spiral cum drought of 1979 (when the output gap reached an all-time high of 18.7 percent) has led to an irreversible jump onto a higher plateau of the output gap. As a result of all these recessions, which have been occurring with distressing frequency of late, the output gap in 1983 amounted to as much as Rs. 7675 crores ($ 6.1 billion) at 1970-71 prices.

All these factors clearly highlight the inescapable conclusion that the Indian economy is currently operating far below its potential. Under the circumstances, it is imperative that urgent <u>optimal</u> policy measures need to be initiated soon enough, seriously enough and effectively enough in order to combat the menace of this 'expanding frontier'. It is here that optimal control theory can help the most.

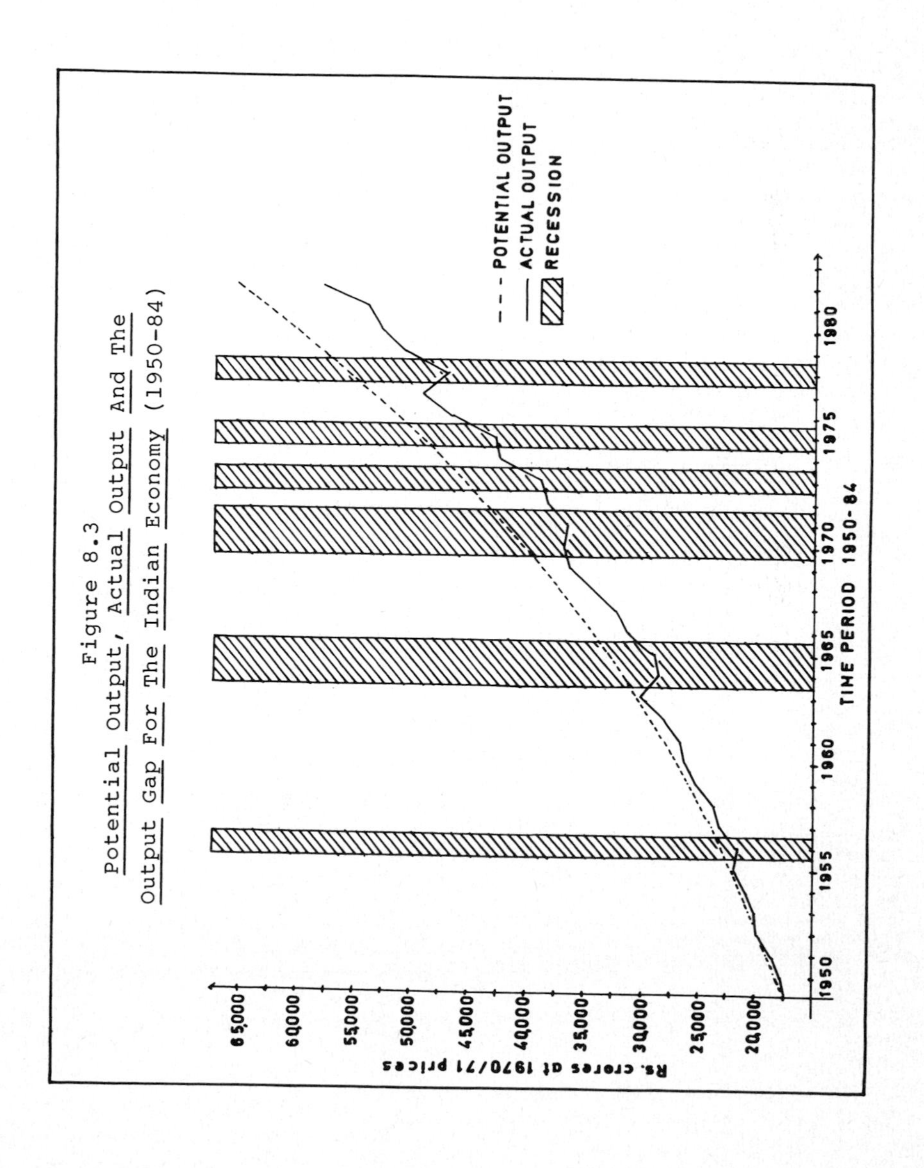

Figure 8.3
Potential Output, Actual Output And The Output Gap For The Indian Economy (1950-84)

8.5 Contributions Of The Study

The essence of the study was to provide an optimal control framework within which short-term Indian macroeconomic countercyclical policy could be prescribed. In the presence of random disturbances affecting the state-transition equation and measurement errors contaminating the observation mechanism, the use of closed-loop control policy was clearly shown to improve short-term macroeconomic performance. Thus, the use of the Kalman filter as a basis on which optimal regulatory policy for the Indian economy can be executed is highly recommended.

The construction of a resolved linearized model with a quadratic cost functional and affected by disturbances resulted in the incorporation of a closed-loop control policy based on noisy observations of the system state. It was seen that the optimal control actions could be designed after predictions of the system state, corrected on the basis of latest available information, had been obtained.

The construction of the Kalman filter for the Indian economy, probably for the very first time, yielded extremely interesting results regarding the updating of predictions based on the latest available observations of the state. These results dovetailed very well into the existing concept of errors in information signalling which is quite endemic within the Indian data base context. The use of such a filter could go a long way in drastically improving the efficiency of Indian plan models which currently have to depend upon scarce, inaccurate and, often, conflicting observations on the state of the economy. The resolution of our model and the subsequent exogenization of the loop variables rendered the original system recursive. That the Kalman gain converged so very rapidly to a unique steady-state solution can be partially attributed to the recursive nature of the state transition matrix.

One very important finding of the study is that although we designed our regulatory policy based upon some weighted average of past and present prediction errors in a scheme referred to as proportional-plus-integral control which ought to have contributed to the overall stability and continuity of the optimal control actions, it was generally noticed that these control actions were not very smooth. In fact, in certain cases, impulse responses were very much evident. We have every reason to believe that this was due to the fact that we were using an annual model to predict policies over a relatively short-term planning horizon and in view of the intensity of the exogenous shocks imparted to the system during the period of experimentation there was not much scope nor time for the control actions to be smooth or stable if they were expected to satisfactorily track the target variables. All this makes the construction of a quarterly model for the Indian economy extremely vital.

We also noticed that optimal policy settings to attain short-term targets could not be maintained indefinitely from the point of view of long-term stability. There were no control settings which could achieve short-term and long-term optimality simultaneously and hence we realize that there has to be some form of trade-off between these two extreme settings of optimal policy. This is a very important finding as it indicates that it is essential to take a broader view of the underlying implications of an optimal short-term policy package. In effect, if the technological, behavioural, institutional or exogenous constraints placed on an economy are too great to allow reasonable optimal values for the policy variables, then a protracted setting of the instrument values around their extreme 'optimal' values should be avoided in order to prevent the economy from 'overheating'.

8.6 Alternative Methodologies

This study does not intend to set up any sort of rivalry between planning by the use of consistency models and macro-econometric control models. The former are extremely important for medium- and long-term planning in India. On the other hand, for short-term planning, they are incapable of forecasting optimal policies under conditions of uncertainty and the Indian economy being extremely vulnerable to such manifestations, Indian planners must realize that they have to adopt control models and accept their predictions as aids that will lead to substantial improvements in plan implementation.

However, of late, prompted by the numerous failures of consistency as well as optimization models, many Indian economists have been advocating the time-series approach to measure and predict the characteristics and contents of the economy. While the forecasting successes of these time-series models have been undisputed, protagonists of econometric models have argued that a successful _ex-post_ prediction of economic variables by a time-series model does not _per se_ vindicate its specifications, since it could be a case of 'spurious' correlations, because these predictions, by extrapolating past trends, may only indicate that econometric models have tried lifting the economy onto a path higher than that indicated by previous experience. But this may be equally consistent with saying that econometric models were designed with scant regard to existing bottlenecks and past shortfalls in the economy. Thus, time-series models being descriptive of these shortfalls, estimate the lower bound of the growth rate of the economy. A successful prediction by a time-series model can then be considered as a flaw in the control design of an econometric model.

While alternative stances on this issue can be taken, it is essential to realize that, both, time-series as well as control methodologies have extremely important roles to play in improving the efficiency of economic planning in India. In this context, the studies by Pethe (1986) who formulated a stochastic control design for the derivation of robust economic policies; Karnik (1986) who applied time-series methodology to ascertain the direction of causality between real government expenditure and real national income; Nachane and Nadkarni (1985) who used time-series analysis to measure the extent of causality between money supply, prices and real output; amongst others, assume great relevance from the viewpoint of designing an optimal interface between control theory and time-series analysis. It is studies of these types that will eventually increase our comprehension regarding the Indian economy.

8.7 Epilogue

The scheme described here could very well be used for actual short-term policy prescription in the Indian economy because of the nature of uncertainty residing in the data base of the Indian planning apparatus. The generalized optimal control trajectories computed in this study are highly indicative of the fact that the current instrument settings in the Indian economy are sub-optimal. Thus, these results provide 'broad-spectrum' policy guidelines to help in the adjustment of instrument values in future. We are well aware that the control variables as designed in this study cannot be fine-tuned or adjusted very rapidly. However, given such a framework, the central planners could be appraised of the general nature of the optimal control trajectories to be selected and such information could be of immense assistance to them in their extremely onerous task of streamlining economic activity in India.

In essence, we suggest that the approach outlined in this study be seriously considered to determine short-run macroeconomic countercyclical policies for the Indian economy because, from a planning angle, the methods promise solution techniques to the main problem perpetually confronted by the Indian planner - that of accurately interpreting tardy, incomplete, biased and often conflicting current statistics. Should he wait to collect and evaluate more evidence before prescribing optimal compensatory policy or would that imply valuable time is lost? This study proposes precise answers to such questions and although, in practice, such answers may be patently unacceptable to those unaccustomed to sophisticated mathematical methods, a beginning has been made and a design has been set up and we hope that Indian, as well as other, planners involved in short-term planning heed the signals this study hopefully sends forth.

APPENDIX A

THE OPTIMUM LINEAR DISCRETE FILTER

To obtain a linear minimum-variance estimator, we must use one of two procedures. One approach is to specify a linear conditional mean and find the best linear form; this is the approach using orthogonal projections. The other approach is to assume gaussian amplitude distributions for the input and measurement noises. Because of the repeatability property associated with linear systems and gaussian distribution, the exact conditional mean in this case will be linear. The linear minimum-variance estimate must be equal to the minimum-variance estimate if the minimum-variance estimate is, in fact, linear. This is the case if gaussian distributions are assumed.

Note that if we require that the estimator be linear in the observations, then the actual distributions of the input and measurement noises are not important. However, if the distributions are, in fact, gaussian, as is often the case, then the conditional mean is actually linear. In other words, the Kalman filter is the best (minimum-error-variance) linear filter for any distribution; it is the best filter of all possible linear and nonlinear estimators if the input and measurement noises are gaussian.

In the derivation of the Kalman filter, we shall assume and require that the observations be processed sequentially. Regardless of whether the estimator is sequential or not, the values of the resulting state estimates are unaltered. In this appendix, we shall obtain the optimum linear minimum-error-variance unbiased estimates of the state of a linear time-invariant dynamic system whose state-transition equation is contaminated by additive white noise of zero mean and known finite variance. To obtain this estimate, we observe a time-invariant linear function of the state whose measurement equation is also contaminated by additive white noise of zero mean and known finite variance. There is no correlation between the input and measurement noises and we wish to determine a sequential form of the estimator. The solution to this problem, based on Sage and Melsa (1971), is the Kalman filter algorithm.

A.1 The Model

Consider a stochastic system which is represented mathematically by the following set of linear difference equations:

$$x(t+1) = Ax(t) + Bu(t) + w(t) \qquad \text{---(1)}$$
$$z(t) = Cx(t) + v(t) \qquad \text{---(2)}$$

where
t: 1,2,...,T; discrete time index with a fixed finite horizon
x(t): an $\underline{n}$ dimensional state vector at time $\underline{t}$
u(t): an $\underline{r}$ dimensional control vector at time $\underline{t}$
z(t): an $\underline{l}$ dimensional observation vector at time $\underline{t}$
w(t): an $\underline{n}$ dimensional input noise vector
v(t): an $\underline{l}$ dimensional measurement noise vector
A,B,C: time-invariant matrices of appropriate dimensions(given).

The random disturbances w(t) and v(t) are supposed to have zero mean and to be independent of each other at all times and uncorrelated amongst themselves over different time periods.In addition, they are assumed to have finite covariance matrices. Thus,

$$\left.\begin{aligned} E(w(t)) &= 0 \\ E(v(t)) &= 0 \end{aligned}\right\} \qquad \text{---(3)}$$

$$E(w(t)w'(t+k)) = \begin{cases} Q, & \text{when } k=0 \\ 0, & \text{otherwise} \end{cases} \qquad \text{---(4)}$$

$$E(v(t)v'(t+k)) = \begin{cases} R, & \text{when } k=0 \\ 0, & \text{otherwise} \end{cases} \qquad \text{---(5)}$$

$$E(w(t)v'(t+k)) = 0 \qquad \text{---(6)}$$

In eqs.(4) and (5), the covariance matrices Q and R are positive-(semi)definite matrices. This representation of a linear stochastic system is fairly typical in control theory. The presence of the random disturbances w(t) in eq.(1) implies that the transitions of the state are known in a probabilistic sense only. The presence of the observation equation (2) recognizes the empirical fact that the system state may not be observed as such; only certain linear combinations perturbed by random disturbances v(t) are measured as observation variables.

Based on a set of sequential observations Z(t) = (z(1),.., z(t)), we wish to determine an estimate of x(t), which we shall symbolically represent by $\hat{x}(t/t+k)$. The estimation error will be denoted by

$$\bar{x}(t/t+k) = x(t) - \hat{x}(t/t+k) \qquad \text{---(7)}$$

Depending upon the value of k, the estimation is referred to as prediction or extrapolation ($k < 0$), filtering ($k = 0$), smoothing or interpolation ($k > 0$). In this appendix, we shall be concerned only with the filtering problem. The estimate will be conditionally and unconditionally unbiased, that is, $E(\hat{x}(t/t+k)/Z(t+k)) = E(x(t)/Z(t+k))$ and $E(\hat{x}(t/t+k)) = E(x(t))$, and a linear function of the observation sequence. Of the many possible unbiased linear estimators, we wish to select the one which gives the minimum error variance, that is, the one for which $\mathrm{var}(\bar{x}(t/t+k)/Z(t+k))$ is minimum.

As orthogonal projection theory is needed for the following derivation, we present here without proof some required results (see Deutsch 1965). The linear minimum-variance estimate of x, given a linear observation space Z, is given by the orthogonal projection of x onto Z, that is $\hat{x} = \hat{E}(x/Z)$. Here we have used the symbol $\hat{E}$ rather than E, because the linear minimum-variance estimator is, in general, not the true conditional mean. We could have assumed gaussian distribution, in which case $\hat{E}(x/Z)$ would become $E(x/Z)$; however, we have chosen not to take that approach in order to emphasize that the gaussian distribution is not necessary as long as one remembers that the resulting estimator may not be the best but only the best linear estimator.

If an orthogonal sequence $(a(1),a(2),\ldots,a(m))$ forms a basis for Z, then $\hat{x}$ may be represented by

$$\hat{x} = \hat{E}(x/Z) = \sum_{i=1}^{m} E(x\,a'(i))\,E(a(i)\,a'(i))\,a(i) \qquad \text{---(8)}$$

A final property is needed to obtain a sequential form for the solution. If B is orthogonal to Z, that is $E(B'a(i)) = 0$ for $i=1,2,\ldots,m$, where $(a(1),a(2),\ldots,a(m))$ is an orthogonal basis for Z, then

$$\hat{E}(x/Z,B) = \hat{E}(x/Z) + \hat{E}(x/B) \qquad \text{---(9)}$$

This result is merely a statement of the linearity property of the linear minimum-variance estimator and the orthogonal projection lemma.

A.2 The Discrete One-Stage Predictor

Although we are interested in the filtered estimate of x, i.e., $\hat{x}(t/t) = \hat{E}(x(t)/Z(t))$, we shall begin the derivation by considering one-stage prediction, i.e., $\hat{x}(t+1/t)=\hat{E}(x(t+1)/Z(t))$ To obtain the desired sequential form of our solution, we shall proceed by induction, and assume that $\hat{x}(t/t-1)$ is known. We shall then compute $\hat{x}(t+1/t)$ in terms of $\hat{x}(t/t-1)$ and the new observation $z(t)$. However, $z(t)$ is, in general, not orthogonal to $Z(t-1)$ and before using eq.(9), we must determine the portion of $z(t)$ which is orthogonal to $Z(t-1)$. This amounts to extracting the 'new information' in $z(t)$.

It can easily be shown that

$$I(t/t-1) = \bar{z}(t/t-1) = z(t) - C\hat{x}(t/t-1) \qquad \text{---(10)}$$

is orthogonal to $Z(t-1)$. Note that $\bar{z}(t/t-1)$ represents the 'new information' contained in $z(t)$ since the best estimate of $z(t)$ given $Z(t-1)$, i.e., $C\hat{x}(t/t-1)$, is subtracted from $z(t)$ to obtain $\bar{z}(t/t-1)$. This is just an alternative way of saying that $\bar{z}(t/t-1)$ is orthogonal to $Z(t/t-1)$. The random variable I is known as the innovation.

By using eq.(10), we can write $\hat{x}(t+1/t)$ in terms of the innovation as

$$\begin{aligned}\hat{x}(t+1/t) &= \hat{E}(x(t+1)/Z(t-1),z(t)) \\ &= \hat{E}(x(t+1)/Z(t-1),\bar{z}(t/t-1))\end{aligned}$$

These two expressions are equivalent, since $z(t)-\bar{z}(t/t-1)$ is contained in the observation space $Z(t-1)$, and hence adds nothing additional to the representation of $\hat{x}(t+1/t)$ other than that contained in $Z(t-1)$.

Because $Z(t-1)$ and $\bar{z}(t/t-1)$ are orthogonal, we may use eq.(9) to write $\hat{x}(t+1/t)$ as

$$\hat{x}(t+1/t) = \hat{E}(x(t+1)/Z(t-1)) + \hat{E}(x(t+1)/\bar{z}(t/t-1))$$

Since $\hat{E}(x(t+1)/Z(t-1)) = \hat{x}(t+1/t-1)$, the last expression becomes

$$\hat{x}(t+1/t) = \hat{x}(t+1/t-1) + \hat{E}(x(t+1)/\bar{z}(t/t-1)) \qquad \text{---(11)}$$

This relation indicates that $\hat{x}(t+1/t)$ is obtained by predicting the value of $x(t+1)$, based on previous observations, i.e., $Z(t-1)$, and then correcting the predictions with the new information in the current sample, i.e., $\bar{z}(t/t-1)$. This concept of prediction and correction is important and is a useful interpretation of the Kalman filter algorithm. The development of the algorithm will therefore be presented in terms of this predictor-corrector concept.

Let us consider each of the two terms on the right-hand side of eq.(11) separately. From eq.(1) we know that $x(t+1)$ is given by

$$x(t+1) = Ax(t) + Bu(t) + w(t)$$

Therefore $\hat{x}(t+1/t-1)$, which is equal to $\hat{E}(x(t+1)/Z(t-1))$ by definition, becomes

$$\begin{aligned}\hat{x}(t+1/t-1) &= \hat{E}(Ax(t)+Bu(t)+w(t)/Z(t-1)) \\ &= A\,\hat{E}(x(t)/Z(t-1)) + \hat{E}(w(t)/Z(t-1)) + Bu(t)\end{aligned}$$

By definition, $\hat{E}(x(t)/Z(t-1)) = \hat{x}(t/t-1)$, and we have

$$\hat{x}(t+1/t-1) = A\hat{x}(t/t-1) + \hat{E}(w(t)/Z(t-1)) + Bu(t)$$

Since $Z(t-1)$ depends only on $w(i)$ for $i \leqslant t-1$, and since w is a white-noise process, the expected value of $w(t)$, given $Z(t-1)$, is just the unconditional expectation $E(w(t)) = 0$. Thus the foregoing result becomes

$$\hat{x}(t+1/t-1) = A\hat{x}(t/t-1) + Bu(t) \qquad \text{---(12)}$$

Here we see that the prediction of $x(t+1)$, based on $Z(t-1)$ is obtained by using the transition mechanism, with no forcing function, i.e., by setting $w(t) = 0$, to translate $\hat{x}(t/t-1)$ forward one stage. It follows that

$$\hat{w}(t/t+k) = E(w(t)/Z(t+k)) = 0, \qquad k \leqslant 0 \qquad \text{---(13)}$$

which says that the best filtered or predicted estimate of zero-mean white-noise is zero.

If we substitute eq.(12) into eq.(11), we obtain

$$\hat{x}(t+1/t) = A\hat{x}(t/t-1) + Bu(t) + \hat{E}(x(t+1)/\bar{z}(t/t-1)) \qquad \text{---(14)}$$

Let us consider the last term in this equation. We may use

eq.(8) to write $\hat{E}(x(t+1)/\bar{z}(t/t-1))$ as

$$\hat{E}(x(t+1)/\bar{z}(t/t-1)) = E(x(t+1)\bar{z}'(t/t-1)) \cdot (E(\bar{z}(t/t-1)\bar{z}'(t/t-1)))^{-1}\,\bar{z}(t/t-1) \qquad ---(15)$$

Let us examine each of the terms on the right-hand side of the foregoing equation individually. By substituting eq.(1) for $x(t+1)$, we obtain for the first term

$$E(x(t+1)\bar{z}'(t/t-1)) = E((Ax(t)+Bu(t)+w(t))\bar{z}'(t/t-1)) \qquad ---(16)$$

Now use the definition of $z(t)$ from eq.(2) and the definition of $\bar{z}(t/t-1)$ from eq.(10) to write $\bar{z}(t/t-1)$ as

$$\begin{aligned}\bar{z}(t/t-1) &= z(t) - C\hat{x}(t/t-1)\\ &= Cx(t) + v(t) - C\hat{x}(t/t-1)\\ &= C\bar{x}(t/t-1) + v(t)\end{aligned}$$

where $\bar{x}(t/t-1) = x(t) - \hat{x}(t/t-1)$. Eq.(16) therefore becomes

$$E(x(t+1)\bar{z}'(t/t-1)) = E((Ax(t)+Bu(t)+w(t))(C\bar{x}(t/t-1)+v(t))')$$

which becomes

$$\begin{aligned}E(x(t+1)\bar{z}'(t/t-1)) &= A\,E(x(t)\bar{x}'(t/t-1))C' + A\,E(x(t)v'(t))\\ &+ E(w(t)\bar{x}'(t/t-1))C' + E(w(t)v'(t))\\ &+ B\,E(u(t)\bar{x}'(t/t-1))C' + B\,E(u(t)v'(t))\end{aligned} \qquad ---(17)$$

Since $x(t)$ depends only on $x(t-1)$, $u(t-1)$ and $w(t-1)$, and since w and v are uncorrelated, $E(x(t)v'(t)) = 0$. Because w is a white-noise process and $\bar{x}(t/t-1)$ depends on $w(i)$ for $i \leqslant t-1$, the third term on the right-hand side of the foregoing equation must be zero, since $E(w(t)\bar{x}'(t/t-1)) = 0$. The fourth term on the right-hand side of the equation is zero, since w and v are uncorrelated. Because $\bar{x}(t/t-1)$ depends on $u(i)$ for $i \leqslant t-1$, the fifth term in the above equation is zero. And, since u and v are uncorrelated, $E(u(t)v'(t)) = 0$. Therefore, only the first term remains and we have

$$E(x(t+1)\bar{z}'(t/t-1)) = A\,E(x(t)\bar{x}'(t/t-1))C' \qquad ---(18)$$

This expression may be further simplified by writing $x(t)$ as $\hat{x}(t/t-1)+\bar{x}(t/t-1)$, so that $E(x(t)\bar{x}'(t/t-1))$ becomes

$$\begin{aligned}E(x(t)\bar{x}'(t/t-1)) &= E((\hat{x}(t/t-1) + \bar{x}(t/t-1))\bar{x}'(t/t-1))\\ &= E(\hat{x}(t/t-1)\bar{x}'(t/t-1)) + E(\bar{x}(t/t-1)\bar{x}'(t/t-1))\end{aligned}$$

But the first term is zero because of the orthogonal projection lemma, so that eq.(18) becomes

$$E(x(t+1)\bar{z}'(t/t-1)) = AV\bar{x}(t/t-1)C' \qquad ---(19)$$

where $V\bar{x}(t/t-1) = \mathrm{var}(\bar{x}(t/t-1))$.

In a similar manner, it is possible to show that

$$E(\bar{z}(t/t-1)\bar{z}'(t/t-1)) = CV\bar{x}(t/t-1)C' + R \qquad ---(20)$$

where R = var(v(t)) from eq.(5). If we substitute eqs.(19), (20) and (10) into eq.(15), we obtain

$$\hat{E}(x(t+1)/\bar{z}(t/t-1)) = AV\bar{x}(t/t-1)C'(CV\bar{x}(t/t-1)C' + R)^{-1}. (z(t) - C\hat{x}(t/t-1)) \qquad ---(21)$$

The expression for $\hat{x}(t+1/t)$ is therefore

$$\hat{x}(t+1/t) = A\hat{x}(t/t-1)+Bu(t) + AV\bar{x}(t/t-1)C'(CV\bar{x}(t/t-1)C'+R)^{-1}(z(t)-C\hat{x}(t/t-1)) \qquad ---(22)$$

This result may be written more conveniently as

$$\hat{x}(t+1/t) = A\hat{x}(t/t-1) + Bu(t) + K(t+1,t)(z(t)-C\hat{x}(t/t-1)) \qquad ---(23)$$

where

$$K(t+1,t) = AV\bar{x}(t/t-1)C'(CV\bar{x}(t/t-1)C'+R)^{-1} \qquad ---(24)$$

The quantity K(t+1,t) is referred to as the Kalman gain for one-stage prediction.

Before we can use the above result, it is necessary to find an expression for $V\bar{x}(t/t-1)$, which is needed to compute K(t+1,t). Alternatively, we may determine $V\bar{x}(t+1/t)$.

By combining eqs.(1) and (23), we obtain

$$\begin{aligned}\bar{x}(t+1/t) &= x(t+1) - \hat{x}(t+1/t)\\ &= Ax(t) + Bu(t) + w(t) - A\hat{x}(t/t-1) - Bu(t)\\ &\quad - K(t+1,t)(z(t) - C\hat{x}(t/t-1))\end{aligned}$$

If we substitute eq.(2) for z(t), we obtain

$$\bar{x}(t+1/t) = (A - K(t+1,t)C)\bar{x}(t/t-1) + w(t) - K(t+1,t)v(t) \qquad ---(25)$$

Since $\bar{x}(t+1/t)$ has zero mean (due to the fact that the estimate is unbiased); and $\bar{x}(t/t-1)$, w(t) and v(t) are all uncorrelated, the expression for $V\bar{x}(t+1/t)$ may easily be obtained directly from its definition and eq.(25) as

$$\begin{aligned}V\bar{x}(t+1/t) &= var(\bar{x}(t+1/t)) = var(x(t+1)/Z(t))\\ &= (A - K(t+1,t)C)V\bar{x}(t/t-1)(A - K(t+1,t)C)'\\ &\quad + Q + K(t+1,t)RK'(t+1,t)\end{aligned}$$

where Q = var(w(t)), from eq.(4). This may be written as

$$\begin{aligned}V\bar{x}(t+1/t) &= AV\bar{x}(t/t-1)A' + K(t+1,t)(CV\bar{x}(t/t-1)C'+R)K'(t+1,t)\\ &\quad - K(t+1,t)CV\bar{x}(t/t-1)A' - AV\bar{x}(t/t-1)C'K'(t+1,t) + Q\end{aligned}$$

Substituting eq.(24) for K(t+1,t) and simplifying the result, we find that the expression for the error variance becomes

$$V\bar{x}(t+1/t) = AV\bar{x}(t/t-1)A' + Q - AV\bar{x}(t/t-1)C'(CV\bar{x}(t/t-1)C'+R)^{-1}. CV\bar{x}(t/t-1)A' \qquad ---(26)$$

This equation, which is referred to as the variance algorithm for one-stage prediction, describes how the error variance propagates. Eq.(26), along with eqs.(23) and (24), completely defines the sequential one-stage linear minimum-error-variance predictor.

A.3 The Discrete Kalman Filter

Let us now turn to the basic problem of interest, which is the filter problem. The one-stage predictor is a prelude to our derivation because the filtering solution involves one-stage prediction, which is then corrected with the current data.

If the filtered estimate of x(t), i.e., $\hat{x}(t/t)$, were known then $\hat{x}(t+1/t)$ could be obtained as

$$\hat{x}(t+1/t) = A\hat{x}(t/t) + Bu(t) \qquad \text{---(27)}$$

Since x(t), and hence z(t), depends on w(i) for $i<t$, the observation space Z(t) contains no information about w(t), w being a discrete white-noise process. Hence, to predict x(t+1) based on Z(t), we could predict $\hat{x}(t/t)$ forward one stage, with $\hat{w}(t/t) = 0$. Such an approach would give eq.(27), which we shall use for the following development.

For notational simplicity, we shall write $\hat{x}(t/t)$ as $\hat{x}(t)$. The conditioning space is implicitly assumed as Z(t), unless specified otherwise. In this notation, eq.(27) becomes

$$\hat{x}(t+1/t) = A\hat{x}(t) + Bu(t) \qquad \text{---(28)}$$

We had previously developed eqs.(23), (24) and (26) as a means of obtaining $\hat{x}(t+1/t)$. Clearly, the two estimates of x(t+1), based on Z(t), must be equal, and hence we may use eq.(28) to develop a sequential algorithm for $\hat{x}(t)$ from eqs. (23), (24) and (26). As the first step, we substitute eq.(28) into eq.(23) to obtain

$$A\hat{x}(t) + Bu(t) = A\hat{x}(t/t-1) + Bu(t) + K(t+1,t)(z(t) - C\hat{x}(t/t-1))$$

which simplifies into the following sequence if we re-substitute eq.(28), with $t = t-1$, for the right-hand side elements representing $\hat{x}(t/t-1)$.

$$A\hat{x}(t) = A(A\hat{x}(t-1) + Bu(t-1)) \\ + K(t+1,t)(z(t) - C(A\hat{x}(t-1) + Bu(t-1)))$$

Pre-multiplying both sides by A^{-1}, which is assumed to exist, we obtain

$$\hat{x}(t) = A\hat{x}(t-1) + Bu(t-1) \\ + A^{-1}K(t+1,t)(z(t) - C(A\hat{x}(t-1) + Bu(t-1)))$$

To simplify this expression, let us define

$$K(t) = A^{-1}K(t+1,t)$$

where K(t) is the Kalman gain for optimal filtering. This may be written as

$$K(t) = V\bar{x}(t/t-1)C'(CV\bar{x}(t/t-1)C' + R)^{-1} \qquad ---(29)$$

by using the definition of K(t+1,t) given in eq.(24). Therefore $\hat{x}(t)$ becomes

$$\hat{x}(t) = A\hat{x}(t-1) + Bu(t-1) + K(t)(z(t) - C(A\hat{x}(t-1) + Bu(t-1))) \qquad ---(30)$$

Eq.(30) is the most convenient form for the estimator equation for the Kalman filter which, along with eqs.(29) and (26), completely specifies the solution to the linear minimum-variance filtering problem.

The Kalman filter algorithm can be put into a more computationally tractable form by finding an alternative and simpler expression for the variance of the filtering error $V\bar{x}(t)$, which can be used to assess the quality of the estimation procedure. The variance $V\bar{x}(t/t-1)$ is referred to as the à priori variance, since it is the variance of the estimate of x(t) before the observation z(t) is received. By the same reasoning, $V\bar{x}(t)$ is called the à posteriori variance.

To find an alternative expression for $V\bar{x}(t)$, let us make use of the fact that $\hat{x}(t/t-1) = A\hat{x}(t-1)+Bu(t-1)$ and therefore eq.(30) can be written as

$$\hat{x}(t) = \hat{x}(t/t-1) + K(t)(z(t)-C\hat{x}(t/t-1)) \qquad ---(31)$$

This may be simplified by using the innovation, eq.(10), to obtain

$$\hat{x}(t) = \hat{x}(t/t-1) + K(t)\bar{z}(t/t-1) \qquad ---(32)$$

Therefore $\bar{x}(t)$ becomes

$$\begin{aligned}\bar{x}(t) &= x(t) - \hat{x}(t)\\ &= x(t) - \hat{x}(t/t-1) - K(t)\bar{z}(t/t-1)\\ &= \bar{x}(t/t-1) - K(t)\bar{z}(t/t-1)\end{aligned}$$

Therefore $V\bar{x}(t)$ is

$$\begin{aligned}V\bar{x}(t) = {} & V\bar{x}(t/t-1) + K(t)E(\bar{z}(t/t-1)\bar{z}'(t/t-1))K'(t)\\ & - K(t)E(\bar{z}(t/t-1)\bar{x}'(t/t-1))\\ & - E(\bar{x}(t/t-1)\bar{z}'(t/t-1))K'(t)\end{aligned}$$

If we substitute eq.(29) for K(t), and eqs.(19) and (20) for $E(\bar{x}(t/t-1)\bar{z}'(t/t-1))=E(x(t)\bar{z}'(t/t-1))$ and $E(\bar{z}(t/t-1).\bar{z}'(t/t-1))$, respectively, into the foregoing equation, then we obtain

$$V\bar{x}(t) = V\bar{x}(t/t-1)-V\bar{x}(t/t-1)C'(CV\bar{x}(t/t-1)C'+R)^{-1}CV\bar{x}(t/t-1)$$

which may be written as

$$V\bar{x}(t) = (I - K(t))V\bar{x}(t/t-1) \qquad ---(33)$$

by using eq.(29) for $K(t)$. This last expression gives the filtered error-variance as a function of the one-period prediction error variance.

An advantage of using $V\bar{x}(t)$ is that it allows the transition expression for $V\bar{x}(t/t-1)$ in eq.(26) to be considerably simplified. Let us rewrite eq.(26) as

$$V\bar{x}(t+1/t) = A(V\bar{x}(t/t-1) - V\bar{x}(t/t-1)C'(CV\bar{x}(t/t-1)C'+R)^{-1}CV\bar{x}(t/t-1))A' + Q$$

By making use of eq.(29) for $K(t)$, we may write this expression as

$$\begin{aligned} V\bar{x}(t+1/t) &= A(V\bar{x}(t/t-1)-K(t)CV\bar{x}(t/t-1))A' + Q \\ &= A((I - K(t)C)V\bar{x}(t/t-1))A' + Q \end{aligned}$$

The quantity within brackets can be recognized as $V\bar{x}(t)$, so that we have

$$V\bar{x}(t+1/t) = AV\bar{x}(t)A' + Q \qquad \text{---(34)}$$

Eqs.(29), (33) and (34) comprise the final form of the discrete Kalman filter algorithm. The entire model is summarized in Table A.1 for easy reference.

Table A.1
<u>Summary Of The Discrete Kalman Filter Algorithm</u>

Message model:	$x(t+1) = Ax(t) + Bu(t) + w(t)$
Observation model:	$z(t) = Cx(t) + v(t)$
Prior statistics:	$E(w(t)) = 0$ $E(v(t)) = 0$ $cov(w(t),w(t+k)) = Q$ $cov(v(t),v(t+k)) = R$ $cov(w(t),v(t+k)) = 0$
Filter estimator:	$\hat{x}(t/t-1) = A\hat{x}(t-1) + Bu(t-1)$ $\hat{x}(t) = \hat{x}(t/t-1) + K(t)(z(t)-C\hat{x}(t/t-1))$
A priori variance:	$V\bar{x}(t+1/t) = AV\bar{x}(t)A' + Q$
Gain equation:	$K(t+1) = V\bar{x}(t+1/t)C'(CV\bar{x}(t+1/t)C'+R)^{-1}$
A posteriori variance:	$V\bar{x}(t+1) = (I-K(t+1)C)V\bar{x}(t+1/t)$
Initial conditions:	$V\bar{x}(0/0) = V\bar{x}(0) = Vx(0)$

It is to be noted that the variance and the gain equations are independent of the observation sequence and can be precomputed. The researcher is cautioned to enforce the symmetry of $V\bar{x}(t)$ and $V\bar{x}(t/t-1)$ at each step when making practical applications of the Kalman filter algorithm. Although eqs.(33) and

(34) are clearly symmetric in theory, finite word length and round-off error in the implementation of the equations can cause $V\bar{x}(t)$ and $V\bar{x}(t/t-1)$ to become non-symmetric, resulting in serious degradation of performance and even instability of the estimation procedure. This difficulty is most pronounced for small (relative to R) values of Q.

An examination of the table reveals that an inbuilt predictor-corrector concept is present in the Kalman filter. The previous estimate $\hat{x}(t-1)$ is predicted forward one stage and used to obtain the "best estimate" of the new observation z(t) based on all previous observations. The error between this "best estimate" of the current observation and the actual observation,namely, I(t/t-1) or $\bar{z}(t/t-1)$, represents new information, i.e., the portion of z(t) which is orthogonal to Z(t-1). The error is weighted by K(t) which is based on a knowledge of the input, measurement and estimation error variances, to obtain a correction which is then added to the predicted value to update the new estimate.

A.4 The Kalman-Bucy Theorem

We now indicate a very important theorem due to Kalman and Bucy (1961) which can be stated as follows.

Given, for the linear stochastic system of equations (1) and (2), that:

(1) the matrices A, B and C are finite and time-invariant,
(2) the error covariance matrices Q and R are finite, positive-definite and time-invariant,
(3) the linear system is completely controllable with respect to w(t), and
(4) the linear system is perfectly observable.

Then, it follows:

(i) the Kalman filter is uniformly asymptotically stable, and
(ii) every solution $V\bar{x}(t)$ of the set of recurrence relationships, given by eqs.(29), (33) and (34), which results from initializing with any symmetric non-negative matrix Vx(0) tends uniformly to a unique finite positive-definite matrix V*.

APPENDIX B

THE DISCRETE MAXIMUM PRINCIPLE

The discrete version of the maximum principle is usually stated in three parts. These parts constitute a set of necessary conditions for minimizing an objective function, which is a sum of functions, for different periods subject to a set of difference equations. For the given optimal control problem, the specific Hamiltonian is first defined, and minimized by the optimal value of the control vector, given the optimal values of the state and co-state variables. Second, the state variables (corresponding to the optimal values of the control variables) and the co-state variables are made to satisfy, respectively, the difference equations obtained by differentiating the Hamiltonian with respect to the co-state and state variables. Third, if the state vector in the terminal period has to satisfy certain end-point restrictions, then the terminal co-state vector must satisfy the transversality condition. The optimal control that results will be seen to depend on the solution of two difference equations: a 'Riccati' equation which depends on both the system itself and the weighting matrices in the cost functionals, and a 'tracking' equation which depends on the solution of the Riccati equation as well as the nominal state and control trajectories that are to be tracked. The solution to this problem, based on Chow (1975), is the maximum principle algorithm.

B.1 The Problem

Our system of interest takes the general form

$$x(t+1) = Ax(t) + Bu(t+1) + Cz(t+1) \qquad \text{---(1)}$$

with the initial condition

$$x(0) = E \qquad \text{---(2)}$$

Here, $x(t)$ is the n-dimensional state vector at time $\underline{t}$, $u(t)$ is the s-dimensional control vector at time $\underline{t}$ and $z(t)$ is an m-dimensional vector representing, at time $\underline{t}$, $\underline{m}$ exogenous variables which are known for all $\underline{t}$ but which cannot be controlled by the policy planner. The existence of such exogenous uncontrollable variables is invariably the case with most econometric models. A, B and C are appropriately dimensioned time-invariant matrices. The elements of A, B and C can be estimated by standard econometric and state-space techniques and are therefore assumed to be known.

Let $\bar{\bar{x}}(t)$ and $\bar{\bar{u}}(t)$ be the nominal state and control vectors that we would like to track. We assume that $\bar{\bar{x}}(t)$, $\bar{\bar{u}}(t)$ and $z(t)$ have been specified for the entire planning period, $t=1,\ldots,N$. It must be noted that the $\bar{\bar{x}}(t)$ need not be the result of substituting the $\bar{\bar{u}}(t)$ and $z(t)$ into eq.(1), i.e., given $z(t)$, the choice of the vectors $\bar{\bar{x}}(t)$ and $\bar{\bar{u}}(t)$ need not be simultaneously consistent with eq.(1), but can be chosen independently. The cost functional is given by

$$J = \frac{1}{2} \sum_{t=1}^{N} ((x(t)-\bar{\bar{x}}(t))'Q(x(t)-\bar{\bar{x}}(t)) + (u(t)-\bar{\bar{u}}(t)'R(u(t)-\bar{\bar{u}}(t))) \qquad \text{---(3)}$$

where Q is an nxn positive-(semi)definite matrix and R is an sxs positive-definite matrix.

The optimal control problem is to find a control sequence, $(u^*(t), t=1,\ldots,N)$, such that

$$x^*(0) = E \qquad \text{---(4)}$$

$$x^*(t+1) = Ax^*(t) + Bu^*(t+1) + Cz(t+1) \qquad \text{---(5)}$$

and the cost functional (3) is minimized.

B.2 The Necessary Conditions

We begin by expressing the necessary conditions set forth by the maximum principle. The Hamiltonian becomes

$$\begin{aligned} H(x(t),u(t),p(t+1)) &= (1/2)(x(t)-\bar{\bar{x}}(t))'Q(x(t)-\bar{\bar{x}}(t)) \\ &+ (1/2)(u(t)-\bar{\bar{u}}(t))'R(u(t)-\bar{\bar{u}}(t)) \\ &+ p'(t+1)(Ax(t)+Bu(t+1)+Cz(t+1)) \end{aligned} \qquad \text{---(6)}$$

The necessary conditions provide the equations describing the optimal trajectories for $x^*(t)$, $u^*(t)$ and $p^*(t)$. The canonical equations for the problem are determined by the first necessary condition

$$x^*(t+1) = \partial H/\partial p(t+1) = Ax^*(t) + Bu^*(t+1) + Cz(t+1) \qquad \text{---(7)}$$

$$p^*(t) = \partial H/\partial x(t) = Q(x^*(t)-\bar{\bar{x}}(t)) + A'p^*(t+1) \qquad \text{---(8)}$$

and these are subject to the split-boundary conditions

$$x^*(0) = E \qquad \text{---(9)}$$

$$p^*(N) = Q(x^*(N)-\bar{\bar{x}}(N)) \qquad \text{---(10)}$$

Note that eq.(10) is the result of the second two-point boundary condition imposed upon the costate variables by the maximum principle. The endpoint cost $K(x^*(N))$ for our problem is given by

$$K(x^*(N)) = (1/2)(x^*(N)-\bar{\bar{x}}(N))'Q(x^*(N)-\bar{\bar{x}}(N)) \qquad \text{---(11)}$$

and therefore

$$p^*(N) = (\partial/\partial x(N))\, Kx^*(N) = Q(x^*(N)-\bar{\bar{x}}(N))$$

Finally, the minimization of the Hamiltonian with respect to the control variables yields

$$\partial H/\partial u(t) = R(u^*(t)-\bar{\bar{u}}(t)) + B'p^*(t) = 0 \qquad ---(12)$$

B.3 <u>The Solution</u>

Eqs.(7), (8) and (12) constitute a set for the unknowns $x^*(t)$, $u^*(t)$ and $p^*(t)$, $t=1,...,N$. The dynamic situation is treated by defining the variables $x^*(t)$, $u^*(t)$ and $p^*(t)$ at different points in time as separate variables. The dynamic nature of the problem, thus, gives a special structure to the simultaneous equations (7), (8) and (12), requiring, thereby, special methods for their efficient solution.

To solve this system of equations, we start with $t = N$, and repeat the following three steps backward in time for $t = N-1,..., 1$. First, eq.(8) is used to express $p^*(t)$ as a function of $x^*(t)$. Second, the result, together with eqs.(7) and (12), is used to solve for $u^*(t)$. Third, the results of the first two steps together with eq.(7) are used to express $x^*(t)$ and $p^*(t)$ as linear functions of $x^*(t-1)$. Using the last linear function, we express $p^*(t-1)$ as a linear function of $x^*(t-1)$ and we are back in step one of the next round.

Step one: For $t = N$, using eq.(10) gives

$$p^*(N) = Qx^*(N) - Q\bar{\bar{x}}(N) \qquad ---(13)$$

We set the following

$$H(N) = Q \qquad ---(14)$$

$$h(N) = Q\bar{\bar{x}}(N) \qquad ---(15)$$

Therefore, we have

$$p^*(N) = H(N)x^*(N) - h(N) \qquad ---(16)$$

Step two: For $t = N$, using eq.(12) gives

$$Ru^*(N) - R\bar{\bar{u}}(N) + B'p^*(N) = 0 \qquad ---(17)$$

Substituting eq.(16) in eq.(17) above, we get

$$Ru^*(N) - R\bar{\bar{u}}(N) + B'(H(N)x^*(N)-h(N)) = 0 \qquad ---(18)$$

From eq.(7) we have without any loss of generality

$$x^*(N) = Ax^*(N-1) + Bu^*(N) + Cz(N) \qquad ---(19)$$

Substituting eq.(19) in eq.(18), we get

$$Ru^*(N) - R\bar{\bar{u}}(N) + B'(H(N)(Ax^*(N-1)+Bu^*(N)+Cz(N)) - h(N)) = 0 \qquad ---(20)$$

Solving eq.(20) uniquely in terms of $u^*(N)$, we obtain

$$u^*(N) = G(N)\, x^*(N-1) + J(N) \qquad ---(21)$$

where

$$G(N) = -(R + B'H(N)B)^{-1} B'H(N)A \qquad ---(22)$$

and

$$J(N) = -(R + B'H(N)B)^{-1} (B'H(N)Cz(N)-B'h(N)-R\bar{\bar{u}}(N)) \qquad ---(23)$$

The matrix $(R+B'H(N)B)$ is assumed to be non-singular.

Step three: Substituting the linear feedback rule given by eq.(21) into eq.(7), we solve for $x^*(N)$ as a function of $x^*(N-1)$ by means of

$$x^*(N) = (A + BG(N))x^*(N-1) + BJ(N) + Cz(N) \qquad ---(24)$$

Substituting eq.(24) above into eq.(16), we express $p^*(N)$ as a function of $x^*(N-1)$ by means of

$$p^*(N) = H(N)\,(A+BG(N))x^*(N-1) + H(N)\,(BJ(N)+Cz(N)) - h(N) \qquad ---(25)$$

Having solved for $p^*(N)$ in terms of $x^*(N-1)$, we substitute eq.(25) into eq.(8) to obtain an equation analogous to eq.(13) in step one, by means of

$$\begin{aligned} p^*(N-1) &= Qx^*(N-1) - Q\bar{\bar{x}}(N-1) + A'p^*(N) \\ &= H(N-1)x^*(N-1) - h(N-1) \end{aligned} \qquad ---(26)$$

where

$$H(N-1) = Q + A'H(N)(A + BG(N)) \qquad ---(27)$$

and

$$h(N-1) = Q\bar{\bar{x}}(N-1) - A'H(N)\,(BJ(N) + Cz(N)) + A'h(N) \qquad ---(28)$$

The development from step two onwards can now be followed with N-1 replacing N, and so on.

B.4 The Algorithm

To apply this solution to the deterministic control problem in the form of an algorithm, we combine the pair of equations, (22) and (27), to form a matrix difference equation in the nxn matrix $H(t)$:

$$\begin{aligned} H(t-1) &= Q + A'H(t)\,(A + BG(t)) \\ &= Q + A'H(t)A - A'H(t)B\,(R + B'H(t)B)^{-1} B'H(t)A \end{aligned} \qquad ---(29)$$

Eq.(29) is known as a matrix Riccati equation and is an example of a nonlinear difference equation. As can be noticed, the solution to this depends upon both the system itself and the weighting matrices in the cost functional. Eq. (29) is solved backward in time with eq.(14) as the initial condition.

Then, having obtained H(N), H(N-1),..., H(1), we use eq.(22) to obtain G(N), G(N-1),..., G(1) forward in time. The matrices H(t) are symmetric because H(N) is symmetric; given symmetric Q and H(N), H(N-1) is also symmetric, and so on.

Similarly, the intercepts J(t) in the optimal control equation (21) are obtained by combining the pair of equations, (23) and (28), to form a vector difference equation in the nx1 vector h(t):

$$\begin{aligned} h(t-1) &= Q\underline{\bar{\bar{x}}}(t-1) - A'H(t)(BJ(t)+Cz(t)) + A'h(t) \\ &= Q\bar{\bar{x}}(t-1) - A'(H(t)Cz(t) - h(t)) \\ &\quad + A'H(t)B(R+B'H(t)B)^{-1}(B'H(t)Cz(t)-B'h(t)-R\bar{\bar{u}}(t)) \end{aligned} \qquad ---(30)$$

Eq.(30) is the tracking equation for the optimal control problem, the solution of which depends on the system, the weighting matrices in the cost functional, the nominal state and control trajectories that are to be tracked, as well as the solution of the Riccati equation. Eq.(30) is solved backward in time with eq.(15) as the initial condition, and in conjunction with the stored H(t), t=N, N-1,..., 1. Having thus obtained h(N), h(N-1),..., h(1), we use eq.(23) to obtain J(N),...,J(1).

Having thus computed G(t) and J(t), t=1,...,N, we set the optimal control u*(t) by the feedback rule given by eq.(21).

B.5 The Discrete Maximum Principle

Table B.1 summarizes a special form of the Pontryagin maximum principle upon which we based our effort.

Table B.1
Summary Of The Discrete Maximum Principle (Special Case)

Transition equation:	$x(t+1) = f(x(t),u(t),t)$
Initial conditions:	$x(t(0)) = E$
Cost functional:	$J = K(x(N),N) + \sum_{t=0}^{N-1} z(x(t),u(t),t)$
Hamiltonian:	$H = z(x(t),u(t),t)+p'(t+1)\, f(x(t),u(t),t)$
First necessary condition:	$x(t+1) = \partial H/\partial p(t+1);\quad p(t) = \partial H/\partial x(t);\quad \partial H/\partial u(t) = 0.$
Two-point boundary conditions:	$p(t(0)) = \partial K(x(t(0),0))/\partial x(t(0))$ $p(t(N)) = \partial K(x(t(N),N))/\partial x(t(N))$
Sufficient condition:	$P = \begin{bmatrix} \partial^2 H/\partial x^2 & (\partial/\partial u)(\partial H/\partial x) \\ ((\partial/\partial u)(\partial H/\partial x))' & (\partial^2 H/\partial u^2) \end{bmatrix}$ P is positive-definite along $x^*(t+1) = f(x^*(t),u^*(t),t)$ where x* and u* are the optimal x and u.

We must not expect this special case of the maximum principle to solve all types of optimization problems. There are many other considerations imposed by control variable inequality constraints, state variable inequality constraints, initial and terminal manifold constraints, nonfixed terminal times and singular solution possibilities which cannot be dealt with by the maximum principle in its present form. Moreover, the maximum principle conditions are, in general, not sufficient, nor do they necessarily yield a unique solution or a global maximum. However, it has been shown that the conditions are necessary and sufficient if the Hamiltonian is linear in the control variables or if the maximized Hamiltonian is a concave function of the state variables.

APPENDIX C

CONTROLLABILITY AND OBSERVABILITY

When we select a particular system in the modelling of economic phenomena, we must be aware of the broader implications behind such a specification and, subsequent, construction of the system. It would be self-defeating to construct and discuss the dynamic behaviour of a macroeconomic model for a certain policy implication study, if the concerned dynamic model were incapable of achieving the desired policy objectives, due to an inherent limitation in the set of differential or difference equations used to represent the macroeconomic system, regardless of the type or number of instruments that were employed. It is, thus, of extreme importance to be able to determine whether or not the dynamic system possesses the necessary properties without actually solving the equations of the model. From the viewpoint of the stability property of dynamic systems, Lyapunov partially answered this question (see LaSalle and Lefschetz 1961) through the specification of his famous 'stability i.s.L.' conditions (where i.s.L. refers to "in the sense of Lyapunov"). However, there are other equally powerful properties besides stability, and two of the most important ones are called controllability (or reachability) and observability (or reconstructibility). These concepts, apart from highlighting the fact that dynamic systems do have properties that are absent in systems of algebraic equations, also establish an important link between systems theory and macroeconomic stabilization theory. For example, it has been shown (see Aoki 1976) that if a dynamic economic model has the proper controllability and observability properties, then under certain technical conditions it is possible to find a stabilization policy for the dynamic model. Thus, the absence of these properties can immediately alert us to the limitations of the system's performance.

Controllability, which ensures that an instrument time-path can be found to achieve a given target value without actually choosing the instrument time-path beforehand, is therefore an existence condition for a dynamic policy. We say that a deterministic system

$$x(t+1) = f(x(t),u(t),t)$$

is (completely) controllable at time $t(0)$ if for each pair of states $x(0)$ and $x(1)$, there exists a feasible control vector $u(.)$ on some finite time interval $t(0) \leqslant t \leqslant t(1)$ such that the system moves from the state $x(0)$ at time $t(0)$ to $x(1)$ at time $t(1)$. Thus, the controllability of a model has to do with the effectiveness of the instruments in influencing or modifying the dynamic behaviour of the system. The problems associated with the existence and design of policy instruments to achieve a set of targets were initially considered by Tinbergen (1955) for a static economic system. Controllability generalizes his approach to dynamic models. Therefore, the concept of controllability ranks equally with that of stability in as far as its importance in ascertaining the capabilities of dynamic models is concerned.

For dynamic models, it is also pertinent to enquire if targets, once attained, can be tracked, i.e., if the state vector can be made to track the target vector if the components of the latter change dynamically over time and also if there is a policy which can accomplish this by means of stable instruments. Yet another important question is whether some outputs or targets can be changed keeping the rest of the target values unchanged. This concept is referred to as the decoupling problem in control literature (see Falb and Wolovich 1967). This turns out to be a generalized version of the so-called assignment problem of Mundell (1962,1968) where subsets of instruments are assigned to subsets of targets in a one-to-one correspondence such that the instruments in each subset influence only a limited number of targets so as to achieve non-interaction of some grouping of targets. For linear dynamic systems, all these questions are closely connected with the concept of controllability.

Observability is also equally important and this concept can be established as the dual to that of controllability. More directly, the observability property has to do with the ability to recover unobservable systems' data from a set of observed data. Such a property is extremely important in giving an operational definition to variables that are not directly available to model builders.

In defining state observability, we have a dynamic state transition equation

$$x(t+1) = f(x(t),u(t),t)$$

as well as an observation equation

$$z(t) = h(x(t),u(t),t)$$

and this combined system is said to be observable at time $t(0)$ if for each feasible instrument path over some time interval $t(0) \leqslant t \leqslant t(1)$, the observation record over the same time interval uniquely determines $x(t(0))$.

Thus, the study of controllability and observability is a necessary preliminary step in the discussion of stabilization and optimization topics in linear (as well as nonlinear) dynamic economic models. Although we do not have as extensive a connection between the controllability or observability property and other desired properties described above for nonlinear systems as in linear ones, these properties are still very important in the intertemporal optimization or stabilization considerations of nonlinear dynamic systems.

C.1 Controllability Of Dynamic Economic Systems

C.1.1 State-Space Controllability

Given an economic system and a set of instruments (also called the control vector) of a specified type generally varying with time (such as piecewise continuous in time $\underline{t}$ or differentiable), the question that needs to be asked is whether or not the economic system in some initial state $x(t(0))$ can be brought to a desired target state $\underline{x}$ in some finite time using admissible controls?

If the answer is in the affirmative, we say that the system is controllable with respect to the admissible set of instrument values. We say that controllability is complete if the system in an arbitrary initial state and initial time can be brought to an arbitrarily specified target state in a finite time. Therefore, we must select a set of instruments so that the system is uniformly completely controllable. We now proceed with providing precise definitions of controllability.

Theorem C.1 The n-dimensional linear time-invariant discrete-time system

$$x(t+1) = Ax(t) + Bu(t) \qquad ---(1)$$

is completely controllable if and only if the rank of the controllability matrix

$$(B, AB, A^2B, \ldots, A^{n-1}B)$$

is equal to n. For a proof see Aoki (1976).

C.1.2 Path controllability

Some economists dismiss the concept of (point) controllability as being of limited interest for the theory of economic policy (see Nyberg and Vioti 1976), while others regard it as crucial (see Buiter and Gersovitz 1979). Be it as it may, policy makers are interested not only in achieving desired targets at a specified point in time, but also keeping targets on some desired time paths once achieved. For example, they would be interested not only in achieving a low rate of inflation but also in manipulating instruments which are either numerous or powerful enough to keep the economy in the desired low state, indeed, as long as possible, once it gets there. Thus, although point controllability is important, it does not contain the notion associated with guiding the economy along an equilibrium or trend or desired growth path. This is because point controllability focuses entirely on moving the economy to a prespecified state at a prespecified time but expresses nothing about the economy once it reaches its target. This type of target variable manipulation requires more than the condition that dynamic systems are completely controllable (see Preston 1974).

Since it is important that instruments have this added capability, it is essential to describe the conditions necessary for policy makers to be able to force the target variables to track their desired trajectories. This condition is known as path controllability. This "perfect controllability" concept asks whether a given set of policy instruments is capable of guiding a given set of target variables along arbitrarily specified time paths. It generalizes the original Tinbergen conditions (Tinbergen 1955) which is the condition for the existence of instrument variables to achieve assigned (or desired) fixed target variable values in static macroeconomic models. Thus path controllability may be considered as the dynamic analogue of Tinbergen's conditions in the theory of economic policy and can be used to analyze dynamic policy trade-off questions in situations where there are no policy coordination problems.

Theorem C.2 The n-dimensional linear time-invariant discrete-time system given by eq.(1) is path controllable if and only if the matrix

$$\begin{bmatrix} B & AB & A^2B & \cdots & \cdots & \cdots & A^{2n-2}B \\ 0 & B & AB & \cdots & \cdots & \cdots & \cdots \\ \cdot & \cdot & \cdot & \cdot & \cdot & \cdot & \cdot \\ \cdot & & & \cdot & & & \\ 0 & 0 & 0 & B & \cdots & \cdots & A^{n-1}B \end{bmatrix}$$

has rank n^2. For a proof see Aoki (1976).

C.1.3 Perfect output controllability

Instead of requiring that a state vector be path controllable, we sometimes may require only that an output vector track its desired trajectories (including stationary values) for all time. Conditions for following desired output time paths have been studied in control theory, using either algebraic or geometric approaches. These conditions are referred to as perfect output controllability (Brockett and Mesarovic 1965) or functional reproducibility (Basile and Marro 1971).

Theorem C.3 The n-dimensional linear time-invariant discrete-time system given by eq.(1) together with its m-dimensional observation (output) equation

$$z(t) = Cx(t) + Du(t) \qquad \text{---(2)}$$

is perfectly output controllable if and only if the matrix

$$\begin{bmatrix} D & CB & CAB & \cdots & \cdots & CA^{n-1}B & \cdots & 0 \\ 0 & D & CB & & & & & \\ \cdot & & \cdot & & \cdot & & & \\ \cdot & & & & \cdot & & & CA^{n-2}B \\ 0 & & & & D & CB & & CA^{n-1}B \end{bmatrix}$$

has rank m(m+1). For a proof see Aoki (1981).

C.1.4 Sufficient conditions for controllability

Since all the rank conditions described above are too cumbersome to actually verify, Aoki and Canzoneri (1979) provided an alternative set of sufficient conditions which, though stronger than necessary, are very convenient to work with and seem to cover most macroeconomic applications.

<u>Theorem C.4</u> Let the state-space representation of an econometric model be given by the following pair of equations:

State equation: $x(t+1) = Ax(t) + B(1)\,u1(t) + B(2)\,u2(t)$
Target equation: $z(t) = Cx(t) + D(1)\,u1(t) + D(2)\,u2(t)$

where <u>x</u> is an n-dimensional state vector, <u>u1</u> is an r-dimensional control vector consisting of the policy variables whose potential is currently being assessed, <u>u2</u> is a vector of exogenous variables, intercept terms and remaining policy variables (if any) and <u>z</u> is an m-dimensional target (observation) vector.

The target vector <u>z</u> is said to be perfectly (or path) controllable by <u>u1</u>, i.e., there exists a path u1(t) that will guide z(t) along any differentiable time path T(t) in m-dimensional space, if any <u>one</u> of the following four conditions is satisfied:

(i) $|D(1)| \neq 0$,
(ii) $D(1) = 0$; $|CB(1)| \neq 0$,
(iii) $|D(1) - CA^{-1}B(1)| \neq 0$,
(iv) rank D(1) is <u>l</u>, which is less than m=dim z, and there exists a nonsingular matrix P, such that

$$\begin{bmatrix} PC & PD(1) \end{bmatrix} = \begin{bmatrix} C(1) & D(3) \\ C(2) & 0 \end{bmatrix}$$

such that

$$\begin{bmatrix} D(3) \\ C(2)B(1) \end{bmatrix}$$

is invertible. The matrix P represents a series of elementary row operations to reduce D(1) to the form

$$\begin{bmatrix} D(3) \\ 0 \end{bmatrix}$$

For proof see Aoki and Canzoneri (1979).

Application:

As we wanted to examine whether our resolved 20-equation model was path controllable or not, we rewrote its state-space form as follows:

State equation: $x(t) = Ax(t-1) + B1\,u1(t) + B2\,u2(t)$
Target equation: $z(t) = Cx(t)$

where <u>x</u> and <u>z</u>, and therefore the matrices A and C, conformed to their specifications. However, our vectors <u>u1</u> and <u>u2</u> were:

u1: (ITR M1 T GDE DEF WPF WPNF XR AF ANF)'

u2: (R MDL CR PM t 1)'

The state-space matrices, B1 and B2, are provided in Table C.1. It is noticed that the second of the Aoki-Canzoneri conditions is indeed satisfied with D(1)=0 (by definition) and $|C\,B1| \neq 0$.(The product of these matrices,denoted by H,is provided in Table C.2).Thus our model is shown to be path controllable.

Table C.1
State-Space Matrices
Of Potential Control Variables
And Exogenous Variables

B1	COL1	COL2	COL3	COL4	COL5	COL6	COL7	COL8	COL9	COL10
	0.3307	0	0	0	0	0	0	0	0	0
	0	0	0	0	0	0	0	0	1.465	0
	0	0	0	0	0	0	0	0	0	11.5406
	0	-1.3856E-19	-1.2962E-14	0	0	-1.8783E-15	-1.0527E-15	0	128.098	472.253
	0.252721	0	0	0	0	0	0	0	0	213.72
	-.000021734	0.000039	3.6484	0	0	0.5287	0.2963	0	0	-0.01838
	0.310276	-.000854763	-79.962	0.1987	0	-11.5875	-6.49401	0	11.6441	43.3306
	0.893697	-.000854763	-79.962	0.1987	0	-11.5875	-6.49401	0	139.742	729.304
	0.858664	-.000821256	-76.8275	0.190911	0	-11.1333	-6.23944	0	134.264	700.715
	0.567319	-.000542604	-50.7599	0.126135	0	-7.35576	-4.1224	0	88.7085	462.963
	0.000007402	-7.0795E-09	-.000662277	.0000016457	0	-.000095972	-.000053786	0	-.000013403	0.00172404
	0.373744	-.000357462	-33.4401	0.0830964	0	-4.8459	-2.71579	5.5511E-17	18.0252	155.999
	0.411396	0.0303085	2835.32	0.0952739	0	410.875	230.267	-1.0582	24.2373	177.546
	0.0376521	0.030666	2868.76	0.0121775	0	415.721	232.983	-0.0582011	6.21201	21.5472
	0.143461	1.34539	125859	0.1987	0	18238.6	10221.5	0	139.742	94.8463
	0.158123	1.48288	173241	0.219007	0	20102.6	11266.1	0	154.024	104.54
	0.00464881	0.0435968	5093.28	0.00643881	0	591.016	331.224	0	4.5283	3.07346
	0.0114482	0.107362	40132.6	0.0158563	0	1455.44	815.675	0	11.1514	7.56874
	0.00289791	0.0271768	2542.35	0.00401374	0	368.42	206.474	0	2.82279	1.9159
	-0.0096973	-0.0909418	-37581.6	0.986569	1	-1232.84	-690.925	0	-9.44592	-6.41117

B2	COL1	COL2	COL3	COL4	COL5	COL6
ROW1	0	0	0	0	0	95.1237
ROW2	0.8031	0	0	0	1.921	-2.3764
ROW3	0.3454	0	0	0	1.224	-0.013
ROW4	84.3564	-5.0677E-15	0	0	218.058	-401.855
ROW5	6.39646	-22.823	0	0	22.6673	277.34
ROW6	-.000550096	0.00196278	0	0	-0.00194938	-0.0355512
ROW7	7.68005	-0.0430182	0	0	19.8642	324.925
ROW8	98.4329	-22.866	0	0	260.589	295.534
ROW9	94.5743	-21.9697	0	0	250.374	387.804
ROW10	62.4853	-14.5154	0	0	165.422	-34.3089
ROW11	.0000442517	-.000189386	0	0	0.000165278	0.0142206
ROW12	14.5502	-9.56257	0	6.3998E-13	40.1812	529.982
ROW13	18.6005	-9.41795	0	-609.989	50.612	708.485
ROW14	4.05026	0.144622	0	-609.989	10.4309	178.504
ROW15	79.4441	44.8871	0	0	193.298	-931.659
ROW16	87.5633	49.4746	0	0	213.053	-1026.87
ROW17	2.57436	1.45455	0	0	6.26377	292.765
ROW18	6.33964	3.58199	0	0	15.4252	420.516
ROW19	1.60477	0.90672	0	0	3.90462	76.6646
ROW20	-5.37005	-3.03416	-1	0	-13.0661	-204.415

Table C.2
The Controllability Criterion Matrix

H

COL1	COL2	COL3	COL4	COL5	COL6	COL7	COL8	COL9	COL10
-.000021265	.0000381576	3.56959	0	0	0.51728	0.2899	0	0	-0.0179829
0.904243	-.000864849	-80.9055	0.201045	0	-11.7243	-6.57064	0	141.391	737.91
0.572369	-.000547433	-51.2117	0.127257	0	-7.42123	-4.15909	0	89.498	467.083
.0000075293	-7.2013E-09	-.000673668	0.000001674	0	-.000097623	-.000054711	0	-.000013633	0.0017537
0.383686	-.000366971	-34.3256	0.0853067	0	-4.9748	-2.78803	5.6988E-17	18.5047	160.149
0.44546	0.0328181	3070.09	0.103163	0	444.895	249.333	-1.14582	26.2441	192.247
0.0374149	0.0304728	2850.69	0.0121008	0	413.102	231.515	-0.0578344	6.17288	21.4115
0.143963	1.3501	126300	0.199395	0	18302.4	10257.3	0	140.231	95.1783
0.158992	1.49104	174194	0.220212	0	20213.2	11328.1	0	154.871	105.115
-0.0080507	-0.0754999	-31200.3	0.819049	0.8302	-1023.51	-573.606	0	-7.842	-5.32255

C.2 Observability Of Dynamic Economic Systems

Reconstructibility refers to the possibility of determining the state at some past date from the observation data currently available, while observability refers to the possibility of determining the current state from future observation data.

C.2.1 Reconstructibility

Consider an n-dimensional linear time-varying discrete-time system

$$x(t+1) = A(t)\,x(t) + B(t)\,u(t) \qquad ---(3)$$

together with its m-dimensional observation (output) equation

$$z(t) = C(t)\,x(t) + D(t)\,u(t) \qquad ---(4)$$

and denote the output at time $\underline{t}$ when $x(t(0)) = x(0)$ by $z(t,t(0),x(0),u)$. The system is said to be completely reconstructible if and only if for all $t(1)$, there exists $-\infty < t(0) < t(1)$ such that

$$z(t,t(0),x(0),0) = 0, \qquad t(0) \leqslant t \leqslant t(1)$$

implies $x(0) = 0$.

Theorem C.5 A linear time-invariant system is completely reconstructible if and only if it is completely observable.

For a proof see Aoki (1976).

C.2.2 Observability

Theorem C.6 The n-dimensional linear time-invariant discrete-time system, given by eq.(1), together with its m-dimensional observation (output) equation, given by eq.(2), is completely observable if and only if the rank of the observability matrix

$$(C', A'C', (A')^2 C', \ldots, (A')^{n-1} C')$$

is equal to n. For a proof see Aoki (1976).

It is thus noticed that the condition for complete observability is that (A',C') is a completely controllable pair. In this sense, these two concepts are dual.

For time-invariant linear discrete-time systems we can define the concept of observability analogously. We illustrate by means of the following derivation.

Suppose that at time $\underline{t}$ we know the past values of the observation (output) and the instrument vectors of a dynamic system represented by eqs.(1) and (2). Eq.(1) can be written in its final form representation as

$$x(t) = A^t x(0) + Bu(t-1) + ABu(t-2) + \ldots + A^{t-1} Bu(0) \qquad ---(5)$$

Substituting eq.(5) in eq.(2), we obtain

$$z(t) = C A^t x(0) + C (Bu(t-1) + \ldots + A^{t-1} Bu(0)) + Du(t) \qquad ---(6)$$

We rewrite this into matrix form as

$$\begin{bmatrix} z^*(0) \\ z^*(1) \\ \cdot\cdot \\ \cdot\cdot \\ z^*(t) \end{bmatrix} = R(t)\, x(0) \qquad ---(7)$$

where

$$z^*(t) = z(t) - C (Bu(t-1) + \ldots + A^{t-1} Bu(0)) - Du(t) \qquad ---(8)$$

$$t=0,1,\ldots,t$$

and

$$R(t) = \begin{bmatrix} C \\ CA \\ \cdot\cdot \\ \cdot\cdot \\ CA^t \end{bmatrix} \quad : m(t+1) \times n \text{ matrix} \qquad ---(9)$$

Since $R(t)$ and all the z^*'s are known in eq.(7), the only unknown vector in that equation (by assumption) turns out to be $x(0)$. Thus, the observability question reduces to that of an existence of a unique solution, in terms of $x(0)$, to

$$z^* = R(t)\, x(0) \qquad ---(10)$$

where z^* is the (known) stacked vector of z^*'s defined in eq.(8). To obtain this, we premultiply both sides of eq.(10) by $R'(t)$. The solution of $x(0)$ now given by

$$x(0) = (R'(t)R(t))^{-1} R'(t)\, z^* \qquad ---(11)$$

will exist if and only if the matrix $(R'(t)R(t))$ is nonsingular. This implies that the initial state has been recovered from future observation data and therefore the system is completely observable. Under the circumstances, this is exactly the same as saying that the observability matrix has full rank. This corollary is often made use of when the actual computation of the rank of the observability matrix becomes too tedious.

C.3 Stochastic Controllability And Observability

We have discussed the concepts of controllability and observability for deterministic dynamic systems in state-space form. We shall now briefly extend the scope of the discussion to stochastic systems.

The concept of observability easily finds a counterpart in statistics since observability is a condition on the behaviour of the estimation error of the state vector as the size of the observation data grows indefinitely. This can be clearly seen from the last section where a stacked set of observation vectors is related to the unknown state vector $x(0)$ by the algebraic eq.(7), from which we have

$$x(0) = R(t)^{+} z^{*} + r \qquad \text{---(12)}$$

where z* is the stacked observation vector and $\underline{r}$ is some vector in the null space of R(t). The estimated or reconstructed x(0) from z* is therefore given by eq.(12) where $R(t)^{+}$ denotes the Penrose pseudo-inverse. The null space of R(t) is nonincreasing with $\underline{t}$ and reduces to $\{0\}$ if and only if the system is observable. When we have exogenous random noises, then the estimated or reconstructed vectors do not converge to the true unknown vector when the observation data is finite. Stochastic observability can therefore be defined as a condition for estimates to converge in some probability sense (see Aoki 1967, Fitts 1972).

In a similar manner, stochastic controllability can be defined by replacing deterministic descriptions by probabilistic ones (see Connors 1967).

It needs to be noted that the deterministic concepts of controllability and observability are useful even in systems with random disturbances. Thus, if a system is observable then we can determine an initial state from a finite data set. Observability can therefore be considered a special case of constructing asymptotically consistent estimators or of statistical hypotheses testing, such as determining which alternative models are consistent with the observed data. While examples of determining alternate economic models consistent with observation were provided by Basmann (1965), it was shown by Aoki and Li (1973) that if a system is observable, then the set of initial states that is consistent with the observed data eventually converges to a singleton.

ANNEXURE

DATA USED IN THE ESTIMATION PROCESS

Endogenous Variables

Year	QF	QNF	QA	YAR	YMR	YTR	YSR	YNAR	YNFR	YGFR
1961:	85.7	91.6	87.6	13685	4681	1009	5811	11501	25186	26440
1962:	83.0	90.3	85.3	13323	5002	1084	6174	12260	25583	27003
1963:	83.5	93.2	86.6	13694	5477	1165	6580	13222	26916	28380
1964:	92.5	105.2	96.5	14932	5872	1215	7007	14094	29026	30617
1965:	74.9	93.5	80.8	12842	6049	1282	7162	14493	27335	29023
1966:	76.8	87.8	80.3	12678	6121	1322	7403	14846	27524	29307
1967:	98.7	99.0	98.8	14633	6319	1397	7644	15360	29993	31868
1968:	97.3	97.4	97.3	14713	6619	1496	7950	16065	30778	32725
1969:	104.0	103.6	103.9	15869	6936	1537	8431	16904	32773	34883
1970:	112.9	108.7	111.6	16980	7117	1574	8848	17539	34519	36736
1971:	111.4	110.9	111.2	16867	7293	1650	9218	18161	35028	37313
1972:	102.3	102.2	102.3	15780	7561	1729	9432	18722	34502	36910
1973:	110.3	117.0	112.4	16955	7724	1760	9764	19284	36203	38646
1974:	104.3	118.3	108.8	16618	7921	1985	10100	20006	36624	38979
1975:	127.2	119.8	124.8	18777	8348	2165	10865	21378	40155	42662
1976:	115.7	117.8	116.4	17532	9161	2329	11543	23033	40565	43208
1977:	133.6	130.9	132.7	19743	9774	2432	12207	24413	44156	46948
1978:	139.3	134.6	137.8	20057	10542	2547	13220	26309	46366	49403
1979:	114.8	122.3	117.2	17466	10198	2714	13502	26414	43880	47024
1980:	137.5	130.1	135.1	19692	10478	2827	14408	27713	47405	50682
1981:	140.8	145.9	142.4	20312	11106	2992	15392	29490	49802	53439

Endogenous Variables

Year	YPDR	CPR	s	SNR	INR	IAR	ITR	KR	DR	FR
1961:	23950	21596	0.0910	2291	2886	463	517	60930	1254	595
1962:	24173	22037	0.1036	2651	3388	525	631	63816	1420	737
1963:	25224	22605	0.1068	2874	3616	541	695	67204	1464	742
1964:	27510	24571	0.1004	2915	3990	648	715	70820	1591	1095
1965:	26025	23836	0.1232	3367	4482	790	721	74810	1688	1115
1966:	26399	24382	0.1292	3557	4892	727	605	79292	1783	1335
1967:	28972	26272	0.1038	3113	4264	664	465	84184	1875	1151
1968:	29648	27056	0.1044	3212	3811	707	449	88448	1947	599
1969:	31327	28082	0.1292	4234	4567	835	435	92259	2110	333
1970:	33062	29838	0.1323	4567	4961	901	609	96826	2217	394
1971:	33444	30714	0.1382	4840	5262	931	638	101786	2285	422
1972:	33239	30070	0.1255	4330	4667	927	726	107048	2408	337
1973:	35011	30880	0.1649	5969	6629	1122	763	111715	2443	660
1974:	35088	31143	0.1530	5604	5850	875	715	118344	2355	246
1975:	38776	33467	0.1736	6971	5915	804	683	124194	2507	-1056
1976:	38889	33422	0.1965	7970	6491	1233	687	130109	2643	-1479
1977:	42813	36800	0.1869	8253	6876	1369	745	136600	2792	-1377
1978:	45076	38086	0.2224	10311	9071	1878	845	143476	3037	-1240
1979:	42316	36589	0.1929	8466	7440	1447	791	152547	3144	-1026
1980:	45556	39077	0.1871	8870	8613	1612	959	159987	3277	-257
1981:	49194	41008	0.1989	9906	9282	1726	1028	168600	3637	-624

Endogenous Variables

Year	MR	WPF	WPNF	WP	P	YNFN	YNMN	CC	SCBR	TBR
1961	1540	0.489	0.613	0.5522	0.5592	14084	15164	2201	125	142
1962	1730	0.521	0.562	0.5732	0.5825	14902	16166	2379	132	155
1963	1891	0.564	0.556	0.6085	0.6349	17089	18654	2606	140	172
1964	2092	0.662	0.733	0.6753	0.6941	20147	21930	2769	168	190
1965	1960	0.707	0.823	0.7267	0.7610	20802	22884	3034	180	197
1966	2080	0.837	0.951	0.8277	0.8748	24078	26259	3197	225	266
1967	2057	1.016	0.851	0.9238	0.9440	28313	30737	3376	229	285
1968	1907	0.963	0.830	0.9130	0.9375	28854	31574	3682	306	387
1969	1588	1.007	0.948	0.9475	0.9753	31964	35023	4010	322	381
1970	1634	1.000	1.000	1.0000	1.0000	34519	38047	4371	364	452
1971	1962	1.034	0.986	1.0560	1.0535	36902	40994	4800	447	580
1972	1925	1.195	1.074	1.1620	1.1778	40636	45260	5420	487	595
1973	2141	1.419	1.466	1.3970	1.4021	50757	55924	6308	856	952
1974	1891	1.958	1.637	1.7490	1.6286	59646	65980	6347	908	1040
1975	1880	1.741	1.398	1.7300	1.5492	62208	69922	6710	912	1098
1976	1825	1.527	1.674	1.7660	1.6571	67220	75750	7893	1500	1905
1977	2418	1.704	1.780	1.8580	1.7234	76098	85017	8631	2143	2310
1978	2621	1.726	1.704	1.8580	1.7491	81099	91634	10220	3191	3862
1979	2539	1.854	1.946	2.1760	2.0120	88287	100488	11699	4250	4873
1980	3705	2.167	2.177	2.5730	2.2434	106348	120269	13463	4858	5989
1981	4020	2.374	2.395	2.8060	2.4304	121039	137924	14512	5675	6467

Endogenous Variables

Year	M1	MS	v	BD	NDE	CR	TR	NTR	AF	ANF
1961:	3046	5516	4.7933	114	1113	957	875	162	117.2	31.0
1962:	3310	5775	4.6758	156	1204	1204	1061	367	117.8	31.0
1963:	3752	6166	4.6027	167	1613	1381	1374	472	117.4	31.1
1964:	4080	6042	5.0674	172	2048	1645	1563	518	118.1	31.3
1965:	4529	6232	4.6571	173	2231	1705	1784	536	115.1	31.3
1966:	4950	5980	4.9008	295	3279	2473	1934	539	115.3	30.9
1967:	5350	5791	5.5030	206	2962	2251	1937	617	121.4	31.1
1968:	5779	6330	5.1698	262	2855	2078	2019	741	120.0	30.6
1969:	6387	6741	5.1748	46	3202	2508	2201	866	123.6	31.1
1970:	7140	7140	5.1451	285	3474	2524	2451	891	124.3	31.5
1971:	8138	7706	4.8421	519	3739	3031	2928	1100	122.6	31.8
1972:	9413	8101	4.5562	869	4398	2974	3443	1135	119.3	31.1
1973:	10848	7765	4.9769	328	4861	3646	3900	1173	126.5	31.7
1974:	11557	6608	5.8988	721	5285	3296	5097	1460	121.1	31.7
1975:	12745	7367	5.7910	366	6722	4697	6010	2066	128.2	32.0
1976:	15073	8535	5.0624	131	7204	5607	6581	2158	124.4	31.8
1977:	17792	9576	4.9027	933	8065	5589	7060	2732	127.5	32.6
1978:	21075	11343	4.3554	1506	10416	6938	8568	2672	128.6	33.0
1979:	23424	10675	4.3682	2700	9291	6153	8219	2958	125.2	32.1
1980:	27256	10593	4.7845	2577	9757	7939	9341	3553	125.8	32.5
1981:	29114	10376	5.1503	1539	10077	9005	10537	3790	128.6	33.3

Exogenous Variables

Year	R	MDL	XR	PM	DFEA	DBFA
1961:	97.4	19.92	945	0.6533	-21	778
1962:	94.7	16.91	993	0.6342	-28	881
1963:	85.9	10.77	1149	0.6584	31	1133
1964:	97.5	19.11	1017	0.6793	-28	1039
1965:	80.0	15.93	845	0.6978	-35	1594
1966:	87.9	14.69	745	0.9573	65	1788
1967:	92.0	17.18	906	1.0042	27	1713
1968:	93.5	17.70	1308	1.0033	151	1215
1969:	97.0	19.44	1255	0.9973	272	1068
1970:	96.5	19.56	1240	1.0000	-36	1681
1971:	99.0	17.55	1540	0.9265	70	2030
1972:	86.0	20.57	1588	0.9717	-45	2272
1973:	99.5	25.52	1481	1.3770	109	237
1974:	85.5	35.68	1645	2.3886	-281	3008
1975:	97.0	19.62	2936	2.8000	687	3785
1976:	94.3	15.88	3304	2.7737	1464	2591
1977:	98.5	24.37	3795	2.4873	1990	6405
1978:	97.5	21.51	3861	2.5971	899	3024
1979:	85.0	43.85	3565	3.5934	-43	7433
1980:	97.0	21.93	3962	3.3840	-613	10418
1981:	99.0	20.52	4644	3.3751	-2069	9982

Exogenous Variables

Year	DEF	IGAR	IGTR	GDE	T	BR
1961:	313	247	417	682	0.0767	4.00
1962:	474	282	526	822	0.0848	4.25
1963:	816	278	602	965	0.0916	4.50
1964:	806	315	601	1044	0.0885	5.25
1965:	885	335	526	1082	0.1001	6.00
1966:	909	269	395	1053	0.0906	6.00
1967:	968	262	314	1081	0.0856	5.50
1968:	1033	313	318	1212	0.0946	5.00
1969:	1101	283	247	1318	0.0957	5.00
1970:	1200	292	428	1477	0.1022	6.00
1971:	1525	308	455	2314	0.1109	6.00
1972:	1652	362	572	2371	0.1138	6.00
1973:	1680	364	503	2506	0.1018	7.00
1974:	2112	322	529	3177	0.1062	9.00
1975:	2472	374	546	3945	0.1240	9.00
1976:	2563	506	532	4710	0.1269	9.00
1977:	2634	567	516	5615	0.1172	9.00
1978:	2868	601	580	6400	0.1299	9.00
1979:	3273	588	589	7466	0.1382	9.00
1980:	3800	622	733	9251	0.1309	9.00
1981:	4200	665	785	10594	0.1395	10.00

Note: In all cases, 'Year' represents the financial year. Thus, for example, 1961 would represent the year 1961-62.

REFERENCES

Aghevli, B.B. and M.S. Khan (1977), Inflationary finance and the dynamics of inflation: Indonesia 1951-72, American Economic Review, June, 390-403.

Aghevli, B.B. and M.S. Khan (1978), Government deficits and the inflationary process in developing countries, IMF Staff Papers, September.

Allen, R.G.D. (1967), Macroeconomic Theory (MacMillan, London).

Anderson, B.D.O. (1967), An algebraic solution to the spectral factorization problem, IEEE Trans. Autom. Control, AC - 12(4), 410-414.

Andrews, A. (1968), A square-root formulation of the Kalman covariance equations, AIAA J., 6(6), 1165-1166.

Aoki, M. (1967), Optimization Of Stochastic Systems (Academic Press, New York).

Aoki, M. (1968), Control of large-scale dynamic systems by aggregation, IEEE Trans. Autom. Control, AC-13(3),246-253.

Aoki, M. (1976), Optimal Control And Systems Theory In Dynamic Economic Analysis (North-Holland, Amsterdam).

Aoki, M. (1981), Dynamic Analysis Of Open Economies (Academic Press, New York).

Aoki, M. (1983), Notes On Economic Time Series Analysis: System Theoretic Perspectives (Springer-Verlag, Berlin).

Aoki, M. and M.B. Canzoneri (1979), Sufficient conditions for control of target variables and assignment of instruments in dynamic macroeconomic models, International Economic Review, 20, 605-616.

Aoki, M. and M.T. Li (1973), Partial reconstruction of state vectors in decentralized systems, IEEE Trans. Autom. Control, AC-18(2), 289-292.

Athans, M. (1968), The matrix minimum principle, Information and Control, 11, 592-606.

Athans, M. (1974), The Importance of Kalman filtering methods for economic systems, Annals of Economic and Social Measurement, 3, 49-64.

Athans, M. and P.L. Falb (1966), Optimal Control Theory - An

Introduction To The Theory And Its Applications (McGraw Hill, New York).

Athans, M. and E. Tse (1967), A direct derivation of the optimal linear filter using the maximum principle, IEEE Trans. Autom. Control, AC-12(6), 690-698.

Ball, R.J., Chairman (1978), Committee On Policy Optimisation: Report (Her Majesty's Stationery Office, London).

Barnett, S. (1971), Matrices In Control Theory (Van Nostrand, London).

Basile, G. and G. Marro (1971), On the perfect output controllability of linear dynamic systems, Richerche di Automatica, 2, 1-10.

Basmann, R.L. (1965), A note on the statistical testability of 'explicit causal chains' against the class of 'interdependent models', Journal of the American Statistical Association, 60, 1080-1093.

Bellantoni, J.F. and K.W. Dodge (1967), A square-root formulation of the Kalman-Schmidt filter, AIAA J., 5(7),1309-1314

Bellman, R.E.(1957), Dynamic Programming (Princeton University Press, New Jersey).

Bellman, R.E.(1961), Adaptive Control Processes: A Guided Tour (Princeton University Press, New Jersey).

Bhattacharya, B.B. (1984), Government budget, inflation and growth: A macroeconometric policy analysis, Technical Report, Institute Of Economic Growth, Delhi.

Booton, R.C. (1952), An optimization theory for time-varying linear systems with nonstationary statistical inputs, Proc. IRE, 40, 977-981.

Brahmananda, P.R. (1980), Growthless Inflation By Means Of Stockless Money (Himalaya Publishing House, Bombay).

Brockett, R.W. and M. Mesarovic (1965), The reproducibility of multivariable systems, Journal of Mathematical Analysis and Applications, 11, 548-563.

Brown, R.G. (1963), Smoothing, Forecasting, And Prediction Of Discrete Time Series (Prentice Hall, New Jersey).

Buiter, W.H. and M. Gersovitz (1979), Issues in controllability and the theory of economic policy, Journal of Public Economics.

Cagan, P. (1956), The monetary dynamics of hyper-inflation, in: M. Friedman (Ed.), Studies In The Quantity Theory Of Money (University of Chicago Press, Chicago).

Chakravarty, S. (1973), Reflections on the growth process in the Indian economy, in: C.D. Wadhwa (Ed.) Some Problems Of India's Economic Policy (Tata McGraw Hill, New Delhi).

Choudhary, N.K. (1963), An econometric model of India: 1930-55, Ph.D. Thesis, Dept. of Economics, University of Wisconsin.

Chow, G.C. (1967), Multiplier, accelerator, and liquidity

preference in the determination of national income in the United States, Review of Economics and Statistics, 49,1-15.

Chow, G.C. (1968), The acceleration principle and the nature of business cycles, Quarterly Journal of Economics,82,403-418

Chow, G.C. (1975), Analysis And Control Of Dynamic Economic Systems (John Wiley, New York).

Chow, G.C. (1981), Econometric Analysis By Control Methods (John Wiley, New York).

Chow, G.C. and P. Corsi, Eds. (1982), Evaluating The Reliability Of Macroeconomic Models (John Wiley, New York).

Christ, C.F. (1975), Judging the performance of econometric models of the U.S. economy, International Economic Review, 16, 54-74.

Connors, M.M. (1967), Controllability of discrete, linear, random dynamical systems, SIAM Journal of Control, 5, 183-210.

Cooley, T.F. and E. Prescott (1976), Estimation in the presence of stochastic parameter variation, Econometrica,44,167-183

Darlington, S. (1958), Linear least squares smoothing and predictions with applications,Bell System Tech J,37,1221-1294

Davis, M.C. (1963), Factoring the spectral matrix, IEEE Trans. Autom. Control, AC-8(4), 296-305.

Desai, M.J. (1973), Macroeconometric models for India: A survey, Sankhya, Series B, 35, 169-206.

Deutsch, R. (1965), Estimation Theory (Prentice Hall, New Jersey).

Dorfman, R. (1969), An economic interpretation of optimal control theory, American Economic Review, 59(5), 817-831.

Dreyfus, S.E. (1968), Introduction to stochastic optimization and control, in: H.F. Karreman (Ed.) Stochastic Optimization And Control (John Wiley, New York).

Dutton, D.S. (1971), A model of self-generating inflation: The Argentina case, Journal of Money, Credit and Banking, 3, 245-262.

Dyer, P. and S. McReynolds (1969), Extension of square-root filtering to include process noise, J. Opt. Theory Appl., 3(6), 444-458.

Falb, P.L. and W.A. Wolovich (1967), Decoupling in the design and synthesis of multivariable control, IEEE Trans. Autom. Control, AC-12(6), 651-659.

Fel'dbaum, A.A. (1953), Optimal processes in automatic control systems, Automat i Telemehan, 14, 712-728.

Fitts, J.M. (1972), On the observability of nonlinear systems with applications to nonlinear regression analysis, Information Sciences, 4, 129-156.

Fox, K.A., J.K. Sengupta and E. Thorbecke (1966), The Theory Of

Quantitative Economic Policy With Applications To Economic Growth And Stabilization (North-Holland, Amsterdam).

Friedland, B. (1969), Treatment of bias in recursive filtering, IEEE Trans. Autom. Control, AC-14(4), 359-367.

Friedman, B.M. (1973), Methods In Optimization For Economic Stabilization Policy (North-Holland, Amsterdam).

Friedman, M. (1969), The Optimum Quantity Of Money (Aldine Publ. Co., Chicago).

Fromm, G. and L.R. Klein (1976), The NBER/NSF model comparison seminar: An analysis of results, Annals of Economic and Social Measurement, 5, 1-28.

Gabay, D., P. Nepomiastchy, M. Rachdi and A. Ravelli (1978), Etude, resolution et optimisation de modeles macroeconomiques, Rapport LABORIA no. 312, IRIA, B.P. Le Chesnay.

Gabay, D., P. Nepomiastchy, M. Rachdi and A. Ravelli (1980), Numerical methods for simulation and optimal control of large-scale macroeconomic models, in: A. Bensoussan, P. Kleindorfer and C. Tapiero (Eds.) Applied Stochastic Control In Econometrics And Management Science (North-Holland, Amsterdam).

Gilbert, E.G. (1969), The decoupling of multivariable systems by means of state variable feedback, SIAM Journal of Control, 7.

Goldberger, A.S. (1959), Impact Multipliers And Dynamic Properties Of The Klein-Goldberger Model (North-Holland, Amsterdam).

Hadley, G. and M.C. Kemp (1971), Variational Methods In Economics (North-Holland, Amsterdam).

Halkin, H. (1964), Optimal control for systems described by differential equations, in: C.T. Leondes (Ed.) Advances in Control Systems: Theory And Applications (Academic Press, New York).

Halkin, H. (1966), A maximum principle of the Pontryagin type for systems described by nonlinear difference equations, SIAM Journal of Control, 4(1).

Harvey, A. and G.D.A. Phillips (1978), The maximum likelihood estimation of autoregressive moving average models by Kalman filtering, Working paper No. 3, University of Kent.

Hestenes, M.R. (1950), A general problem in the calculus of variations with application to the paths of least time, RAND Corporation, RM-100, ASTIA Document No. AD-112382.

Ho, Y.C. and R.C.K. Lee (1964), A Bayesian approach to problems in stochastic estimation and control, IEEE Trans. Autom. Control, AC-9(4), 333-339.

Holt, C.C. (1960), A general solution for linear decision rules, Unpublished report, Graduate School of Industrial Administration, Carnegie Mellon University.

Holt, C.C. (1962), Linear decision rules for economic stabilization and growth, Quarterly Journal of Economics,76,20-45

Holt, C.C., F. Modigliani, J.F. Muth and H.A. Simon (1960), Planning Production, Inventories And Work Force (Prentice Hall, New Jersey).

Holtzman, J.M. (1966), Convexity and the maximum principle for discrete systems, IEEE Trans. Autom. Control, AC-11.

Howrey, E.P. (1966), Stabilization policy in linear stochastic systems, Report No. 83, Econometric Research Program, Princeton University.

Intriligator, M.D. (1971), Mathematical Optimization And Economic Theory (Prentice Hall, New Jersey).

Intriligator, M.D. (1978), Econometric Models, Techniques, And Applications (Prentice Hall, New Jersey).

Jazwinski, A.H. (1968), Limited memory optimal filtering, IEEE Trans. Autom. Control, AC-13(5), 558-563.

Jazwinski, A.H. (1969), Adaptive filtering, Automatica, 5(4), 475-485.

Jordan, B.W. and E. Polak (1964), Optimal control of aperiodic discrete-time systems, SIAM Journal on Control,2, 332-346.

Kalman, R.E. (1960), A new approach to linear filtering and prediction problems, Journal of Basic Engineering, 82D, 35-45.

Kalman, R.E. (1963), New methods in Weiner filtering theory, in: J.L. Bogdanoff and F. Kozin (Eds.) Proc. Symp. Eng. Appl. Random Functions Theory And Probability (John Wiley, New York).

Kalman, R.E. (1964), When is a linear control system optimal? Journal of Basic Engineering, 86D, 51-60.

Kalman, R.E. and R.S. Bucy (1961), New results in linear filtering and prediction theory, Journal of Basic Engineering, 83D, 95-108.

Kalman, R.E. and B.L. Ho (1966), Spectral factorization using the Riccati equation, Aerospace Corp. Report, TR-1001-1.

Kalman, R.E., P.L. Falb and M. Arbib (1961), Topics In Mathematical Systems Theory (McGraw Hill, New York).

Kalman, R.E., Y.C. Ho and K.S. Narendra (1962), Controllability of linear dynamical systems, Contributions to Differential Equations, 1(2), 189-213.

Karchere, A.J. and E. Kuh (1982), Some initial thoughts about model reliability, in: G.C. Chow and P. Corsi (Eds.) Evaluating The Reliability Of Macroeconomic Models (John Wiley, New York).

Karnik, A.V. (1986), Estimating the direction of causality between public expenditure and national income: The Indian case, Unpublished report, Dept. of Economics, University of Bombay.

Katz, S. (1962), A discrete version of Pontryagin's maximum

principle, Journal of Electronics and Control, 13, 179-184

Kau, S. and K.S.P. Kumar (1969), Successive linearization and nonlinear filtering, in: M. Beckman and H.P. Kunzi (Eds.) Computing Methods In Optimization Problems (Springer-Verlag, Berlin).

Kendrick, D. (1972), On the Leontief dynamic inverse, Quarterly Journal of Economics, 86, 653-696.

Klein, L.R. (1982), Economic theoretic restrictions in econometrics, in: G.C. Chow and P. Corsi (Eds.) Evaluating The Reliability Of Macroeconomic Models (John Wiley,New York).

Kleinman, D.L. (1969), Optimal control of linear systems with time delay and observation noise, IEEE Trans. Autom. Control, AC-14(4).

Kleinman, D.L. and M. Athans (1966), The discrete minimum principle with application to the linear regulation problem, MIT Electronic System Laboratory, Technical Report: ESL-R-260.

Knoll, A. and M. Edelstein (1965), Estimation of local,vertical and orbital parameters for an earth satellite using horizon sensor measurements, AIAA J., 3(2), 338-345.

Krasnakevich, J.R. and R.H. Haddad (1969), Generalized prediction correction estimation, SIAM Journal on Control, 7, 469-511.

Krishnamurty, K. (1983), Inflation and growth: A model for India, 1961-1980, Technical Report, Institute Of Economic Growth, Delhi.

Kuh, E. and J. Neese (1982), Econometric model diagnostics, in: G.C. Chow and P. Corsi (Eds.) Evaluating The Reliability Of Macroeconomic Models (John Wiley, New York).

Kuhn, H.W. and A.W. Tucker (1951), Nonlinear programming, in: J. Neyman (Ed.) Proc. Second Berkeley Symp. On Mathematical Statistics And Probability (University of California Press, Berkeley).

Kwakernaak, H. and R. Sivan (1972), Linear Optimal Control Systems (Wiley Interscience, New York).

LaSalle, J. and S. Lefchetz (1961), Stability By Liapunov's Direct Method (Academic Press, New York).

Leamer, E. (1978), Specification Searches (John Wiley,New York)

Luenberger, D.G. (1969), Optimization By Vector Space Methods (John Wiley, New York).

Mahalanobis, P.C. (1953), Some observations on the process of growth in national income, Sankhya, 12(4).

Mahalanobis, P.C. (1955), The approach of operational research to planning in India, Sankhya, 16, 1-46.

Mammen, T. (1967), An econometric study of the money market in India, Ph.D. Thesis, Dept. of Economics, University of Pennsylvania.

Mangasarian, O.L. (1966), Sufficient conditions for the optimal control of nonlinear systems, SIAM Journal on Control, 4, 139-152.

Mariano, R.S. and S. Schleicher (1972), On the use of Kalman filters in economic forecasting, Discussion Paper No. 3, University of Pennsylvania.

Marwah, K. (1964), An econometric model of price behaviour in India, Ph.D. Thesis, Dept. of Economics, University of Pennsylvania.

Meditch, J.S. (1967), Orthogonal projection and discrete optimal linear smoothing, SIAM Journal on Control, 5(1),74-80.

Mujumdar, N.A., T.R. Venkatachallam and M.V. Raghavachari(1980) The high saving phase of the Indian economy: 1976-79 - An exploratory interpretation, Occasional Papers of the Reserve Bank of India, 1(1).

Mundell, R.A. (1962), Appropriate use of monetary and fiscal policy for internal and external stability, IMF Staff Papers, 9, 70-79.

Mundell, R.A. (1968), International Economics (MacMillan, New York).

Muth, J.F. (1960), Optimal properties of exponentially weighted forecasts of time series with permanent and transitory components, Journal of American Statistical Association, 55, 299-306.

Muth, J.F. (1961), Rational expectations and the theory of price movements, Econometrica, 29, 315-335.

Nachane, D.M. and R.M. Nadkarni (1985), Empirical testing of certain monetarist propositions via causality theory: The Indian case, Indian Economic Journal, 33(1), 13-42.

Narasimham, N.V.A. (1956), A Short-Term Planning Model For India (North-Holland, Amsterdam).

Nepomiastchy, P., A. Ravelli and F. Rechenmann (1978), An automatic method to get an econometric model in a quasitriangular form, Communications at NBER Conference on Control and Economics, Austin.

Nerlove, M. (1972), Lags in economic behaviour, Econometrica, 40, 221-252.

Nerlove, M. and S. Wage (1964), On the optimality of adaptive forecasting, Management Science, 10, 207-224.

Nikaido, H. (1972), Introduction To Sets And Mapping In Modern Economics (North-Holland, Amsterdam).

Nyberg, L. and S. Viotti (1976), Controllability and the theory of economic policy: A critical note, Seminar Paper No. 61, IIE, University of Stockholm.

Olivera, J.H.G. (1967), Money, prices and fiscal lags: A note on the dynamics of inflation, Banca Nazionale De Lavoro Quarterly Review, 82, 258-267.

Pagan, A. (1975), A note on the extraction of components from time series, Econometrica, 43, 163-168.

Pagan, A. (1980), Some identification and estimation results for regression models with stochastically varying coefficients, Unpublished manuscript, Australian National University.

Pandit, V.N. (1980), Macroeconomic structure and policy in a less developed economy, Working Paper No. 235, Delhi School Of Economics, University of Delhi.

Pani, P.K. (1984), A macro model of the Indian economy with special reference to output, demand and prices: 1969-82, Occasional Papers of the Reserve Bank of India, 5(2), 113-239.

Pearson, J.B. and R. Sridhar (1966), A discrete optimal control problem, IEEE Trans. Autom. Control, AC-11(1), 171-174.

Pethe, A.M. (1986), Optimization dynamics: Algorithmic applications within a macroeconometric control system, Ph.D. Thesis, Dept. of Economics, University Of Bombay.

Phillips, A.W. (1957), Stabilization policy and the time-form of lagged responses, Economic Journal, 67, 265-277.

Pindyck, R.S. (1973), Optimal Planning For Economic Stabilization (North-Holland, Amsterdam).

Pindyck, R.S. and D.L. Rubinfeld (1976), Econometric Models And Economic Forecasts (McGraw Hill, New York).

Pontryagin, L.S. (1961), Optimal regulation processes, American Mathematical Society Transactions, Series D, XVIII,295-339

Pontryagin, L.S., V. Boltyanskii, R. Gamkrelidze and E. Mishchenko (1962), The Mathematical Theory Of Optimal Processes (Wiley Interscience, New York).

Porter, B. and R. Crossley (1972), Modal Control (Taylor and Francis, London).

Potter, J.E. (1967), A formula for updating the determinant of the covariance matrix, AIAA J., 5(7), 1352-1354.

Preston, A.J. (1974), A dynamic generalization of Tinbergen's theory of economic policy, Review of Economic Studies, 41, 65-74.

Raj, K.N. (1962), The marginal rate of savings in the Indian economy, Oxford Economic Papers, 14(1).

Rao, M.J.M. (1984a), Planning And Control Theory (Himalaya Publishing House, Bombay).

Rao, M.J.M. (1984b), Control systems and quantitative economic policy, in: R. Trappl (Ed.) Cybernetics And Systems Research 2 (Elsevier Science Publishers, North-Holland, Amsterdam).

Rao, M.J.M. and C.J. Fernandes (1986), Measurement of potential output and the output gap for the Indian economy, Journal of the University of Bombay, 54(1).

Rao, V.K.R.V. (1980), Savings, capital formation and national income, Economic and Political Weekly, XV(22).

Riddle, A.C. and B.D.O. Anderson (1966), Spectral factorization - Computational aspects, IEEE Trans. Autom. Control, AC-11(4), 764-765.

Rosen, J.B. (1967), Optimal control and convex programming, in: J. Abadie (Ed.) Nonlinear Programming (North-Holland, Amsterdam).

Rosenberg, B. (1973), A survey of stochastic parameter regression, Annals of Economic and Social Measurement, 2,399-428

Roy, J. and M.J.M. Rao (1980), Short-term macroeconometric forecasting model used for Indian annual plans, in: A. Ghosal (Ed.) Cybernetics And Applied Planning (South Asian Publishers, New Delhi).

Rozonoer, L.I. (1966), L.S. Pontryagin's maximum principle in optimal control theory, in: R. Oldenburger (Ed.) Optimal And Self Optimizing Control (MIT Press, Cambridge, Massachusetts).

Rutishauser, H. (1958), Solution of eigen value problems with the LR transformation, NBS Applied Mathematics Series, 49.

Sage, A.P. (1968), Optimum Systems Control (Prentice Hall, New Jersey).

Sage, A.P. and J.L. Melsa (1970), System Identification (Academic Press, New York).

Sage, A.P. and J.L. Melsa (1971), Estimation Theory With Applications To Communications And Control (McGraw Hill, New York).

Salmon, M. and K.F. Wallis (1982), Model validation and forecast comparisons: Theoretical and practical considerations, in: G.C. Chow and P. Corsi (Eds.) Evaluating The Reliability Of Macroeconomic Models (John Wiley, New York)

Salmon, M. and P. Young (1979), Control methods and quantitative economic policy, in: S. Holly, B. Rustem and M.B. Zarrop (Eds.) Optimal Control For Econometric Models (MacMillan, London).

Sant, D. (1977), Generalized least squares applied to time-varying parameter models, Annals of Economic and Social Measurement, 6, 301-314.

Schlee, F.H., C.J. Standish and N.F. Toda (1967), Divergence in the Kalman filter, AIAA J., 5(6), 1114-1120.

Schultz, D.G. and J.L. Melsa (1967), State Functions And Linear Control Systems (McGraw Hill, New York).

Simon, H.A. (1956), Dynamic programming under uncertainty with a quadratic criterion function, Econometrica, 24, 74-81.

Sims, C.S. and J.L. Melsa (1970), A fixed configuration approach to the stochastic linear regulator problem, Preprints, 11th Joint Automatic Control Conference, Atlanta, Georgia.

Smith, G.L. (1967), Sequential estimation of observation error variances in a trajectory estimation problem, AIAA J., 5(11), 1964-1970.

Smith, V.K. (1973), Monte Carlo Methods (Lexington Books, Lexington).

Swamy, P.A.V.B. (1974), Linear models with random coefficients, in: P. Zarembka (Ed.) Frontiers In Econometrics (Academic Press, New York).

Taylor, L. (1970), The existence of optimal distributed lags, The Review of Economic Studies, 37, 95-106.

Theil, H. (1957), A note on certainty equivalence in dynamic planning, Econometrica, 25, 346-349.

Theil, H. (1958), Economic Forecasts And Policy (North-Holland, Amsterdam).

Theil, H. (1964), Optimal Decision Rules For Government And Industry (North-Holland, Amsterdam).

Theil, H. (1965), Linear decision rules for macrodynamic policy problems, in: B.G. Hickman (Ed.) Quantitative Planning of Economic Policy (The Brookings Institution, Washington, D.C.).

Theil, H. and S. Wage (1964), Some observations on adaptive filtering, Management Science, 10, 198-206.

Tinbergen, J. (1939), Statistical Testing Of Business Cycle Theories (League Of Nations, Geneva).

Tinbergen, J. (1956), On The Theory Of Economic Policy (North-Holland, Amsterdam).

Tustin, A. (1953), The Mechanism Of Economic Systems: An Approach To The Problem Of Economic Stabilization From The Point Of View Of Control-Systems Engineering (Harvard University Press, Cambridge, Massachusetts).

Valentine, F.A. (1937), The problem of Lagrange with differential inequalities as added side conditions, in: Contributions To The Calculus Of Variations 1933-37 (University of Chicago Press, Chicago).

Vishwakarma, K.P. (1970), Prediction of economic time-series by means of the Kalman filter, International Journal of System Sciences, 1, 25-32.

Vishwakarma, K.P. (1974), Macroeconomic Regulation (Rotterdam University Press, Rotterdam).

Vishwakarma, K.P., P.M.C. de Boer and F.A. Palm (1970), Optimal prediction of inter-industry demand, Econometric Institute Report No. 7022, Netherlands School Of Economics,Rotterdam

Waelbroeck, J. (1973), Survey of short run model research outside the U.S.A., Paper No. 5, Economic Research Unit, University of Pennsylvania.

Waelbroeck, J. (1982), Discussant of 'Econometric model diagnostics' by E. Kuh and J. Neese in: G.C. Chow and P. Corsi

(Eds.) Evaluating The Reliability Of Macroeconomic Models (John Wiley, New York).

Wang, S.H. and C.A. Desoer (1972), The exact model matching of linear multivariable systems, IEEE Trans. Autom. Control, AC-17.

Weiner, N. (1949), The Extrapolation, Interpolation And Smoothing Of Stationary Time Series With Engineering Applications (John Wiley, New York).

Westcott, J.H., M.B. Zarrop, S. Holly, B. Rustem and R. Becker (1979), A control theory framework for policy analysis, in: S. Holly, B. Rustem and M.B. Zarrop (Eds.) Optimal Control For Econometric Models (MacMillan London).

Wishart, D.M.G. (1969), A survey of control theory, Journal of Royal Statistical Society, Series A, 293-319.

Wong, E. and J.B. Thomas (1961), On the multidimensional prediction and filtering problem and the factorization of spectral matrices, J. Franklin Inst., 272(2), 87-99.

Wonham, W.M. (1974), Linear Multivariable Control (Springer-Verlag, Berlin).

Young, P.C. (1977), A general theory of modelling for badly defined systems, in: G.C. Vansteekiste (Ed.) Modelling Of Land, Air And Water Resource Systems (Academic Press, New York).

Zadeh, L.A. and C. Desoer (1963), Linear Systems Theory (McGraw Hill, New York).

Zadeh, L.A. and J.R. Ragazzini (1950), An extension of Weiner's theory of prediction, J. Applied Physics, 21, 645-655.

Zadeh, L.A. and J.R. Ragazzini (1952), Optimum filters for the detection of signals in noise, Proc. IRE, 40, 1223-1231.

AUTHOR INDEX

SUBJECT INDEX